JOURNAL OF CHROMATOGRAPHY LIBRARY — volume 39B

selective sample handling and detection in high-performance liquid chromatography

part B

JOURNAL OF CHROMATOGRAPHY LIBRARY – volume 39B

selective sample handling and detection in high-performance liquid chromatography

part B

edited by
K. Zech

Byk Gulden Pharmaceuticals, Byk Guldenstrasse 2, P.O. Box 6500, 7750 Konstanz, F.R.G.

and

R.W. Frei [†]

Department of Analytical Chemistry, Free University, De Boelelaan 1083, 1081 HV Amsterdam, The Netherlands

ELSEVIER
Amsterdam — Oxford — New York — Tokyo **1989**

ELSEVIER SCIENCE PUBLISHERS B.V.
Sara Burgerhartstraat 25
P.O. Box 211, 1000 AE Amsterdam, The Netherlands

Distributors for the United States and Canada:

ELSEVIER SCIENCE PUBLISHING COMPANY INC.
655, Avenue of the Americas
New York, NY 10010, U.S.A.

ISBN 0-444-88327-4 (Vol. 39B)
ISBN 0-444-41616-1 (Series)

CONTENTS

LIST OF CONTRIBUTORS

Prof. Dr. W. Haerdi,
Dr. J.L. Veuthey, Dr. M.A. Bagnoud
Department of Inorganic,
Analytical and Applied Chemistry
University of Geneva
30 q.E. Ansermet
1211 Geneva 4
SWITZERLAND

Prof. Dr. R.W. Frei
Dr. H. Jansen
Philips Lighting b.v.
P.O.Box 8 00 20
5600 JM Eindhoven
THE NETHERLANDS

Associate Prof. P.R. Haddad
Department of Analytical Chemistry
University of New South Wales
P.O. Box 1
Kensington N.S.W. 2033
AUSTRALIA

Prof. Dr. N.H. Velthorst,
Dr. C. Gooijer,
Prof. Dr. R.W. Frei
Department of General and
Analytical Chemistry
Free University at Amsterdam
De Boelelaan 1083
1081 HV Amsterdam
THE NETHERLANDS

Prof. Dr. K.-Fr. Sewing and
Dr. U. Christians
Medizinische Hochschule Hannover
Zentrum Pharmakologie und Toxikologie
Abteilung Allgemeine Pharmakologie
P.O. Box 61 01 80
3000 Hannover
F.R.G.

Dr. M. Valcárcel,
M.D. Luque de Castro
Departamento de Quimica Analitica
Universidad de Cordoba
Facultad de Ciencias
Dr. M. Valcárcel
14004 Cordoba
SPAIN

Dr. A.C. Veltkamp
E C N
Stichting Energieonderzoek
Centrum Nederland
Researchcentrum
P.O. Box 1
1755 ZG Petten
THE NETHERLANDS

PREFACE

On 29 January 1989, during the completion of this second volume, Professor Roland Frei died after an illness of several months. Through his death I have lost a dear fried, to whom I am grateful for many scientific stimuli. Roland Frei's contributions to the field of selective sample handling and detection in HPLC will be missed not only by myself but also by many colleagues with similar interests. It was fortunate that we shared an interest in sample preparation, with, in Amsterdam, the emphasis on environmental samples and, in Constance, on biological samples arising from pharmaceutical research.

This book is the second and provisionally last part of a two-volume project. It follows the previously expressed view that the handling, separation and detection of complex samples should be considered as an integrated, interconnected process. On the basis of this philosophy we choose the contributions, which we hope will convince the reader - perhaps even more so than in Part A - that optimal sample preparation leads to a simplification of detection or reduced demands on the separation process. The reverse of this is also shown in detail.

In accordance with the aims of this book, we have tried once again to put the emphasis on chemical principles and have, therefore, suppressed, as far as possible, a discussion of the equipment required. This reflects our opinion that the limiting factor in the analysis of complex samples is incomplete knowledge of the underlying chemistry rather than the available hardware. This lack of knowledge is becoming more evident as the demands for lower detection limits grow, as resolving complex matrix problems requires even more understanding of the chemical interaction between the substance to be analysed and the stationary phase.

Thus, apart from one chapter dealing with chemically modified silicas, the main theme of this book is developed in three chapters on sample preparation and three on detection.

The first chapter outlines concentration and chromatography on chemically modified silicas with complexing properties. Examples of the use of these phases with organic and inorganic compounds are given.

Chapter II is the first of three contributions dealing with sample preparation. In particular, ion chromatography clearly exposes the following critical questions regarding sample preparation. Is the prepared sample representative of the material to be analysed? How can contamination be avoided, which results from the ubiquitous nature of the compound to be analysed? What is the best separation procedure to use in order to avoid tedious sample preparation?

In Chapter III the processing of whole blood (rather than plasma or serum) for drug analysis is described. The analysis of cyclosporine and its metabolites, an especially difficult case, demonstrates how comprehensive the optimisation of sample preparation must be in order to successfully perform the analysis. Several other examples are also described.

Chapter IV deals with radio-column liquid chromatography and introduces the other theme of this book: selective detection methods. The widespread use of radioisotopes requires a high degree of purification during the manufacture of the compounds, as well as highly accurate detection methods in biological and biochemical studies.

Chapter V continues the theme of selective detection with an overview of post-column reaction detection. The use of immobilised enzymes in post-column reactors or so called 'pumpless' reactor systems for on-line reagent generation after the chromatographic separation step is discussed in detail. Various examples of the separation of biological compounds show how the production of electrochemical reagents and photochemical reaction detection have increased the selectivity of the detection. This has led to more economical analytical systems.

Selective detection employing luminescence detection techniques is outlined in Chapter VI. The use of immobilised fluorophores or the coupling to photochemical reactions leads to highly selective detection systems, which again greatly simplify the sample handling.

The volume concludes with a review of the use of continuous separation techniques in flow injection analysis. It demands that a strong interdisciplinary dependence between sample handling and separation in this area is essential.

The completion of this second volume marks the achievement of the aims first set out by Professor Frei in the preface to Part A. It is thus a fitting conclusion to this preamble to quote these words:

"By the nature of its content, and written as it is by experienced practitioners, the book should be useful to investigators in many areas of application. Each chapter includes sufficient references to the literature to serve as a valuable starting point for more detailed investigation. The strong emphasis on sample handling makes the book unique in many ways and it should prove useful to the environmental scientist as well as to investigators from the clinical, pharmaceutical and bioanalytical fields."

Finally, I would like to thank the authors for their contributions, many friends for stimulating discussions, Dr. M. Galvan for linguistic assistance and Mrs. G. Bader and C. Jantke for the preparation of the camera ready manuscript. In particular, I am extremely grateful to my wife, who took over most of the editorial work following Roland's death.

September 1989   K. Zech
                                                 (Constance, F.R.G.)

CHAPTER I

PRECONCENTRATION AND CHROMATOGRAPHY ON CHEMICALLY MODIFIED SILICAS WITH COMPLEXATION PROPERTIES

J.L. VEUTHEY, M.A. BAGNOUD and W. HAERDI

1. INTRODUCTION

In the past 20 years numerous studies have been made to develop solid surfaces containing complexing sites for various applications. The first complexing surfaces found applications in preconcentrating transition metals from natural media (ref. 1). Through improved cross-linking techniques, complexing groups with faster exchange kinetics were bound to the support. It then became possible to use these supports in chromatography (ref. 2). More recently, other applications include:

- the use of these supports in heterogeneous catalysis at the complexing metal sites (ref. 3)
- retaining organics from air or water (ref. 4)
- peptide synthesis (ref. 5)
- immobilizing enzymes (ref. 6).

These applications have only been made possible after having tested a number of different types of supports. For example, the first supports employed for metal retention were divinylbenzene polystyrene copolymer resins. The development of other supports such as silicates and cellulose, have recently replaced these resins. Cross-linked silicate supports have been highly developed and have found a wide range of applications, especially in chromatography, due to their low cost and the facility of binding a wide range of functional groups.

This chapter will discuss the use of these supports in liquid-solid extractions, the types of supports available and a certain number of considerations which should be made concerning their application.

1.1   PRECONCENTRATION USING LIQUID-SOLID EXTRACTION

The analysis of inorganic, organic, or organometallic species necessitates, for the most part, a purification step to eliminate a large proportion of unwanted products. Extraction techniques seem to be the most appropriate in fulfilling this need. These techniques, in addition to clean-ups, offer the advantage of reducing the initial sample volume, thus concentrating it.

Selectively extracting a compound or class of compounds from a complex mixture while reducing the final sample volume before analysis is called "preconcentration" or "enrichment". This technique is very useful when compounds are found at trace levels.

Highly measured precautions must be taken by the analyst during trace analysis. A slight contamination of the extracting solvent, for example, may contribute to the errors revealed in the final analysis since contaminations may be preconcentrated as much as the compound of interest. Consequently, it is necessary to find techniques which reduce contamination risks keeping the number of operations to a strict minimum. Liquid-solid extraction, which consists of retaining solutes in a liquid phase on a solid support (filter, column, suspension) fulfill these pre-requisites better than any other technique. These extraction principles are not exclusive to liquid phases alone but are also applicable to gas phases as well.

1.2   STATIONARY PHASES FOR LIQUID-SOLID EXTRACTION

Four different types of supports exist for liquid-solid extraction, namely (ref. 7):

1.)   synthetic and foamed plastic resin

2.)   silica and alumina

3.) cellulose

4.) activated carbon

All of these supports have advantages and disadvantages, hence the choice of the support depends on the type of analysis needed. Several criterion are necessary for this choice. For example, the purity of the support is a very important factor. Consideration should also be given to certain properties such as the rigidity of the support or the ease of analysis or even the cost of the support, all of which play an important role in this choice.

1.2.1   PURITY OF PRECONCENTRATION SUPPORTS

Trace analysis necessitates the use of high purity preconcentration supports. Although silicates have been obtained with a high degree of purity, special attention should be given to those which have been chemically modified as impurities may have been introduced from reagents used during the chemical modification (for example, impurities due to alkoxysilane treatment of silicate gels). Consequently they require a pretreatment step to eliminate these trace compounds. Stationary phases may be pretreated using either a Soxhlet extraction or simply prewashed with an organic solvent or an appropriate mineral acid.

1.2.2   OTHER FACTORS INFLUENCING THE CHOICE OF SUPPORTS

One important physical property of these supports is their rigidity. Elution is usually performed under applied pressures deforming the support thus changing their physicochemical properties. In general, supports having higher rigidity allow higher elution flow rates to be obtained thus making analysis time shorter.

For instance, divinylbenzene polystyrene resins (Chelex 100) tend to swell, retract or become deformed upon contact of the mobile phase. These problems may be minimized when they are mixed with silica gel of approximately the same particle size.

Foamed plastics, on the other hand, are rigid and remain so even at high elution flow rates. At lower elution flow rates, silicates, controlled pore glasses (CPG) and cross-linked celluloses containing chelating substitutents are not deformed either and may be utilized for preconcentration as well as for high performance chromatography.

Another factor affecting the choice of a support is the type of analysis which can be performed. After preconcentration, analysis of retained products using either neutron activation or X-ray fluorescence may be done directly on silicas, resins or cellulose. However, direct analysis of CPG or foamed plastics is not recommended and elution is necessary.

Ligand exchange kinetics also plays a very important role in the choice of a support. Resins, for example, have very slow exchange kinetics as opposed to those of modified silicates. This difference may be attributed to the structure itself of the supports, since in fact, the cross-bonded functional groups of the resin are practically inaccessible whereas those of the modified silicates lie at the surface.

1.3   CHOICE OF A SUPPORT

Among the four different types of solid supports employed for the extraction or preconcentration of trace organics or inorganics, silicas offer the most attractive properties and consequently are receiving increasing interest for the following reasons:

1.)   their low cost
2.)   they may be used directly or modified through the physical or chemical adsorption of chelating agents
3.)   they are applicable to the extraction of organic as well as inorganic constituents from liquid or gas phase systems
4.)   they are used not only for preconcentration purposes but for chromatographic separation of organic or mineral components as well
5.)   they have fast ligand exchange kinetics.

The following discussion will focus on complexing silicas and will cover a number of their applications. With respect to chronological order, it will be necessary to first mention the synthesis of these complexing silicas, subsequently their use for preconcentration applications, and finally, their use as chromatography supports. Certain silicas, such as the cross-linked alkyl-chain silicas, which are readily employed in liquid chromatography, will not be discussed here since they have been thoroughly described in numerous articles (refs. 8-10).

2.   CHEMICALLY MODIFIED SILICAS

Silicas can easily be modified to obtain a wide range of supports of differing properties. Numerous factors may influence the chemical modification of silicates making their quantitative synthesis difficult. An entire monography (ref. 11) treats the subject of silicas, their physico-chemical properties and their reaction mechanisms.

Chemical modification of silicas is principally accomplished by one of two methods:

a.)   a reaction called "surface modification" between an organosilane and the silicate

b.)   or by hydrolytic polycondensation of organosilanes.

Surface modification is by far the most often employed as it is a much simpler procedure. In contrast, the second method is found to be difficult, uncontrollable, and gives rise to undefined supports. Consequently, this discussion will focus on surface modification.

2.1   SURFACE MODIFICATION

Several types of reactions between various organic groups -R and silicas are employed to modify these supports, namely:

1.)   -R may be directly bonded to the silicium atom, Si-R

2.)   -R may be bonded to the silicate via a heteroatom

    a. Si-O-R

    b. Si-NH-R

    c. Si-O-Si-R

Among the four different procedures of cross-linking organics, -R, onto silicas, Si-O-Si-R is the most often utilized. The siloxane bond is, in effect, the most stable one and consequently is the most suitable for applications in complex systems such as natural aquatic systems. Problems such as hydrolysis and synthesis are encountered with the three other types of bonds.

Cross-linking of the organic -R group through a siloxane bond may be accomplished using organosilanes such as RnSiX4-n (1<n<3). The trifuctional silane, R-SiX3, is the most often used and reacts according to the following mechanism:

Scheme 1

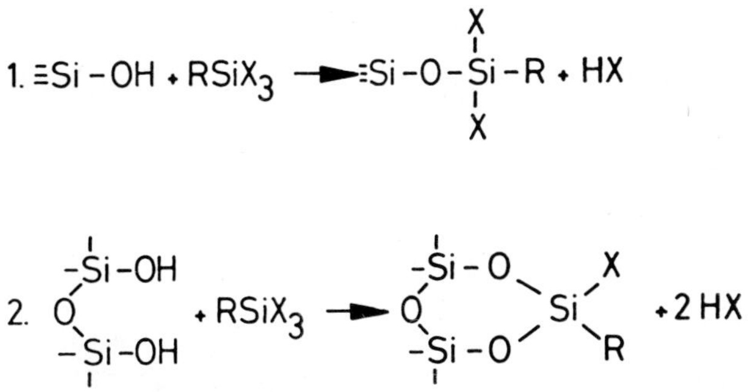

$$1. \equiv Si-OH + RSiX_3 \longrightarrow \equiv Si-O-\underset{\underset{X}{|}}{\overset{\overset{X}{|}}{Si}}-R + HX$$

$$2.\ \begin{matrix} -Si-OH \\ O \\ -Si-OH \end{matrix} + RSiX_3 \longrightarrow \begin{matrix} -Si-O \\ O \quad\quad Si \\ -Si-O \end{matrix} \begin{matrix} X \\ R \end{matrix} + 2\,HX$$

A maximum number of two of the three functional groups may be bonded to the silicate owing to steric hindrance. X is usually an alkoxy (methoxy or ethoxy) or halide functional group.

Another important factor to consider during chemical modification is the change in the surface porosity. Pore size as well as its specific surface decrease as a result of silanization. This factor should be considered when silica surface capacities are calculated.

2.2   CROSS-LINKED FUNCTIONAL GROUPS TO SILICAS

The types of silanes which are likely to bond with silicas have been discussed in a recent review (ref. 12). Only those organosilanes which form non-hydrolyzable cross-linked bonds are suitable for analytical applications.

The silicas 2, 4 - 7, described in Table I may be used directly as chelating surfaces for metals or chemically modified.

The advantage of working with cross-linked silicas is that the desired functional group may eventually be added by: hydrolysis, substitution, oxidation or azo-coupling.

TABLE I   Cross-linked functional groups to silicas by surface modification

No	Organosilane	Modified silica	Ref.
1	$(EtO)_3Si(CH_2)_3NH_2$	$Si-O-Si(CH_2)_3NH_2$	13-17
2	$(EtO)_3Si(CH_2)_3NH(CH_2)_2NH_2$	$Si-O-Si(CH_2)_3NH(CH_2)_2NH_2$	1,13, 14,16, 18-28
3	$(EtO)_3Si(CH_2)_3NHCH_3$	$Si-O-Si(CH_2)_3NHCH_3$	14,28
4	$(MeO)_3Si(CH_2)_3NH(CH_2)_2NH(CH_2)_2NH_2$	$Si-O-Si(CH_2)_3NH(CH_2)_2NH(CH_2)_2NH_2$	13
5	$(EtO)_3Si(CH_2)_3N\begin{smallmatrix}CH_2COOCH_3\\CH_2COOCH_3\end{smallmatrix}$	$Si-O-Si(CH_2)_3N\begin{smallmatrix}CH_2COOCH_3\\CH_2COOCH_3\end{smallmatrix}$	13
6	$(NaO)_3Si(CH_2)_3N\begin{smallmatrix}CH_2COONa\\CH_2COONa\end{smallmatrix}$	$Si-O-Si(CH_2)_3N\begin{smallmatrix}CH_2COONa\\CH_2COONa\end{smallmatrix}$	13

TABLE I   (continued)

No.	Organosilane	Modified silica	Ref.
7	$(NaO)_3Si(CH_2)_3N$ $\begin{array}{c} CH_2COONa \\ CH_2COONa \\ (CH_2)_2N \\ CH_2COONa \end{array}$	$Si-O-Si(CH_2)_3N$ $\begin{array}{c} CH_2COONa \\ CH_2COONa \\ (CH_2)_2N \\ CH_2COONa \end{array}$	13
8	$(MeO)_3Si(CH_2)_3OCH_2CHCH_2$ $\quad O$	$Si-O-Si(CH_2)_3OCH_2CHCH_2$ $\quad O$	27, 29-34
9	$ClR_2Si(CH_2)_nC_6H_4CH_2Cl$	$Si-O-Si(CH_2)_nC_6H_4CH_2Cl$	28, 35-41
10	$(MeO)_3SiC_6H_4NH_2$	$Si-O-SiC_6H_4NH_2$	42

3.   PRECONCENTRATION OF INORGANIC COMPOUNDS USING CHELATING SILICAS

Commercially prepared complexing resins (Chelex, Dowex, etc.) were the first supports that were used for preconcentration purpose. Selectivity is a major problem with these supports since they indifferently retain alkaline and alkaline earth metals as well as the desired transition metals. Secondly, during elution when pressures are applied, these supports undergo swelling or shrinking. Consequently their physical and chemical properties change during the course of the analysis. Thirdly, chelating kinetics are very slow due to steric hindrance.

Complexing silicas, on the other hand, overcame these problems and are currently the most widely used for (refs. 7, 12, 28):

- metal preconcentration from complex natural systems
- analysis and separation of metals in liquid chromatography.

Understanding the mechanism of metal complexing is of outmost importance. Although these mechanisms have been thoroughly studied in homogeneous solutions, they have not been sufficiently elucidated in heterogeneous liquid-solid systems. Retention properties are dependent upon a number of different phenomena. For instance, the formation of a stable complex may be prevented by steric hindrance. Other important factors are:

- adsorption
- interactions between the silica and the solution.

Hence, the retention behavior of these systems is difficult to predict since many unknown factors are important.

3.1 METAL PRECONCENTRATION

Table II lists the principal complexing functional groups as well as their precursors cross-bonded to silicas, employed for the purpose of preconcentration of metals.

Retention selectivity between transition metals and alkaline or alkaline earth metals was the key property evoking the development of chelating silicas. Functional groups such as dithiocarbamates, diamines, diketones, oxines and others are widely used because they form very stable complexes with transition metals even in the presence of commonly encountered interfering species.

As far as possible the preconcentration step needs to be kept simple. This necessitates the use of a minimum number of operations as well as reagents. Most metal preconcentrations require only a pH adjustment as a preliminary step.

TABLE II   Complexing functional groups cross-bonded to silicas for metals preconcentration.

No.	Precursors	Complexing silica	Ref.
1	$Si-O-Si(CH_2)_3NH(CH_2)_2NH_2$	$Si-O-Si(CH_2)_3NH(CH)_2NH_2$	14,16,21-24 43,44
2	$Si-O-Si(CH_2)_3NH_2$	$Si-O-Si(CH_2)_3NHCS_2^-$	1, 14-21
3	$Si-O-Si(CH_2)_3NHCH_3$	$Si-O-Si(CH_2)_3N(CH_3)CS_2Na$	1, 14-21
4	$Si-O-Si(CH_2)_3NH(CH_2)_2NH_2$	$Si-O-Si(CH_2)_3N(CH_2)_2NHCS_2Na$ $\quad\quad\quad CS_2Na$	14,16,20, 21,45-49
5	$Si-O-Si(CH_2)_3NH_2$	$Si-O-Si(CH_2)_3NHCOCH_2COCH_3$	19

TABLE II   (continued)

No.	Precursors	Complexing silica	Ref.
6	Si-O-Si(CH$_2$)$_3$NH(CH$_2$)$_2$NH$_2$	Si-O-Si(CH$_2$)$_3$NCH$_2$CH$_2$ NH (crown ether structure with $N(C_2H_5)_2N(C_2H_5)_2$)	50
7	Si-O-Si(CH$_2$)$_3$NH$_2$	Si-O-Si(CH$_2$)$_3$NH—(cyclopentene ring)—CS$_2$H	51
8	Si-O-Si(CH$_2$)$_3$NHCOC$_6$H$_4$NH$_2$	Si-O-Si(CH$_2$)$_3$NHCO (quinoline azo structure with OH)	17,52-57
9	Si-O-SiC$_6$H$_4$NH$_2$	Si-O-SiC$_6$H$_4$N=N (quinoline structure with OH)	42
10	Si-O-Si(CH$_2$)$_3$NH(CH$_2$)$_2$NH$_2$	Si-O-Si(CH$_2$)$_3$NH(CH$_2$)$_2$NHC(O)CH$_2$C(O)C$_6$H$_5$	16
11	Si-O-Si(CH$_2$)$_3$NH(CH$_2$)$_2$NH$_2$	Si-O-Si(CH$_2$)$_3$NH(CH$_2$)$_2$NHC(O)CH$_2$C(O)CH$_3$	16

TABLE II   (continued)

No.	Precursors	Complexing silica	Ref.
12	$Si-O-Si(CH_2)_3NH(CH_2)_2NH_2$	$Si-O-Si(CH_2)_3NH(CH_2)_2NHC(O)CH_2C(O)CF_3$	16
13	$Si-O-Si(CH_2)_3NH(CH_2)_2NH_2$	$Si-O-Si(CH_2)_3NH(CH_2)_2NHC(O)C_6H_4NH_2$	16
14	$Si-O-Si(CH_2)_3NH_2$	$Si-O-Si(CH_2)_3NH$ $CH_3-N-C_4H_9$	58
15	$Si-O-Si(CH_2)_3NH(CH_2)_2NH_2$	$Si-O-Si(CH_2)_3NHCH_2CH_2N=CHC_6H_4OH$	28
16	$Si-O-Si(CH_2)_3NH(CH_2)_2NH_2$	$Si-O-Si(CH_2)_3NHCH_2CH_2N=CH$	28
17	$Si-O-Si(CH_2)_3NH(CH_2)_2NH_2$	$Si-O-Si(CH_2)_3\overset{+}{N}H_2CH_2CH_2NHCOC_6H_4COO^-$	28
18	$Si-O-Si(CH_2)_3NH(CH_2)_2NH_2$	$Si-O-Si(CH_2)_3\overset{+}{N}H_2CH_2CH_2NHCOCH=CHCOO^-$	28
19	$Si-O-Si(CH_2)_3NH(CH_2)_2NH_2$	$Si-O-Si(CH_2)_3NHCH_2CH_2NHCH_2$	28

TABLE II   (continued)

No.	Precursors	Complexing silica	Ref.
20	$Si-O-Si(CH_2)_3NH(CH_2)_2NH_2$	$Si-O-Si(CH_2)_3NHCH_2CH_2N=C$ (with CF_3, and $HO-$, CF_3)	28
21	$Si-O-Si(CH_2)_3NH(CH_2)_2NH_2$	$Si-O-Si(CH_2)_3NHCH_2CH_2N=C$ (with CH_3, and $HO-$, CH_3)	28
22	$Si-O-Si(CH_2)_3NH(CH_2)_2NH_2$	$Si-O-Si(CH_2)_3NCH_2CH_2N$ (with CH_2COO^-, CH_2COO^-, CH_2COO^-)	28
23	$Si-O-SiC_6H_4CH_2Cl$	$Si-O-SiC_6H_4CH_2$ (dibenzo crown ether structure)	28-35
24	$Si-O-SiC_6H_4CH_2Cl$	$Si-O-SiC_6H_4CH_2$ (with CF_3, $=O$, $-OH$, CF_3)	28
25	$Si-O-SiC_6H_4CH_2Cl$	$Si-O-SiC_6H_4CH_2$ (with CH_3, $=O$, $-OH$, CH_3)	28

The pH of the analytical solution may have an important influence on the stability of the complexing functional group. For instance, siloxane bonds are particularly unstable in either acid (pH < 2) or basic (pH > 8) solutions. Other groups such as dithiocarbamates are also unstable in acid solutions (pH < 5) for they become oxidized or decompose.

Other factors such as stability of the functional group in aqueous solution should be taken into account. For example, diamino-silicas used as precursors for numerous complexing silicas (refs. 13, 59) are unstable over long periods of time in aqueous solutions. Therefore, before using a given silica the physico-chemical properties should be well known.

3.2 ADVANTAGES AND DISADVANTAGES OF CHELATING SILICAS IN PRECONCENTRATING METALS

They offer an important advantage: rapid metal complexing kinetics (ref. 12). Consequently, preconcentration flow rates of up to 50 ml/min may be achieved (refs. 21-23).

Another advantage is the possibility of preconcentrating directly at the sampling site, thus transportation, contamination and sample storage problems may be eliminated (ref. 45).

However, care should be taken to avoid metal impurities, inate to silicas or introduced by reagents used in crossbinding, since they may cause interference with the preconcentration (ref. 60).

Although chelating silicas usually have lower retention capacities than do complexing resins, metals may be preconcentrated from several liters of solution and therefore this is not a major drawback.

3.3 ANALYSIS OF PRECONCENTRATED METALS

Metals which have been preconcentrated on the complexing silicas may be analyzed directly using the following methods: X-ray fluorescence, ESCA, photoacoustics spectroscopy, neutronic activation and IR spectrometry. Alternatively, analyses may be performed by classical techniques once the metal has been eluted from the support: atomic absorption with or without flame, complexometry and colorimetry. The drawbacks of this technique are the difficulty in eluting the strongly retained metal and secondly these procedures are time-consuming.

Analytical techniques used for metal determinations are listed in Table III. X-ray fluorescence is often used since low detection limits may be achieved. Concentrations at ppb levels may be determined in the absence of interferences.

TABLE III   Analysis of preconcentrated metals

Complexing silica	Analysis	Metals (detection)	Ref.
	RFX	-Zn, Cu, Co (10 ppm)	21
	RFX	-Zn, Cu, Hg, Cr, Pb, Mn (.005-10 ppm)	22
$Si-O-Si(CH_2)_3NH(CH_2)_2NH_2$	RFX	-AsO_4, SeO_4, MnO_4, Cr_2O_7, MoO_4, WO_4, VO_4 (1-10 ppb)	23
	RFX	-phosphates (20 ppb)	44
$Si-O-Si(CH_2)_3N(CH_3)CS_2Na$	RFX	-Co, Cu, Ni, Pb, Hg Zn, Mn (.5-5ppm)	21
	RFX	-Cu, Zn, Ni, Mn, Pb, Fe(.5-10ppb)	21
$Si-O-Si(CH_2)_3N(CH_2)_2NHCS_2Na$ CS_2Na	ESCA	-Pb, Hg(10ppb)	1
	ESCA	-Pb(.02-20 ppm)	20
$Si-O-Si(CH_2)_3NHCO$	Colorimetry	-Fe, Cu(1ppb)	52
	AAS	-Cd, Pb, Zn, Cu, Fe, Mn, Ni, Co(.03-1ppb)	57
$Si-O-Si(CH_2)_3NH(CH_2)_2NH_2$	Photoacous. spectr.	-MoO_4, WO_4, Cu	17-24

4. THE PRECONCENTRATION OF ORGANIC COMPOUNDS USING CHELATING SILICAS

A recent application of chelating silicas consists of preconcentrating organic compounds from complex natural systems followed by a chromatographic analysis. In fact, a large number of these organic products are found at trace levels necessitating a preconcentration step. Purification and enrichment of these compounds have usually been accomplished by liquid--liquid extraction techniques.

However, these techniques are laborious, time-consuming and may be the origin of product loss or sample contamination. Most of these problems have been eliminated by replacing liquid-liquid extraction with liquid-solid extraction where silicates are employed as the solid surface (cf. 1.1). The use of nonpolar silicas for the preconcentration of nonpolar compounds from aqueous solutions was first mentioned in 1974 (ref. 61). The strong hydrophobic interaction between the nonpolar organic products and the nonpolar silicas made it possible to preconcentrate them from large volumes of polar solvents such as water (from 1 to 100 ml). By simply changing the polarity of the mobile phase, the preconcentrated compounds could then be eluted in a small volume of solvent. Due to the high retention capacity of these silicas, only a small quantity was necessary to make precolumns (only a few millimeters long) for preconcentration purposes (ref. 62). These small precolumns offer many advantages:
- negligeable pressure drops
- requires small phase volume
- low packing costs
- low dead volumes resulting from coupling the pre-column to the analytical column.

Thus, it is possible to couple a precolumn with an analytical column (on-line) by merely using a high pressure valve as shown in Figure 1.

Using this technique, alkyl-chain cross-bonded silicas were first employed to retain organic compounds. The use of this technique has been reviewed by several authors (refs. 63-65).

Two major problems may be encountered upon using apolar silicas for preconcentration purposes:
1.) lack of selectivity
2.) poor retention of polar organic products.

To overcome these two problems, the utilisation of complexing silicas loaded with metal has proven to be promising. These silicas selectively retain compounds containing an electron-pair donor heteroatom (Lewis base) according to the order given by the ligand exchange principle.

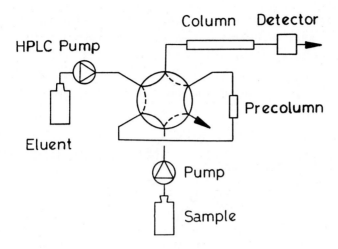

Fig. 1   On-line coupling of the enrichment with the chromatographic
         separation. (——) Enrichment step; (--) elution and chromato-
         graphic separation.

Iminodiacetate (IDA) silica loaded with copper was the first
complexing silica loaded with metal to be studied for its retention
properties of organic compounds (ref. 66). This silica strongly retains
catecholamines from aqueous solutions and then releases them when eluted
with acidic methanol solutions. Separation may then be directly performed
by ion-pair chromatography using a C18 silica analytical column.

Subsequently, other metal loaded complexing silicas have also been
developed. A silica containing the functional group 2-amino-1-cyclopen-
tene-1-dithiocarboxylic acid loaded with platinum irreversibly retains
anilines from water (ref. 67). In this case, this silica serves as a
filter, retaining aniline which would interfere with the ensuing
herbicide analysis. The mercury loaded oxine surface is another example
of a metal loaded support. It preconcentrates 2-mercaptobenzimidazole
(ref. 68) from aqueous solutions following two mechanisms:

1.)   complexation with the mercury

2.)   hydrophobic interaction between the aromatic rings of the
       complexing group and the solute.

The solute is then eluted by exchanging it with a stronger mercury
complexing ligand. A C18 silica column is then employed for the
subsequent separation.

Other examples include the preconcentration of amino and carboxylic acids from complex environmental samples using bisdithiocarbamate, dialkyldithiocarbamate, cyclam and oxine copper-loaded chelating silicas (refs. 68-72) as shown in Fig. 2. These acids are in fact strongly complexed with copper which explains their retention on the support.

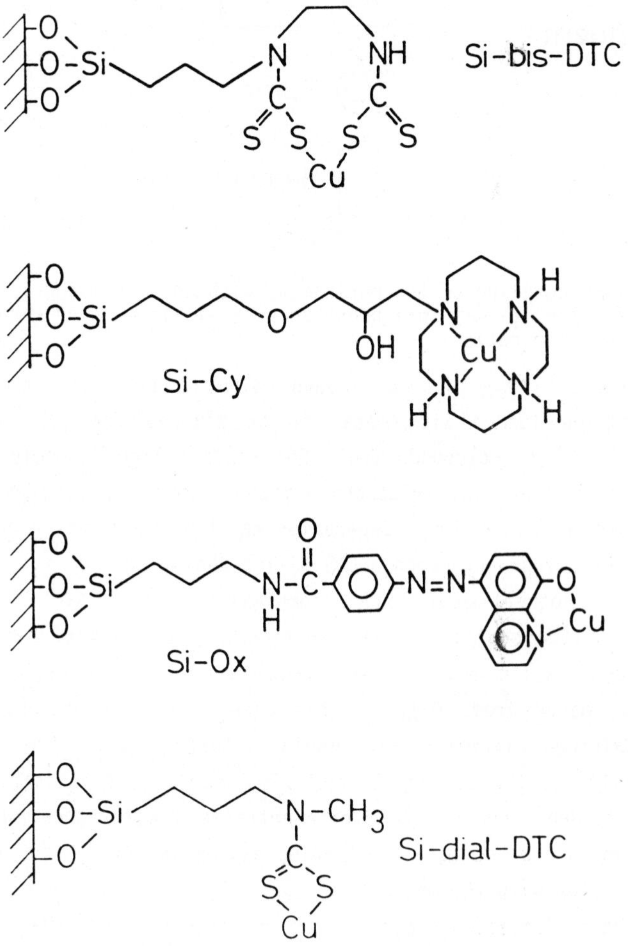

Fig. 2   Copper-loaded chelating silicas: bisdithiocarbamate (Si-bis-DTC), cyclam (Si-Cy), oxine (Si-Ox) and dialkyl-dithiocarbamate (Si-dial-DTC).

Acidification of the eluent allows the copper complexed compounds to be released onto the analytical column. Separation of these compounds is then acomplished by ion-pair chromatography using C18 silica columns. Fig. 3 shows chromatograms obtained after selectively preconcentrating amino acids onto copper loaded dithiocarbamate silicas from different natural systems. Enrichment factors of up to a thousandfold may be obtained from this step when compared with the direct injection technique. Using this technique, amino acid concentrations as low as the ppb range may be detected.

5. CHELATING SILICAS IN CHROMATOGRAPHY

Apart from preconcentration, chelating silicas may find applications in metal chromatography or, if they are saturated in metal, they can be used in ligand exchange chromatography.

5.1 METAL SEPARATION CHROMATOGRAPHY

Numerous articles dealing with the separation of organic compounds by liquid chromatography have been published, but little information is available regarding the separation of metals. This technique is suitable for metal separation whereas other chromatographic techniques, such as gas chromatography, are inapplicable for this purpose.

Among the chromatographic techniques, ionic and ion-exchange chromatography are the most often employed for metal separation. In these techniques, the stationary phase consists of resins. More recently, ionic-exchange functional groups attached to silicas have been substituted for the resins. Several review articles have been published concerning this subject (refs. 73-75).

A less commonly employed chromatographic method involves the use of chelating silicas as the stationary phase. One of the drawbacks of this technique is the difficulty in eluting the metal since strong bonds are formed between the metal and the ligand. Provided that this problem is overcome better metal selectivity may be obtained. One way of circumventing this problem is by varying the pH of the mobile phase. The pH plays an important role in the metal binding strength of the ligand.

Oxine silicas are prime examples of complexing silicas influenced by pH. These silicas were first mentioned in 1979 (ref. 53) and are given in Table II (silica 8). pH adjustment is necessary for a proper separation. If the solution is too acidic, leaching occurs, and the metal is not retained whatsoever. Adversely, if the solution is too basic the metals are not eluted. Hence an intermediary pH gradient has been frequently recommended. In fact, different metals (Mn, Cd, Pb, Zn, and Co) in aqueous solutions have been separated using this method.

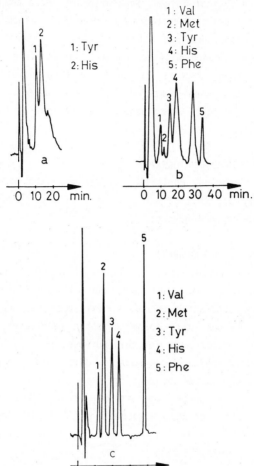

Fig. 3   Enrichment-chromatographic separation of natural solutions
containing amino acids at ppb level. a: urine sample; b: serum
sample; c: river water sample.
Preconcentration step: a) 5 ml of 2000 times diluted urine (>100
ppm for each amino acid) in distilled water. b) 5 ml of 250 times
diluted serum (Monitrol I-E, Merz + Dade , 10-30 ppm) c) 5 ml of
spiked river water (60 ppb of Val, Met and Phe, 50 ppb of Tyr
and His), all enriched on Si-diol-DTC-Cu precolumn (10 x 2 mm
I.D.; particle size 10 μm).
Separation step: Column: Lichrosorb RP-18, 5 μm (125 x 4.6 mm
I.D.) Gradient elution of solvents A and B.
A: 50/50 (v/v) mixture of methanol and water containing
dinatrium citrate 0.04 M, lauryl sulfate 4.10^{-4} M. pH of the
mixture adjusted at 2.25 with concentrated HCl.
B: 10/90 (v/v) of the same mixture adjusted at pH = 2.25 with
concentrated HCl.
Fluorimetric detection λex = 345 nm; λem > 460 nm.

Another factor affecting metal separation in this chromatographic technique is metal-ligand complex formation kinetics. Slow kinetics are reflected by broadening of elution peaks. This is observed in the case of nickel with oxine silica stationary phases.

Before the development of HPLC, thin-layer chromatography was used for the separation of metal ions. Eight different types of silicas were tested for the separation of iron, nickel, copper and zinc. Best results were obtained with a diacetone silica (see No. 12 in Table II). The eluant in this case was a mixture of acetone and trifluoroacetyl-acetone (ref. 16). Diamine silicas (ref. 76) were also found to be suitable for the separation of numerous metals (Cu, Cd, Fe, Mn, Ni, Pd, Zn).

Before discussing the development and application of these chelating silicas in detail, a brief mention should be made concerning high performance liquid chromatography (HPLC) of metal complexes. The metal separation by means of this technique is achieved by forming organo-metallic complexes in the analytical solution which is then introduced into a reversed-phase silica column. Details concerning this technique are described in several reviews (refs. 73, 77, 78).

5.2   LIGAND EXCHANGE CHROMATOGRAPHY

The application of chelating silicas to ligand exchange chromatography is currently being widely developed. Although this technique has been known for the past 20 years, it has only recently been applied to the separation of enantiomers. Several chelating silicas containing different metals have been developed for this purpose.

5.2.1   LIGAND EXCHANGE PRINCIPLES

The term "ligand exchange" was first coined in 1961 by Helfferich (ref. 79). It refers to the process occurring between the metal and the ligand i.e. organo-metallic complexes are first formed and then dissociated by the mobile phase. In most cases, the metal is bound to the stationary support while the ligand is found in the mobile phase. Sometimes, the metal may be found both in the stationary phase as well as the mobile phase. In this case, complexing silicas do not offer any particular advantage.

Transition metals such as Cu, Zn, Cd, Co, etc. bind with these chelating silicas. These metals usually form complexes with compounds containing electron donor functional groups (Lewis bases). In contrast to ion exchange where the interaction takes place in the outer coordination sphere, ligand exchange takes place in the inner coordination sphere of the metal.

This technique offers a number of advantages as opposed to other chromatographic procedures. First of all, the possibility of varying the metal in the stationary phase makes it possible to easily optimize the separation of a large number of solutes. Secondly, the chromatography may be performed at very low solute concentrations, due to the high stability of the ligand-metal complex. This is particularly useful in separating ligands in natural aquatic systems. Thirdly, ligands having different coordination numbers as well as isomers and isotopes may be separated.

This technique has been optimized to separate amines, carboxylic acids, alcohols as well as other compounds. A review, published in 1977 (80), may serve as a standard reference for all practical purposes.

5.2.2   TYPES OF STATIONARY PHASES

Choice of the support should be such that the metal remains irreversibly bound i.e. either by forming an ionic or covalent bond. Sulfonic ion exchange resins were among the first of these supports tested but the bonds were not strong enough to retain the metals. Subsequently, other complexing resins (eg. iminodiacetate, phosphonate, etc.) showing higher metal retention were synthesized. However, as in the case of preconcentration, these resins have two major disadvantages:
- they are deformed under applied pressures
- their exchange kinetics are very slow.
Consequently, they have received little interest.

More recently, metal and chelating silicas have been employed as supports. Chelating silicas have proved to be the most promising since they offer several advantages, namely:
- high metal retention capacity
- high mechanical stability
- fast exchange kinetics
- relatively low cost.
These latter two types of supports are discussed in some detail.

5.2.2.1 METAL SILICAS

A number of metals have been bound to silicas for ligand exchange purposes. Silver was the first metal to be bound with silicas (refs. 87-90). It was used for separating several isomers (olefins, unsaturated esters, unsaturated organic acids and prostaglandins). Other examples include the use of rhodium for separating butene isomers (ref. 87); cadmium silicas for separating chloroaniline and toluidine isomers (refs. 91-92); and copper-silicas for separating amino acids and peptides (refs. 81-84).

Copper silicas have received growing interest due to their ability of separating analytically important compounds such as amino acids and peptides. The preparation of these copper-silicas is therefore schematized below.

SCHEME II

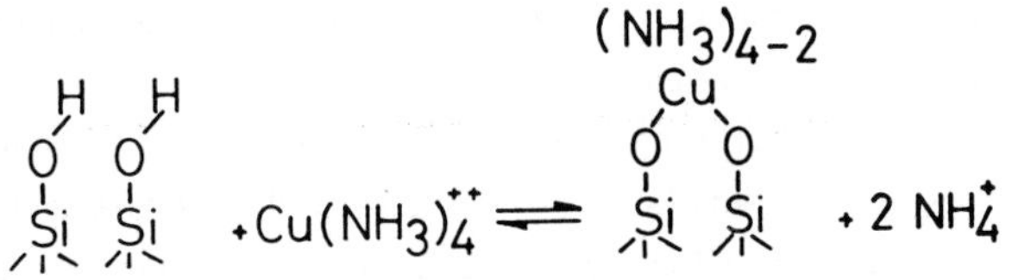

$$\begin{array}{c} H \quad H \\ O \quad O \\ Si \quad Si \end{array} + Cu(NH_3)_4^{++} \rightleftharpoons \begin{array}{c} (NH_3)_4-2 \\ Cu \\ O \quad O \\ Si \quad Si \end{array} + 2\ NH_4^+$$

The metal silica is simply prepared by percolating an ammoniacal solution of copper salt through the silica (ref. 75). The stability, structure and retention mechanisms of these silicas have been thoroughly studied (ref. 85).

The retention capacity of these silicas is approximately 0.6 mmole of copper per gram of silica. This kind of silica was first used in a 15 cm long column to separate 15 natural amino acids (ref. 81). Subsequently, they were used to separate purine and pyrimidine bases found in biological media and for the separation of osamines (ref. 85). More recently, these supports have found applications in the separation of hydrolysis products of proteins, amino acids and dipeptides (ref. 86).

These silicas have one main disadvantage in that they lose part of the bonded metals during elution. Consequently, in order to replace the lost metal, a small quantity of copper is added to the eluent.

5.2.2.2 COMPLEXING METAL SILICAS

The first complexing silica saturated with metal to be used for ligand exchange chromatography was first mentioned in 1977 (ref. 93). It consisted of a propylamine functional group which was saturated with copper. However, the complexing bond was not strong enough to be employed in aqueous solutions.

To resolve this problem, it was necessary to find another functional group capable of forming a stronger bond with the metal. Bidentated ligands, eg. diamines, were chosen for this purpose. Diamine silicas had already been used for preconcentration applications (ref. 14). These silicas were loaded with cadmium for chromatographic ligand separation. In spite of the fact that copper or nickel form stronger bonds with the diamine, faster separations as well as finer chromatographic peaks may be obtained with cadmium. Sulfurcontaining drugs have been separated by this means (ref. 94).

Nevertheless, copper-loaded diamine silicas have been employed for the separation of amino acids (ref. 95). Copper loss may be incurred if proper pH conditions are not selected. Minimum copper loss occurs over the pH range 5 to 9. In addition, this loss may be compensated by adding a small quantity of copper to the mobile phase.

Subsequently, other chelating agents (ref. 96) have been used which bond more strongly with metals. For instance, dithiocarbamate silicas (silica 2 in Table II) and diketone silicas have been synthesized to bind copper. These copper-loaded silicas enable the separation of aromatic amines (ref. 96). The metal loss in this system is considerably lower than that incurred with the diamine silicas.

The preparation of metal chelating silicas, their stability and their application to ligand exchange chromatography has been described in detail (ref. 13). For example, copper-loaded silicas 6 and 7 (Table I) are suitable for separation of carboxylic acids, amino acids as well as triphenylphosphonium salts. These separations are achieved both through metal-ligand complexation and through electrostatic interactions.

Other examples of metal-loaded silicas include iron(III)-loaded oxine silicas which have been utilized to separate phenols in aqueous solutions (ref. 56). In this case iron is irreversibly retained by the silicas and therefore no metal losses are incurred in the mobile phase. The phenols are retained through complex formation with the iron and by hydrophobic interactions with the cross-bonded hydrocarbons of the silica.

More recently new chelating silicas (ref. 97) are being developed to separate dialkyl sulfides. These copper-loaded chelating silicas include the following functional groups: diamines and monoamines, oxines as well as two other special functional groups (Fig. 4).

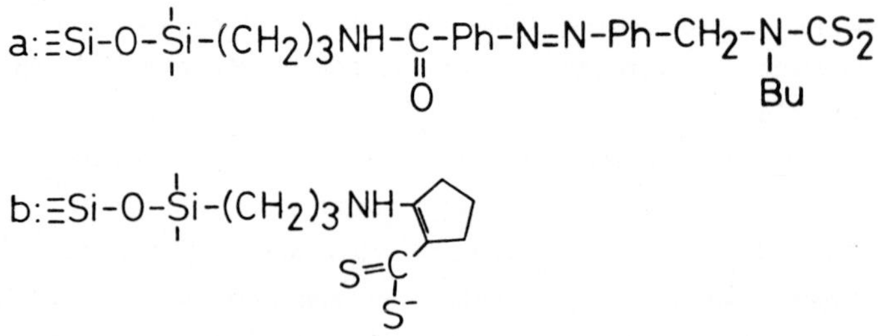

$$a: \equiv Si-O-\overset{|}{\underset{|}{Si}}-(CH_2)_3NH-\overset{O}{\overset{\|}{C}}-Ph-N=N-Ph-CH_2-\overset{|}{\underset{Bu}{N}}-CS_2^-$$

$$b: \equiv Si-O-\overset{|}{\underset{|}{Si}}-(CH_2)_3NH-\overset{}{\underset{S=\overset{}{\underset{S^-}{C}}}{}}$$

Fig. 4   Chelating silicas; a: N-n-butyl-N-(1-naphthylmethyl)carbo-
dithioate (BNMDTC), b: 2-amino-1-cyclopentene-1-dithiocar-
boxylic acid (ACDA) (ref. 97).

An excellent separation of sulfur compounds in hexane is achieved with the copper-loaded 2-amino-1-cyclopentene-1-dithiocarboxylic acid (ACDA) silicas. The separation is based on:
- copper complexation
- steric hindrance
- hydrophobic interactions

By replacing copper with silver, separation of geometric sulfur isomers is possible using this support.

All of these metal complexing silicas have two major problems:
- low column efficiency
- asymetric peaks.

These problems are due to slow complexing and dissociation kinetics in the internal coordination sphere. To overcome this problem, a separation based on outer sphere complex formation seems to be more promising (ref. 94) as the formation and dissociation kinetics are much more rapid. Cobalt-loaded diamine silicas in which all of the coordination sites of the metal are occupied were the first species to be used for outer sphere coordination (ref. 98) (see Fig. 5). Negatively charged species such as phosphates, nucleotides, etc. may be separated with this phase. This silica has high stability in aqueous solutions. Using the same reasoning, a macrocyclic chelating group was cross-bonded to a silica. Copper was then trapped within the cycle and therefore was inaccessible to the ligands. Hence, only the outer coordination sphere was available (ref. 71).

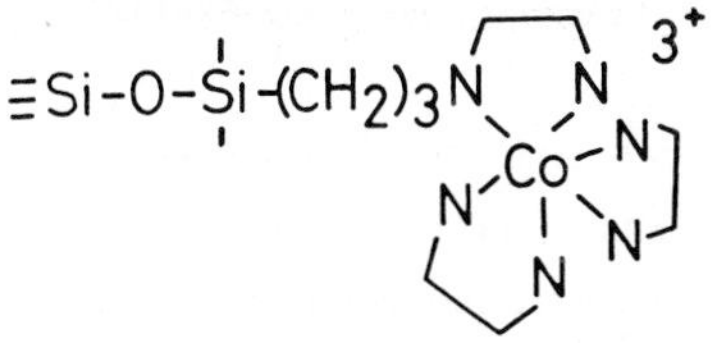

Fig. 5   Bonded silica containing a $Co(en)_3^{3+}$ moiety (ref. 98).

5.3   ENANTIOMER SEPARATION

Recently, there has been an enormous development for the application of metal complexing silicas in separating enantiomers by ligand exchange chromatography. Polystyrene resins were first employed to separate enantiomers (refs. 99-103). Later, in 1979 (ref. 104), chelating silicas containing a chiral center (see Fig. 6) were found to be suitable for enantiomer separation. Once loaded with copper this silica was able to separate L and D isomers of tryptophane as well as tyrosine. A similar type of silica was synthesized in the same year (ref. 83) for separating amino acid enantiomers.

$$a: \equiv Si-O-\overset{|}{\underset{|}{Si}}-(CH_2)_3OCH_2\overset{|}{\underset{OH}{C}}HCH_2\overset{*}{N}-COOH$$

$$b: \equiv Si-O-\overset{|}{\underset{|}{Si}}-(CH_2)_3NH\overset{*}{\underset{O}{C}}-NH$$

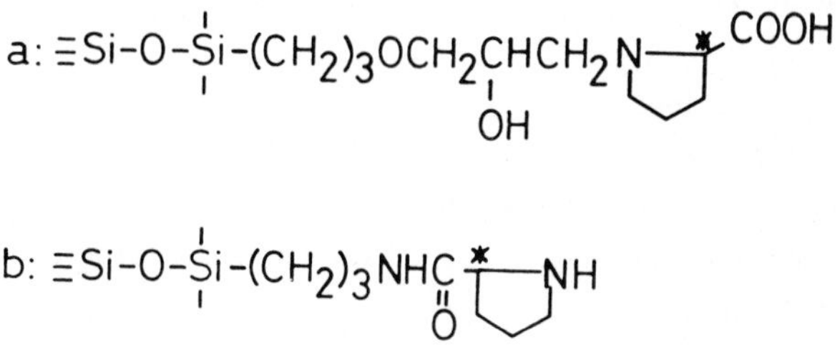

Fig. 6   Chelating silicas containing a chiral center (*). L-Proline is the functional group bonded to the silica. a: (ref. 104), b: (ref. 83).

These two silicas differ in the way in which the functional group, L-proline, is bonded to the silica structure.

Several reviews concerning enantiomer separation, particularly amino acids, using ligand exchange chromatography (refs. 105-107) have recently been published.

Enantiomer separation necessitates the presence of a cross-bonded functional group containing a chiral center (asymetric carbon) in the chelating silica. Optically active amino acids containing a chiral center are then bonded strongly with transition metals. Consequently, they are suitable for cross-linking to the silicas.

The separation of two enantiomers is based on the three point rule (ref. 108), i.e. a three-point bond between the complexating silica and the solute to be analyzed.

Numerous chiral complexing silicas are now commercially available, namely: L-proline, L-hydroxyproline and L-valine (ref. 109). Other amino acids as well as hydroxy-carboxylic acids have also been cross-bonded to silicas (refs. 110-112). All of these silicas when loaded in copper permit the separation of amino acid enantiomers, catecholamines and hydroxycarboxylic acids.

More recently, copper-loaded silicas containing L-proline functional groups have been applied to the separation of primary amine alcohols. However, it requires a derivatization step with a Schiff base (ref. 113).

Finally, a more recent development is the synthesis of a polystyrene silica containing two different amino acid functional groups, L-proline and L-hydroxyproline (ref. 114). This type of silica is more hydrophobic than those without polystyrene functional groups. Like the other silicas, they are also loaded in copper.

6. CONCLUSION

As this discussion has shown, a wide variety of applications have been found for chelating silicas. These silicas offer many advantages, such as:
- good physico-chemical properties
- modification is relatively simple
- low cost

Consequently, they have been chosen for the purpose of preconcentration, liquid-solid extraction as well as high performance chromatography.

They have received increasing interest in the field of enantiomer separation using ligand exchange chromatography. This technique has become essential to the purification and the analysis of optically active compounds in the areas of organic chemistry, pharmacy and biochemistry.

The preconcentration of organic products by liquid-solid extraction is a field which has not been thoroughly studied as of yet. New silicas have recently been developed for this purpose. These silicas contain a metal which selectively retains Lewis bases such as amino acids. Hence, for example, on-line amino acid analyses can be carried out by simply coupling the metal silica preconcentration columns to an analytical column. This technique offers two advantages, namely:
- analyses of trace organics in natural systems.
- automation.

The use of chelating silicas as stationary phase for the chromatography of metals, however, does not seem to be suitable owing to too many interferences. Morover, simpler techniques such as ion chromatographic techniques using silicas as well as complex chromatography using reversed phases are available for metal analysis.

7. ACKNOWLEDGEMENTS

The authors wish to thank M.L. Pelaprat and N. Parthesarathy for their comments.

REFERENCES

1. D.M. Hercules, L. E. Cox, S. Onisick, G.D. Nichols, J.C. Carver, Anal. Chem., 45 (1973) 1973.
2. E. Grushka (Editor) Bonded Stationary Phases in Chromatography, Ann Arbor Sci. Publ., Ann Arbor, Mich., 1974.

30

3. G.V. Lisichkin, A. Ya. Yuffa, Heterogeneous Complex-Metal Catalyst, Khimiya, Moscow (1981).
4. W.A. Aue, P.M. Teli, J. Chromatogr., 62 (1971) 15.
5. M.D. Mattenci, M.H. Caruthers, Tetrahedron Lett. 21 (1980) 719.
6. A.I. Kestner, Usp. Khim., 43 (1974) 1480.
7. R.S.S. Murthy, J. Holzbecher, D.E. Ryan, Rev. Anal. Chem., 6 (1982) 113.
8. K.K. Unger, N. Becker, P. Roumeliotis, J. Chromatogr., 125 (1976) 115.
9. H. Engelhardt, G. Ahr, Chromatographia, 14 (1981) 227.
10. G.E. Berendsen, L. De Galan, J. Chromatogr., 196 (1980) 21.
11. K.K. Unger, Porous Silica, Elsevier (1979).
12. G.V. Lisichkin, G.V. Kudryavtsev, Zh. Anal. Khim., 38 (1983) 1684.
13. M. Gimpel, K.K. Unger, Chromatographia, 16 (1982) 117.
14. D.E. Leyden, G.H. Luttrel, Anal. Chem., 47 (1975) 1612.
15 M. Okamoto, J. Chromatogr., 202 (1980) 55.
16. K.T. Den Bleyker, T.R. Swett, Chromatographia, 13 (1980) 114.
17. J.M. Hill, J. Chromatogr., 76 (1973) 455.
18. M.Okamoto, J. Chromatogr., 212 (1981) 251.
19. T. Seshadri, A. Kettrup, Fresenius Z. Anal. Chem., 296 (1979) 247.
20. H. Jenett, J. Knecht, G. Stork, ibid. 304 (1980) 362.
21. D.E. Leyden, G.H. Luttrel, A.E. Sloan, N.J. de Angelis, Anal. Chim. Acta, 84 (1976) 97.
22. D.E. Leyden, G.H. Luttrel, T.A. Patterson, Anal. Letters 8 (1975) 51.
23. D.E. Leyden, G.H. Luttrel, W.K. Nonidez, D.B. Werho, Anal. Chem. 48 (1976) 67.
24. D.W. Leyden, M.L. Steele, B.B. Jablonski, R.B. Somoano, Anal. Chim. Acta, 100 (1978) 545.
25. S.E. Northcott, D.E. Leyden, Anal. Chim. Acta, 126 (1981) 117.
26. L.W. Burrgraf, D.S. Kendall, D.E. Leyden, F.J. Pern, Anal. Chim Acta, 129 (1981) 19.
27. R.J. Kvitek, M.W. Watson, J.F. Evans, P.W. Carr, ibid. 129 (1981) 269.
28. T.G. Waddel, D.E. Leyden, D.M. Hercules, Middl. Macrom. Monogr., 7 (1980) 55.
29. G. Gubitz, F. Juffmann, W. Jellenz, Chromatographia, 16 (1982) 103.
30. G. Gubitz, W. Jellenz, W. Santi, J. Liq. Chrom. 4 (1981) 701.
31. S.H. Chang, K.M. Gooding, F.E. Regnier, J. Chromatogr., 120 (1976) 321.
32. M. Glad, S.Ohlson, L. Hansson, M.O. Mansson, K. Mosbach, ibid 200 (1980) 254.
33. F.E. Regnier, R. Noel, J. Chrom. Sci., 14 (1976) 316.
34. D.E. Schmidt, J.R.W. Grese, D. Couron, B.L. Karger, Anal. Chem., 52 (1980) 177.
35. T.G. Waddel, D.E. Leyden, J. Org. Chem., 46 (1981) 2406.
36. P.A. Asmus, C.E. Low, M. Novotny, J. Chromatogr. 119 (1976) 25.
37. P.A. Asmus, C.E. Low, M. Novotny, ibid., 123 (1976) 109.
38. M. Novotny, S.L. Bektesh, K.B. Denson, Anal. Chem., 45 (1973) 971.
39. M. Novotny, S.L. Bektesh, K. Grohmann, J. Chromatogr., 83 (1973) 25.
40. E. Grushka, E.J. Kikta, Anal. Chem., 46 (1974) 1370.
41. E. Grushka, R.P.W. Scott, ibid., 45 (1973) 1626.
42. M.A. Marshall, H.A. Mottola, ibid., 55 (1983) 2089.
43. D.E. Leyden, Middl. Macrom. Monogr., 7 (1980) 321.
44. D.E. Leyden, W.K. Nonidez, P.W. Carr, Anal. Chem., 47 (1975) 1449.
45. J.C. Meranger, K.S. Subramanian, C.H. Langford, Rev. Anal. Chem., 5 (1980) 29.
46. A.T. Ellis, D.E. Leyden, W. Wegsheider, B.B. Jablonski, W. Bodnar, Anal. Chim. Acta., 142 (1982) 73.
47. J. Smits, J. Nelissen, R. Van Grieken, ibid., 111 (1979) 215.
48. T. Yao, M. Akino, S. Muska, Bunseki Kaguku, 31 (1982) 409.

49. T. Yao, M. Akino, S. Muska, ibid., 30 (1981) 740.
50. P. Grossmann, W. Simon, J. Chromatogr., 235 (1982) 351.
51. T. Seshadri, A. Kettrup, Fresenius Z. Anal. Chem., 310 (1982) 1
52. K.F. Sugawara, H.H. Weetall, G.D. Schucker, Anal. Chem., 46 (1974) 489.
53. J.R. Jezorek, H. Freiser, ibid., 51 (1979) 366.
54. C. Fulcher, M.A. Crowell, R. Bayliss, K.B. Holland, J.R. Jezorek, Anal. Chim. Acta, 129 (1981) 29.
55. J.R. Jezorek, C. Fulcher, M.A. Crowell, R. Bayliss, B. Greenwood, J. Lyon, ibid., 131 (1981) 223.
56. G.J. Shahwan, J.R. Jezorek, J. Chromatogr., 256 (1983) 39.
57. R.E. Sturgeon, S.S. Berman, S.N. Willie, J.A.H. Desaulniers, Anal. Chem., 53 (1981) 2337.
58. J. Chmielowiec, Chromatographia, 11 (1978) 99.
59. J.R. Jezorek, K.H. Faltynski, L.G. Blackburn, P.J. Henderson, H.D. Medina, Talanta, 32 (1985) 763.
60. M. Verzele, M. De Potter, J. Ghysels, HRC & CC 2 (1979) 151.
61. J.J. Kirkland, Analyst 99 (1974) 859.
62. H.P.M. Van Vliet, T.C. Bootsman, R.W. Frei, U.A.T. Brinkman, J. Chromatogr., 185 (1979) 483.
63. W.A. Saner, Trace Anal., 2 (1982) 151.
64. R.W. Frei, U.A.T. Brinkman, TrAC, 1 (1981) 45.
65. R.W. Frei, Swiss Chem., 6 (1984) 55.
66. C.E. Goewie, Thesis, Free University Amsterdam (1983).
67. C.E. Goewie, P. Kwakman, R.W. Frei, U.A.T. Brinkman, W. Maasfeld, T. Seshadri, A. Kettrup, J. Chromatogr., 284 (1984) 73.
68. M.W.F. Nielen, R. Bleeker, R.W. Frei, U.A.T. Brinkman, ibid., 358 (1986) 393.
69. J.L. Veuthey, M.A. Bagnoud, W. Haerdi, Int. J. Env. Anal. Chem., 26 (1986) 157.
70. J.L. Veuthey, M.A. Bagnoud, W. Haerdi, Chimia, 40 (1986) 353.
71 M.A. Bagnoud, J.L. Veuthey, W. Haerdi, Chimia, 40 (1986) 432.
72. J.L. Veuthey, M.A. Bagnoud, W. Haerdi, J. Chromatogr., 393 (1986) 51.
73. R.M. Cassidy, Trace Anal., 1 (1981) 121.
74. H.F. Walton, J. Chrom. Libr., 22 (1983) A225.
75. R. Rosset, M. Caude, A. Jardy, Manuel Practique de Chromatographie en Phase Liquide, Masson (1982).
76. J.B. Henry, T.R. Sweet, Chromatographia, 17 (1983) 79.
77. G. Schwedt, ibid., 12 (1979) 613.
78. B.R. Willeford, H. Veening, J. Chromatogr., 251 (1982) 61.
79. F. Helfferich, Nature (London) 189 (1961) 1001.
80. V.A. Davankov, A.V. Semechkin, J. Chromatogr., 141 (1977) 313.
81. M. Caude, A. Foucault, Anal. Chem., 51 (1979) 459.
82. E. Schmidt, A. Foucault, M. Caude, R. Rosset, Analysis, 7 (1979) 366.
83. A. Foucault, M. Caude, L. Oliveros, J. Chromatogr., 185 (1979) 345.
84. F. Guyon, A. Foucault, M. Caude, ibid., 186 (1979) 677.
85. F. Guyon, M. Caude, R. Rosset, Analysis, 12 (1984) 321.
86. A. Foucault, R. Rosset, J. Chromatogr., 317 (1984) 41.
87. F. Mikes, V. Schurig, E. Gil-Av, ibid., 83 (1973) 91.
88. R. Aigner, H. Spitzy, R.W. Frei, Anal. Chem., 48 (1976) 2.
89. R.R. Heath, P.E. Sonnet, J. Liq. Chromatogr., 3 (1980) 1129.
90. R. Vivilecchia, M. Thiebaud, R.W. Frei, J. Chrom. Sci., 10 (1972) 411.
91. D. Kunzru, R.W. Frei, ibid., 12 (1974) 191.
92. C. R. Vogt, T.R. Ryan, J.S. Baxter, J. Chromatogr., 136 (1977) 221.
93. F.K. Chow, E. Grushka, Anal. Chem., 49 (1977) 1756.
94. N.H. Cooke, R.L. Viavattene, R. Eksteen, W.S. Wong, G. Davies, B.L. Karger, J. Chromatogr., 149 (1978) 391.

32

95. R.G. Masters, D.E. Leyden, Anal. Chim. Acta, 98 (1978) 9.
96. F.K. Chow, E. Grushka, Anal. Chem., 50 (1978) 1346.
97. H. Takayanagi, O. Hatano, K. Fujimara, T. Ando, Anal. Chem., 57 (1985) 1840.
98. F.K. Chow, E. Grushka, J. Chromatogr., 185 (1979) 361.
99. V.A. Davankov, S.V. Roghozhin, J. Chromatogr., 60 (1971) 280.
100. V.A. Davankov, S.V. Roghozhin, A.V. Semechkin, T.P. Sachkova, ibid., 82 (1973) 359.
101. V.A. Davankov, Y.A. Zolotarev, ibid., 155 (1978) 285.
102. V.A. Davankov, Y.A. Zolotarev, ibid., 155 (1978) 295.
103. V.A. Davankov, Y.A. Zolotarev, ibid., 155 (1978) 303.
104. G. Gubitz, W. Jellenz. G. Löfler, W. Santi, HRC & CC 2 (1979) 145.
105. W. Lindner, Chimia, 35 (1981) 294.
106. V.A. Davankov, A.A. Kurganov, A.S. Bochkov, Adv. Chrom., Vol. 22, M. Dekker (1983).
107. G. Gubitz, G.I.T. Fachz. Lab. Suppl. Chrom., 4 (1985) 6.
108. C.E. Dalgliesh, J. Chem. Soc., (1952) 3940.
109. R. Däppen, H. Arm, V. Meyer, ibid 373 (1986) 21.
110. G. Gubitz, S. Mihellyes, Chromatographia, 19 (1984) 257.
111. N. Watanabe, J. Chromatogr., 260 (1983) 75.
112. H.G. Kicinski, A. Kettrup, Fresenius Z. Anal. Chem., 320 (1985) 51.
113. L.R. Gelber, B.L. Karger, J.L. Neumeyer, B. Feibush, J. Am. Chem. Soc., 106 (1984) 7729.
114. A.A. Kurganov, A.B. Tevlin, V.A. Davankov, J. Chromatogr., 261 (1983) 223.

CHAPTER II

SAMPLE HANDLING IN ION CHROMATOGRAPHY

P. R. HADDAD

1.   INTRODUCTION

The term "Ion Chromatography" was first introduced in 1975 and was originally interpreted to mean the chromatographic analysis (by ion-exchange) of inorganic ions. Throughout the early years of its development, ion chromatography was concerned chiefly with the determination of inorganic anions because it provided the most convenient means of analysis for these species. However, the technique now embraces a much wider range of solutes and separation methods and some overlap with alternative liquid chromatographic methods currently exists. It is fair to say that ion chromatography can be considered to include any liquid chromatographic procedure used for the determination of ionic and ionisable solutes. Fig. 1 shows the main separation modes applicable to ion chromatography.

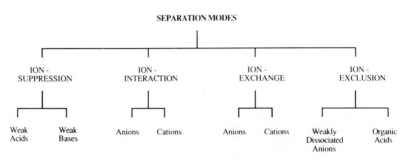

Fig. 1   Separation modes for ion chromatography.

Ion suppression involves adjusting the pH of the mobile phase so that the solutes of interest are non-ionised (or at least only partially ionised). In this form, these solutes can be separated by conventional reversed-phase liquid chromatography and the technique of ion suppression is most successful for weak acids or bases. Ion-interaction methods are applied to the separation of fully ionised solutes by reversed phase HPLC and involve addition to the mobile phase of a relatively hydrophobic ionic species with a charge opposite to that of the solutes of interest. This species is called the "ion-interaction reagent" and serves to convert the non-polar stationary phase into an ion-exchanger suitable for the separation of the desired ionic solutes. Ion-exchange chromatography is a well established technique in which the stationary phase contains ionic functionalities and the solute ions are separated using an eluent which contains a competing ion of the same charge sign as the solute ions. Classical ion-exchangers are not suitable for ion chromatography because their high ion-exchange capacities require the use of high ionic strength eluents, and this creates problems with detection of the eluted ions. For this reason, a new generation of high efficiency, low capacity ion-exchangers has evolved for ion chromatography. It has been common to subdivide ion-exchange methods into those using chemical means to sup-press the background conductance of the eluent, and those using electronic means to achieve the same result. Since this classification is somewhat arbitrary, no distinction between these two approaches will be made in this chapter. The final separation mode, ion-exclusion chromatography, involves the separation of solute ions using an ion-exchanger with functionalities having the same charge sign as the solutes. Donnan exclusion, size exclusion and hydrophobic interactions all play a role in the separation of the injected solutes, and the technique is most successful for partly ionised species. Full details of all of the above separation modes can be found elsewhere (refs. 1, 2).

There exists a great variety of detection methods for ion chromatography and the main approaches are listed in Fig. 2. Conductivity detection is most popular because it is universally applicable, sensitive and convenient to use, however alternative universal methods such as indirect UV absorption are becoming widely used. In many ion chromatographic detection modes, the eluent provides a significant detector signal and it is therefore helpful to differentiate between direct detection (in which the detector signal for the solute exceeds that of the eluent) and indirect detection (in which the reverse situation applies). Indeed, most of the detection methods shown in Fig. 2 can be employed in either a direct or indirect mode with an appropriate choice of eluent. A detailed discussion of the theory and applications of ion chromatographic detection methods is available elsewhere (ref. 3).

In the context of sample handling in ion chromatography, a prime consideration is the diversity of separation and detection methods described above. A judicious selection of the separation and detection modes, and the eluent used, can often mean that sample preparation is minimal. For example, ion-exchange provides good separation of charged species, with uncharged or partially charged species eluting as a group at the column void volume. In contrast, the reverse applies to ion-exclusion chromatography where fully ionised species are usually totally excluded and elute at the void volume, whereas partially charged and uncharged solutes are retained. In other words, the two techniques are best suited to different sample matrices. The same can be said for the detection mode used, since many solutes can be detected selectively and this greatly reduces the requirement for sample preparation. For example, nitrate and nitrite can be determined in cured meats using a simple aqueous extraction coupled with ion-exchange separation and direct UV absorption detection at low wavelength. Under these conditions other ions in the sample, particularly the very high levels of chloride present, are not detected (Fig. 3). Finally, the eluent for the ion chromatographic separation should be selected on the basis of eluotropic strength, so-

36

lubility of the sample, pH, compatibility with the detection mode used, and chemical reactivity with the sample.

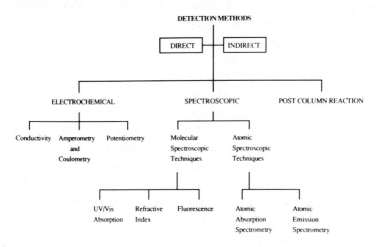

Fig. 2   Detection modes for ion chromatography.

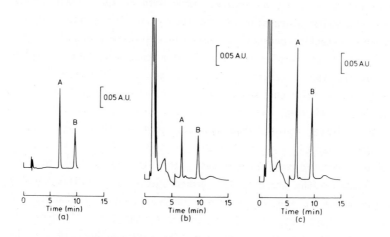

Fig. 3   Separation of nitrite (A) in nitrate (B) in standards (a), bacon sample (b), and spiked bacon sample (c). The sample for (c) was spiked with 0.5 μg of both nitrite and nitrate in a 25 μl injection. Column: Vydac 302 IC 4.6. Eluent: 11.0 mM methanesulphonic acid at pH 5.0. For chromatogram (a), 10 μl of a 50 ppm solution of nitrite and nitrate were injected. Reproduced with permission from ref. 4.

It is thus pertinent to begin this discussion of sample handling in ion chromatography by emphasising that the correct choice of separation and detection modes is imperative if laborious sample preparation procedures are to be avoided. The chromatographer must therefore be fully conversant with the alternatives available. The sample handling methods discussed in this chapter should be viewed as a secondary means of ensuring the success of the ion chromatographic analysis.

2.   SAMPLE COLLECTION AND DISSOLUTION
2.1   SAMPLE COLLECTION

The main concerns when collecting a sample for any analytical method are that the sample taken is representative of the material to be analysed and that no contamination occurs during the sampling process. Statistically based procedures for acquisition of a representative sample have been treated extensively in numerous texts and further discussion is beyond the scope of this chapter.

Contamination is a very important issue and is discussed separately in section 4. Some sampling procedures used for ion chromatography are detailed below.

Gas samples can be conveniently collected using solid sorbents packed into suitable sampling tubes. A known volume of gas is drawn through the tube and the desired sample components are strongly adsorbed. One example of this approach is the adsorption of formaldehyde onto charcoal impregnated with an oxidising solution of proprietary composition (ref. 5). The adsorber tube (Fig. 4) comprises a main adsorbent and an auxiliary or back-up adsorbent designed to detect breakthrough of formaldehyde from the main adsorbent. Formaldehyde reacts with the oxidising solution to produce formate ion, which is desorbed using dilute hydrogen peroxide solution, prior to ion chromatographic analysis. An important aspect of gas analysis is the procedure used to provide calibration standards. Some standards are available commercially, or alternatively a sample generator can be employed. Fig. 5 shows one type of sample generator in which standard solutions are injected into a flowing stream of air for subsequent evaporation and deposition onto an adsorber tube.

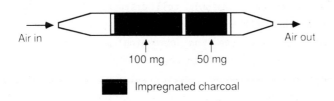

Fig. 4   Solid sorbent tube for sampling of formaldehyde in air.
        Reproduced with permission from ref. 5.

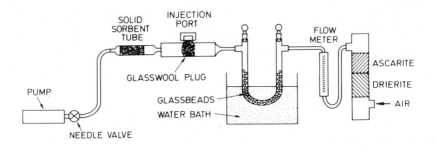

Fig. 5   Sample generator for calibration of sorbent tubes. Reproduced with permission from ref. 5.

Direct absorption of a sample gas into a liquid film has been reported for trace levels of ambient nitrogen dioxide (ref. 6). Here the absorber solution was guaiacol which was deposited as a methanolic solution into the annular cavity between two coaxial glass tubes, before evaporation of the methanol to leave a thin film of guaiacol on the frosted surface of the glass. After passage of a known volume of air, the apparatus was dismantled and the absorbing solution removed with water and analysed for nitrite by ion chromatography. Complete absorption of nitrogen dioxide was observed and there was no detectable oxidation to nitrate ion in the absorbing solution. In a more simple approach, pyruvic acid and methane sulphonic acid have been determined in air by passage of the sample through a microimpinger, using either distilled water or dilute potassium hydroxide as the absorbing solution (ref. 7).

2.2   EXTRACTION METHODS

The removal of ionic species from solid samples prior to ion chromatographic analysis can often be achieved by aqueous extraction of the homogenised sample. This process relies on the high solubility of ionic species in water. Generally a weighed amount of the dry sample is mixed with a known volume of water, extractant solution or eluent, and homogenised in a blender or an ultrasonic cell disrupter for a specified time. The digest is then filtered, subjected to further cleanup where required and injected onto the ion chromatograph. The choice of extracting solution is very dependent on the sample matrix and the nature of the solute ions to be extracted, however water is the prefered extractant whenever possible because alternative extractants often introduce extraneous peaks into the chromatogram. Use of eluent as the extractant is successful only when small injection columes are to be used in the final analysis since the presence of eluent ions in the sample precludes band compression at the head of the column, with subsequent loss of chromatographic efficiency through solute dispersion.

Some samples require extraction with organic solvents before they are suitable for analysis. For example, commercial bromine solutions produced from seawater contain high levels of chloride ion and the analysis of the chloride can be performed after dissolution of the sample in potassium bromide solution, followed by extraction with carbon tetrachloride (ref. 8). Free bromine is extracted and the remaining aqueous solution can be analysed directly by ion chromatography using conductivity detection. Methanol extraction of tetramethylammonium ion from shellfish has been reported (ref. 9), but the methanol must be evaporated and the sample re-dissolved in hydrochloric acid before injection.

2.3. SAMPLE DIGESTION

When samples (particularly solids) are not amenable to simple aqueous extraction, it becomes necessary to digest the sample to obtain a quanti-tative measure of the ionic components.

Traditionally, sample digestion prior to analysis has been performed using concentrated acids, used either alone or in mixtures. This approach is generally inapplicable to ion chromatography because of the large excess of the acid anion(s) introduced and the resulting low pH of the sample digest. The excess anion results in column overloading and the appearance of a major peak in the final chromatogram, whilst the low pH of the digest can cause disruption of the multiple equilibria existing between the eluent species and the column, leading to severe baseline perturbations. For these reasons, acid digestion has been rarely used. One successful application however is the dissolution of geological samples in phosphoric acid prior to the determination of fluorine (as fluoride) by ion chromatography (ref. 10). The fluorosilicic acid produced in the digest is volatilised and collected on a simple condenser apparatus inserted into the digestion tube. The condensed fluorosilicic acid is removed with sodium hydroxide and is converted to fluoride ion which is then determined by ion chromatography.

An attractive alternative to acid digestion of samples is the use of fusion techniques. In this process the sample is mixed with a suitable flux material and is heated until the flux becomes molten. The mixture is then allowed to cool and the fusion cake dissolved in a suitable solvent and then analysed. Typical flux materials include sodium hydroxide, sodium carbonate and lithium tetraborate (ref. 11). Once again, the main problem with this method is the compatibility of the final digest solution with the ion chromatographic eluent, but in this case some of the fluxing materials are identical to eluent components. For example, sodium hydroxide (ref. 12) and sodium carbonate-bicarbonate (ref. 1) are common eluents in ion chromatography. Thus fluoride (ref. 13) and chloride (ref. 14) have been successfully determined in geological materials after fusion with sodium carbonate and injection onto an ion chromatograph using a carbonate-bicarbonate eluent, and boron and fluoride have been determined in glasses after fusion with sodium hydroxide (ref. 15).

Fusion methods are generally quite time-consuming because of the necessity to redissolve the fusion cake and in some cases the high pH of the digest presents a problem. Further disadvantages are the limited applicability of the method, possibile interference from the high level of sodium or lithium present, and the loss of nitrate from the sample during fusion, presumably due to the formation of volatile oxides of nitrogen.

2.4   COMBUSTION METHODS

Analysis of some non-metallic elements in organic samples can be achieved by total combustion of the sample, conversion of the desired elements into gaseous compounds, collection of these gases in a suitable absorber and finally, ion chromatographic analysis of the absorber solution. This approach is suited to the determination of halides (to form such products as HF, HCl, HBr and HI) and sulphur and phosphorus (which form SO_2 and P_2O_5, respectively). The experimental conditions employed for the combustion determine the nature of the final products and in some cases, multiple products are formed for the same element. When this occurs, the composition of the absorber solution should be carefully chosen to convert all forms of an element to a single species suitable for ion chromatographic determination.

The simplest apparatus for combustion of organic samples is a Schoeniger flask, as shown in Fig. 6. A Pyrex glass or quartz vessel containing absorber solution and a small amount of sample (about 0.1 g) in a paper cup is filled with oxygen and inverted. The sample is then electrically ignited and the gases produced are trapped in the absorber solution which provides a gas-tight seal at the mouth of the flask. After an appropriate amount of time has elapsed, the absorber solution is re-moved and analysed. The advantages of this method are that it is inexpensive, rapid and simple, whereas the major disadvantage is that the

oxygen pressure is limited to atmospheric pressure. This in turn limits
the size of the sample which can be analysed and ultimately renders the
method fairly insensitive.

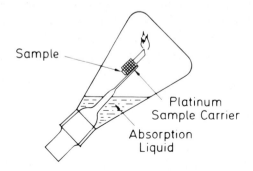

Fig. 6   Schoeniger combustion flask.

Larger samples (up to 1 g) can be accommodated in a bomb combustion
apparatus such as that shown schematically in Fig. 7. Here high pressures
of oxygen (e.g. 40 atm) are used to facilitate complete combustion of the
sample. Because of the high pressure generated within the bomb, obvious
safety considerations apply and analysis is relatively lengthy because of
the time taken to achieve complete absorption of combustion products in
the absorber solution. Absorption may be monitored by a pressure gauge or
using a collapsible expansion bag which inflates during combustion and
deflates during absorption.

As stated earlier, the absorber solution must be carefully selected to
ensure that each element is present in a single form suitable for ion
chromatographic analysis. Elements such as fluorine and chlorine are
converted quantitatively to hydrogen fluoride and hydrogen chloride,
respectively, and so can be absorbed with water or dilute sodium
hydroxide. Hydrogen halides are also produced from bromine and iodine,
but other more oxidised products such as $HBrO_3$ and HIO_3 are also formed.
For these species, a reducing agent such as hydrazine sulphate should be
added to the absorber solution so that only bromide and iodide are
present in the final solution. For sulphur and phosphorus, it is
desirable that they be quantitated as sulphate and phosphate, respectively, and therefore the absorber solution should contain an oxidant
such as dilute hydrogen peroxide.

Some samples, particularly those of a geological origin, may be
combusted using furnace techniques. In this method, the sample is mixed
with a suitable combustion accelerator (such as a mixture of iron and

copper, or iron, tin and vanadium pentoxide) and heated in a ceramic crucible in an induction furnace whilst oxygen is passed over the sample. The combustion products are collected in a suitable absorbing solution. Furnace combustion is very rapid due to the high temperature used and large numbers of samples can be handled with ease. In addition, results are very precise and calibration does not require the use of a large number of geochemical standard materials. Fig. 8 shows a furnace combustion apparatus.

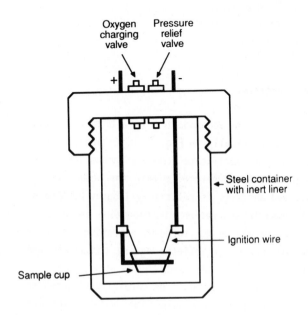

Fig. 7   High pressure combustion bomb.

Table I lists some applications of ion chromatographic analysis of samples prepared by combustion techniques.

3.   SAMPLE CLEANUP METHODS

3.1   INTRODUCTION

When the sample has been dissolved, it is often necessary that some modification of the sample digest be performed before an injection can be made onto the ion chromatograph. This modification may involve a simple filtration step, or it may be more extensive and involve selective removal of the analyte from the sample or removal of interfering matrix components. Alternatively, it may be necessary to change the chemical form of the analyte to improve its separation or detection in the final analysis.

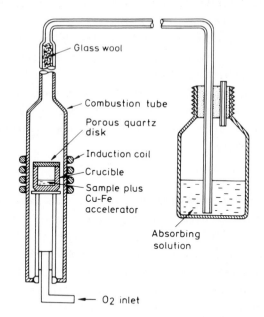

Fig. 8   Furnace combustion apparatus. Reproduced with permission from ref. 19

TABLE I   Applications of combustion methods

Matrix	Species determined	Reference
Plant materials	Cl, S	16
Biological samples	F, Cl, S	17
Fuels, oils and coal	S	17, 23
Foods	I, F, Cl, S	18, 21
Geological samples	F, Cl, S	19, 20
Organic reagents	F, Cl, Br, I, S, P	22

These sample cleanup procedures often take the majority of the total analysis time and contribute significantly to the final cost of the analysis, both in terms of labour and the consumption of materials. In addition, manipulation of the sample can often introduce a major source of imprecision which can greatly outweigh any variables in the chromatographic process itself. Often, the degree of success achieved in the sample cleanup step determines the ultimate success of the analysis.

Sample cleanup can be performed off-line, prior to the chromatographic analysis, or can be incorporated as an on-line process linked with the chromatographic hardware. The goals of cleanup are to achieve: (ref. 1) reduction of the overall loading of sample on the column in order to prevent peak distortion and loss of chromatographic efficiency, (ref. 2) removal of matrix interferences, (ref. 3) concentration or dilution of the analyte, and (ref. 4) preparation of the sample in the solution most appropriate to the analysis. With the exception of sample preconcentration which is discussed in section 5, the achievement of these goals is discussed below.

3.2   FILTRATION

As with all other liquid chromatographic methods, ion chromatography requires that the sample be free from particulate matter to prevent fouling of capillary tubing, column end frits and other hardware components. Many samples, such as water samples, are obtained in a fairly clean form which might appear to require no further treatment prior to injection. Despite appearances, all samples must be filtered through a membrane filter of porosity 0.45 μm or less. Failure to perform this simple step will invariably decrease the column lifetime.

Fortunately, sample filtration is very straightforward if disposable filter units are employed. Careful attention must be paid to sample contamination (see section 4.3), particularly by nitrate ion released from the filter membrane. Ultrafiltration devices wherein the sample is forced under pressure through a membrane, can also be applied to difficult samples, for example, the removal of free calcium and magnesium ions from protein material in biological samples such as serum, milk and egg white (ref. 24).

3.3   CHEMICAL MODIFICATION OF THE SAMPLE
3.3.1   BATCH METHODS USING ION-EXCHANGE RESINS

Perhaps the most common chemical modification of the sample performed in ion chromatography is adjustment of pH of strongly acidic or alkaline samples. Injection of such samples without pH adjustment usually produces an unacceptable chromatogram because of baseline disturbances. Some ion chromatographic detection modes are particularly prone to the formation of system peaks which appear as spurious peaks in the chromatogram. These peaks appear as a direct result of the fact that with some detection methods, such as indirect UV absorption (ref. 3) (or "indirect photo-metric chromatography"), (ref. 25), a characteristic of the eluent (such as absorbance) is being monitored by the detector. Any change in the form of the eluent can therefore lead to a baseline disturbance or system peak. For example, an eluent comprising potassium hydrogenphthalate at pH 4 is commonly employed with indirect UV detection for the determination of anions. This eluent is favoured because it is an effective buffer and provides excellent separation of many inorganic anions. At pH 4, the phthalate is present as a mixture of phthalic acid and hydrogenphthalate ions, both of which exhibit different UV absorption behaviour. It follows therefore that injection of a strongly acidic or alkaline sample will temporarily alter the ratio existing between the two eluent species and will ultimately produce a system peak (ref. 26). The same can be said for other detection modes.

It is usually not possible to adjust the sample pH by simple addition of acid or base because of contamination of the sample by the acid anion or base cation, since these species may be of interest in the sample. In such cases it is often possible to use an ion-exchange resin in the hydrogen form can be added to an alkaline sample in order to lower the pH. The usual procedure is to stir a known weight of resin (e.g. 1 g) with a known volume of sample (e.g. 5 ml) and to monitor the pH of the solution, noting the time required for the sample to reach the desired pH (which is usually that of the eluent to be used). When this reaction time is determined, the process is repeated with a second sample aliquot but with the pH electrode removed. This prevents contamination of the sample by chloride from the electrode filling solution.

Whilst this approach is simple and relatively effective, it suffers from a number of drawbacks. First the sample volume required is large and the reaction time must be adjusted whenever the composition of the sample changes. Second, the resin used must be cleaned thoroughly to prevent contamination of the sample by ions leached from the resin material. Third, the sample volume may change due to uptake or release of solvent from the resin. Finally, some loss of sample components may occur due to adsorption on the resin.

3.3.2   Dialytic techniques

Dialytic techniques in which selected sample components are transfered across a membrane may be subdivided into passive dialysis and active (or Donnan) dialysis procedures. Passive dialysis involves diffusion of particles of a specified molecular weight range through a neutral membrane. On the other hand, active or Donnan dialysis is the transfer of ions of a specified charge sign through an ion-exchange membrane. Both approaches have been applied to the cleanup of samples for ion chromatography.

Passive dialysis is a very slow process, requires appreciable volumes of sample (e.g. 5 ml) and normally results in severe dilution of the sample. These factors have mitigated against its widespread use. Nordmeyer and Hansen (ref. 27) have described an automated device for the rapid dialysis of very small samples (e.g. 40 ml) which enables direct injection of the dialysate onto an ion chromatograph. This device is shown schematically in Fig. 9, from which it can be seen that the sample is introduced into the annular cavity formed between a hollow dialysis fibre and an external concentrically mounted small diameter PTFE tube. The eluent is contained inside the fibre and flow is stopped whilst solute components from the sample dialyse into the interior of the hollow fibre. Because of the small volumes involved, dialysis time is very short (typically less than 1 min), and the sample is then injected directly onto the ion chromatograph. When applied to the removal of free calcium from human serum, linear calibration curves were obtained and peak heights showed a relative standard deviation of less than 5% over a two-week period.

The process of Donnan dialysis can be illustrated by reference to a dialysis system comprising 0.1 M NaCl (solution 1) separated from 0.001 M KCl (solution 2) by a cation-exchange membrane. This experimental arrangement is shown in Fig 10.

Cations can diffuse rapidly through the membrane, according to the following equilibrium:

$$Na^+ + K^+_m \rightleftharpoons Na^+_m + K^+ \qquad \ldots\ldots\ldots(1)$$

where the subscript m refers to the membrane phase

The equilibrium constant for this exchange is given by:

$$K_{Na, K} = \frac{(Na^+_m) (K^+)}{(Na^+) (K^+_m)} \qquad \ldots\ldots\ldots(2)$$

where the brackets denote the activity of the species. Since the equilibrium must exist at both surfaces of the membrane, then:

$$\frac{(Na^+_m)_1 (K^+)_1}{(Na^+)_1 (K^+_m)_1} = \frac{(Na^+_m)_2 (K^+)_2}{(Na^+)_2 (K^+_m)_2} \qquad \ldots\ldots\ldots (3)$$

where the subscripts 1 and 2 refer to the two solutions on either side of the membrane. There can be no concentration gradients for the same ion across the membrane, therefore

$$(Na^+_m)_1 = (Na^+_m)_2 \qquad \ldots\ldots\ldots (4)$$

and

$$(K^+_m)_1 = (K^+_m)_2 \qquad \ldots\ldots\ldots (5)$$

Eqn (3) can be simplified to give eqn (6), or eqn (7) if the activity coefficient is assumed to be unity.

$$\frac{(Na^+)_1}{(Na^+)_2} = \frac{(K^+)_1}{(K^+)_2} \qquad \ldots\ldots\ldots (6)$$

$$\frac{[Na^+]_1}{[Na^+]_2} = \frac{[K^+]_1}{[K^+]_2} \qquad \ldots\ldots\ldots (7)$$

In the system under consideration, there is a strong tendency for the sodium ions to diffuse from the high concentration zone (solution 1) to the low concentration zone (solution 2). As this process occurs, corresponding transfer of potassium ions from solution 2 to solution 1 proceeds in order to preserve electroneutrality. Thus diffusion of 1% of the sodium into solution 2 is accompanied by transfer of 99% of the potassium into solution 1. If the volume of solution 1 is less than that of solution 2, then the concentration of potassium in solution 1 is greater than that originally present in solution 2. In this way, sample preconcentration can be accomplished. Eventually the system will attain chemical equilibrium, but this state is achieved only slowly because transfer of chloride across the membrane is hindered. In the short term therefore, sample modification occurs.

48

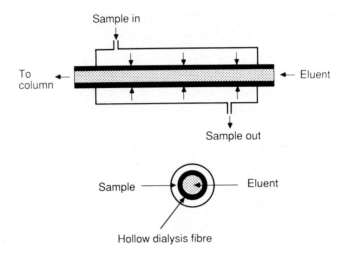

Fig. 9   Schematic representation of a passive dialysis-injection device.
         Adapted with permission from ref. 27.

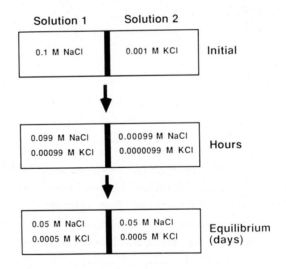

Fig. 10   Schematic representation of Donnan dialysis.

In terms of ion chromatographic sample cleanup, Donnan dialysis can be used to remove a selected species from a sample, or alternatively to selectively add an ion to a sample. An example of the first alternative is the removal of metal cyano complexes from a plating bath by Donnan dialysis into a sodium chloride receiver solution (ref. 28). The second alternative can be illustrated by the dialysis of sodium hydroxide solution using sulphuric acid as the receiver solution. Here hydrogen ions from the sulphuric acid solution diffuse into the sodium hydroxide through a cation-exchange membrane. The pH of the sample is therefore lowered, whilst the anion content is theoretically unaltered, allowing subsequent determination of these anions by ion chromatography.

The second of the above alternatives suffers from a practical limitation which seriously detracts from its routine use. This limitation is that the cation-exchange membrane is not entirely impervious to sulphate ions from the receiver solution, which means that the sample ultimately becomes contaminated with sulphate during dialysis. This problem could be minimised by increasing the permselectivity of the membrane (ie. its ability to permit the transfer of ions of only one charge sign) however this requires careful control over the manufacturing process. A more attractive procedure has been reported by Cox and Tanaka (ref. 29) wherein a slurry of ion-exchange resin in the hydrogen form is used to replace the sulphuric acid employed as the receiver solution in the above example. This method is called "dual ion-exchange" and since the counter anion in the receiver solution is the resin bead itself, transfer across the membrane is eliminated for physical reasons. It should be noted that the ion-exchange membrane may also be used in the form of tube inserted into the resin slurry (ref. 30). Table II shows some applications of Donnan dialysis sample cleanup in ion chromatography.

3.3.3 DISPOSABLE CARTRIDGE COLUMNS

One of the most versatile and convenient means available for sample cleanup is the use of commercially available disposable cartridge columns. These devices offer rapid sample treatment and can usually be employed in tandem with disposable filters so that filtration and sample cleanup can be performed in a single operation. Table III lists some of the common stationary phases available as cartridge column packings from a range of manufacturers.

TABLE II   Applications of donnan dialysis cleanup

Sample	Membrane	Reference
Metal cyano complexes in a plating bath	anex	28
Cl^-, NO_3^-, SO_4^{2-} in conc. NaOH	catex	30
Anions in Na_2CO_3	catex	30
SO_4^{2-} in NaCl	anex	30
Cl^- in polyelectrolytes e.g.. polyacrylic acid	anex	31

Table III   Typical stationary phases for cleanup cartridge columns

1. Silica
2. C18
3. Alumina (acidic, basic and neutral)
4. Anion-exchange
5. Cation-exchange (H^+ or metal form)
6. Polymer (e.g. styrene divinylbenzene and polyvinylpyrrolidine)
7. Activated carbon
8. Chelating agents
9. Amino

Cartridge columns can be employed in one of two ways. The first method is selective removal of the solute ions from the sample matrix and the solvent used to elute the sample through the cartridge should provide chromatographic conditions giving very strong retention of the solute ions. That is, the capacity factors for these solutes should be very large. The alternative operational mode for cartridge columns is to selectively retain matrix components under conditions where the solute ions are unretained. That is, their capacity factors approach zero. It is generally inadvisable to use a cartridge column to achieve chromato-graphic separation of solutes which have capacity factors intermediate between the abovementioned extremes. The reasons for this are that experimental factors are very variable (e.g.. column efficiency, flow-rate, and packing reproducibility) and in most cases the passage of solutes along the column cannot be visually monitored. Thus even if a chromato-graphic separation is optimised on a particular stationary phase, it is probable that the separation would be irreproducible due to changes in the experimental conditions used.

Keeping in mind that we wish the solute to be either well retained or not retained at all, then several possibilities emerge from the stationary phases listed in Table III. Stationary phases which show some ion-exchange ability (such as silica, alumina, anion and cation exchangers, and amino phases), and stationary phases which show chelation ability should be suitable for the selective retention of ionic solutes from a matrix composed largely of neutral, organic species. Alter-natively, hydrophobic stationary phases such as octadexylsilane and the polymeric phases should be useful for the removal of neutral organic components while showing little retention of ionic solutes. A further potential application of cartridge columns is their use for adjusting the pH of a sample in the same manner as that described earlier for ion-exchange resins used in the batch mode. Most of the abovementioned possibilities have been realised in practice and Table IV lists some examples of successful applications.

Several practical aspects should receive attention when using cartridge columns, namely column pretreatment, flow-rate, method of sample application, and sample pH. First, the columns almost invariably require pretreatment in order to remove very fine particles of the packing material, to elute any contaminants, or to condition the stationary phase to improve the efficiency of sample binding. Significant levels of inorganic contaminants are commonly encountered in cartridge columns (see section 4.3), generally as a result of residual reagents from the manufacturing process. Hydrophobic stationary phases usually require pretreatment with an organic solvent such as methanol in order to wet the stationary phase surface so that effective binding of hydrophobic solutes is achieved from aqueous sample solutions.

The flow rate of sample or flushing solution through the precolumn should be kept as low as practicable so that mass transfer effects are minimised, and as reproducible as possible. Most column cartridges are designed for use with disposable syringes and the low packing density of the stationary phase permits very high flow-rates (e.g. 50 ml/min) to be easily achieved. Experience with analytical chromatographic columns suggests that such a high flow-rate is unlikely to produce the degree of selective separation required, so it is advisable to use flow-rates less than 10 ml/min.

TABLE IV   Applications of cleanup with cartridge columns

Matrix	Solute ions	Stationary phase	Reference
Plant extract	NO_2^-, NO_3^-, SO_4^{2-}	C18	4,34
Urine	thiosulphate	C18	32
Urine	oxalate	C18	33
Soil extract	SO_4^{2-}	C18	34
Plasma	NO_2^-, NO_3^-,	C18	36
Plant extract	Cl^-, NO_3^-, SO_4^{2-}	silica	35
High chloride	anions	AG^+ catex	37
NaOH	anions	H^+ catex	38
Leachate	As(III), As(V)	catex	39
Brine	sodium	H^+ catex	43
Air samples	anions	charcoal	40
Digests	metal oxo-anions	anex	41
Natural waters	anions	amino	42
Surfactants	anions	polymer	43
Aromatics	anions	polymer	43

The third important practical consideration is the manner in which the sample is applied to and eluted from the cartridge column. It is possible to apply a known volume of the sample to the head of the column and to elute the sample band through the column with a suitable eluent. However, this method is difficult in practice because of the difficulty in applying an accurate volume of sample using the syringes compatible with the cartridge column, and is recommended only when the sample volume is small or the concentration of the sample is high enough to quickly saturate the cartridge. It is generally more appropriate to pass sample continuously through the column, discarding the first two or three column volumes and then collecting sufficient effluent for analysis.

Finally, the sample pH has an important bearing on the selection of a suitable stationary phase. Apart from the obvious consideration that some stationary phases are intolerant of acidic or alkaline solutions, the sample pH is often a very useful indicator of the ionic strength. In cases where the ionic strength is unacceptably high, it may be necessary to use a second cartridge column, or an alternative cleanup procedure, to remove some of the ionic components from the sample.

In conclusion, it should also be noted that cartridge columns packed with hydrophobic stationary phases can also be used to retain ionic solutes (rather than neutral, organic solutes) if they are first conditioned with an ion-interaction reagent. The success of this approach is dependent on retention of the ion-interaction reagent on the stationary phase during sample elution, thus it is desirable that relatively hydrophobic ion-interaction reagents be used and the sample volume be limited.

Tetramethylammonium hydroxide and pentanesulphonic acid have been employed as ion-interaction reagents for the removal of anionic and cationic surfactants, respectively, using a cartridge column packed with a polymeric divinylbenzene stationary phase (ref. 43).

3.3.4   CHEMICAL REACTION OF SOLUTES

For some samples, cleanup can be best achieved using an appropriate chemical reaction to eliminate a matrix component. Alternatively, it may be necessary to derivatise a solute in order to enhance its detectability or to convert it into a form suitable for separation. Much has been written on the principles of chemical derivatisation of organic solutes (ref. 44), and the same principles apply here to inorganic solutes. Table V lists some reactions which have been employed as sample treatment methods for ion chromatography, or as mobile phase reactions designed to modify the nature of the solute in an ion chromatographic determination.

4.   CONTAMINATION EFFECTS
4.1   INTRODUCTION

One of the most important considerations in sample handling is the possibility of contamination arising from various sources such as the handling procedures used, the volumetric ware employed, filtration or cleanup devices and the chromatographic hardware itself. Such contamination may alter the true concentration of solutes of interest, either directly by contributing detectable levels of the analytes to the final solution, or by promoting chemical reactions which cause levels of analytes to alter. Further the sample itself may be a source of contamination of the chromatographic system, causing column poisoning or memory effects resulting from adsorption of sample constituents on chromatographic components.

In this section, the chief sources of contamination are discussed.

TABLE V   Chemical modification of the sample

Additive	Effect	Reference
Boric acid	prevent oxidation of ascorbic acid (borate $\Longrightarrow H_3BO_3$ in suppressor)	45
N, N dimethyl-phenylenediamine	reacts with H_2S to form methylene blue	46
Boric acid	fluoride $\Longrightarrow BF_4^-$ to eliminate interference of F^- on silica analysis	47
Iodine	$I_2 + HCN \Longrightarrow H^+ + I^- + ICN$ iodide used as a measure of CN^-	48
EDTA	complex interfering metal ions	49
Formaldehyde	$SO_3^{2-} \Longrightarrow$ hydroxymethane-sulphonate	50
Methanol	oxidised to formate by CrO_4^{2-}	51
Barium ions	precipitate sulphate	52

4.2   CONTAMINATION FROM PHYSICAL HANDLING OF THE SAMPLE

The prime sources of sample contamination in physical operations such as weighing and volumetric manipulations are contact of the sample or apparatus with the skin, and leaching of contaminants from volumetric ware. Contact with the skin introduces detectable levels of sodium and chloride to the sample and in cases where trace determination of these solutes is desired, high background levels will invariably occur unless protective gloves are worn.

Volumetric ware should be made from polyethylene or some other inert material and should be washed in non-ionic detergent (sulphate-free) and rinsed thoroughly before use. Standard solutions used for calibration of the ion chromatograph should be stored in inert containers. There is ample evidence to show that even brief exposure of aqueous solutions to conventional laboratory glassware results in significant contamination, particularly by sodium and silicate. It is also noteworthy that samples for anion analysis can be readily contaminated by bicarbonate ion produced by absorption of carbon dioxide from the atmosphere, particularly under alkaline conditions. Care should therefore be taken to exclude carbon dioxide wherever possible.

4.3   CONTAMINATION FROM FILTRATION DEVICES AND CARTRIDGE COLUMNS

As mentioned in the earlier discussion on the use of disposable filtration devices and cartridge columns for the clarification and chemical cleanup of samples for ion chromatography, contamination from these devices must be considered. In most cases, these devices have been manufactured for the general HPLC market where sample contamination by inorganic ions would be a minor problem unless the particular contaminants involved were capable of participating in chemical reactions with the sample components. For this reason, it is not uncommon for inorganic reagents to be employed during the manufacturing process.

Disposable filters and both C18 and alumina cartridge columns produced by the major HPLC supplier have been evaluated for contamination effects (refs. 53, 54) and some of the results obtained are summarised in Tables VI and VII. Whilst the results shown are specific to one brand of product, it is expected that similar contamination levels would exist in alternative products, unless specific means were employed by the manufacturer to remove inorganic contaminants.

Table VI shows that disposable filtration devices release appreciable quantities of nitrate, and lesser amounts of chloride and sulphate, into the initial fraction of solution passed through them. However, the leachable ions are very labile and are essentially removed completely if the filter is pre-washed with 20 ml of water. Care should therefore be taken that such filters are adequately washed before they are used on samples to be subsequently analysed by ion chromatography. Detectable levels of chloride, nitrate, sulphate and lead are leached from C18 cartridge columns by water (Table VII), and a reduced, but still detectable, level of these ions persists after the column has been washed with 20 ml of water. The levels of ions leached from the cartridge are sufficiently low that that they would present a problem only for ultra-trace analyses using sample preconcentration methods. In such cases, it would be necessary to run a blank solution. Alumina columns produce much more severe contamination, undoubtedly due to residues of the reagents used to modify its surface properties during manufacture.

TABLE VI   Contamination from filtration devices
Data taken from ref. 53

ION	Conc. (ppb) in successive 20 ml fractions			
	HA filters		HA filters	
	Fraction 1	Fraction 2	Fraction 1	Fraction 2
F^-	<0.2	<0.2	23.4	<0.2
Cl^-	84.6	13.6	73.2	13.2
NO_3^-	698.8	<0.4	409.5	<0.4
SO_4^{2-}	17.8	2.2	111.9	8.7

TABLE VII   Contamination from cartridge columns

Ion	Conc. (ppb) in successive 20 ml fractions	
	Fraction 1	Fraction 2
F^-	<0.2	<0.2
Cl^-	101.2	29.4
NO_3^-	69.9	36.4
SO_4^{2-}	99.4	39.0
Pb^{2+}	76.3	21.4

Notes
1. No contamination observed for Cd^{2+}, Cu^{2+}, Mn^{2+}, Ni^{2+} and Zn^{2+}
2. Other stationary phases (particularly alumina) contain significant quantities of leachable material
3. Data taken from ref. 53.

4.4   CONTAMINATION FROM CHROMATOGRAPHIC HARDWARE COMPONENTS

Over recent years there has been considerable discussion relating to the suitability of conventional HPLC hardware components for use in ion chromatographic applications. It is clear that the major instrumental components of an ion chromatograph, namely the pump, injector, data management system, and often also the detector, are identical to those used in a typical HPLC system. The chief difference between the two techniques lies in the column used for each method. In view of this, it has been common for HPLC hardware to be used in ion chromatographic systems and this raises the question of the suitability of stainless steel components for use with the aqueous eluents and sample types employed in ion chromatography.

Types 304 and 316 stainless steels are typically used in the construction of solvent-wetted HPLC components. Studies with reversed-phase eluents (ref. 55) have shown that components with small diameter openings, such as capillary tubing, are susceptible to corrosion resulting from mechanical erosion of the protective surface oxide layer due to high fluid velocity. There is also evidence that sample components such as proteins and metal chelates can undergo complexation or ligand exchange reactions at stainless steel surfaces (refs. 56, 57). Aqueous buffers used for the analysis of anions and cations in ion chromatography provide a suitable environment for corrosion and in some cases also exhibit strong complexation properties. It is therefore possible that contamination effects could arise from the use of these eluents on stainless steel hardware components.

Several possibilities exist where contamination of the eluent by metal ions leached from metallic components in the chromatographic system could present serious problems. The first of these is the direct elevation of detector baseline levels in cation analyses using post-column reaction detection. Secondly, interference effects resulting from complexation reactions with solutes can be expected to occur in some anion-exchange separations. For example, iron (III) forms a strong complex with many common inorganic anions and this complex would have different chromatographic and detection characteristics to those exhibited by the free anions. A further effect resulting from eluent contamination occurs in the ion-exchange separation of monovalent cations where the presence of multiply charged cations in the eluent would lead to rapid column deterioration caused by irreversible binding of these ions onto the exchange sites of the low capacity cation-exchange columns used. In such methods, it is common practice to include an ion-exchange guard column between the pump and injector in order to remove any multivalent cations from the eluent. However, this approach would be ineffective against ions produced as corrosion products within the injector or the column itself.

Table VIII shows the compositions of types 304 and 316 stainless steel. These materials are corrosion resistant by virtue of a protective coating of chromium-rich oxides which forms on the surface (ref. 55). This coating can develop gradually during usage or can be formed rapidly by exposing the surface to relatively strong nitric acid solutions. If the latter method is used, the surface is said to be "passivated".

Consideration of the composition of the steel suggests that the species most likely to be produced by corrosion reactions are iron, chromium, manganese, molybdenum and nickel. These metal ions could be leached from the metallic surface through either direct oxidation or by complexation reactions with eluent components. The latter mechanism could be expected to be most prevalent with eluents containing strong complexing agents such as citrate, tartrate, phthalate and ethylenediamine. Side-reaction coefficients for complexation of the above metal ions with the eluents typically used for ion chromatography suggest that significant complexation of iron, chromium and, to a lesser extent, nickel can be expected (ref. 58).

TABLE VIII   Typical composition of types 304 and 316 stainless steel
Data taken from ref. 58

Element	Percentage Type 304	Type 316
Carbon	0.08 (max)	0.08 (max)
Manganese	2.00	2.00
Phosphorus	0.045	0.045
Sulphur	0.030	0.030
Silicon	1.00	1.00
Chromium	18.00 - 20.00	16.00 - 20.00
Nickel	8.00 - 12.00	10.00 - 14.00
Molybdenum	-	2.00 - 3.00

In a recent study (ref. 58), corrosion products were allowed to accumulate to detectable levels by recirculating aqueous eluents through a HPLC system, and the eluents were then analysed by inductively coupled plasma atomic emission spectrometry. The experiment was performed on the chromatographic hardware alone and with a stainless steel column included in the flow-path. The results of this study are summarised in Table IX, and show clearly that the levels of contaminant metal ions (especially iron) found in the eluent were extremely low when the eluent passed only through the chromatographic hardware (pump and injector), but increased markedly when the column was incorporated into the flow-path.

It is interesting to speculate on the source of the corrosion products observed with the stainless steel column. The most probable source is the column frits which have a very high surface area in comparison to the rest of the column and the external chromatographic system and inevitably also contain stress points at which the rate of corrosion would be accelerated. Calculations show that if the eluent wets the entire surface of the column frits, then these frits account for approximately 96% of the total metal surface in contact with the eluent and should therefore be the prime source of eluent contamination.

The levels of metal ions given in Table IX should be viewed from a standpoint which considers the total levels of these species from all sources. The reagents used for the preparation of eluents can be expected to contain residual levels of metal ions: examination of the manufac- turers' specifications shows that the levels of iron and nickel would be in the range of 1-5 ppb in the eluents tested. It can therefore be concluded that oxidising or complexing eluents used for cation-exchange separations can be contaminated with detectable levels of iron, chromium and nickel from stainless steel column frits. Accordingly, cation- exchange columns designed for use with these eluents should be fitted with non-metallic frits, or alternatively, metallic frits should be deactivated by passivation or silanisation reactions. Eluents typically used for ion chromatography of anions do not show any contamination from the metallic components of the chromatographic system and this result indicates that metallic frits are suitable for use in anion-exchange columns.

Chromatographic hardware components such as pumps and injectors do not contribute any significant levels of the metal ions. It should be noted that the contamination levels shown in Table IX were obtained on an un- passivated HPLC system and lower levels occurred when metallic surfaces were passivated by treatment with nitric acid (ref. 58). Corrosion of these components therefore does not represent a problem in ion chromato- graphic methods.

4.5   CONTAMINATION OF THE COLUMN

In the preceding section it was pointed out that metal ions introduced into the eluent from the chromatographic hardware could cause poisoning of cation-exchange columns through irreversible binding of exchange sites. This example serves to illustrate the possibility of column contamination, and in the following discussion such contamination from the sample will be considered.

TABLE IX   Contamination from stainless steel

Data taken from ref. 58.

Eluent[a]	Metal concentration (ppb) per eluent cycle					
	FE	Cr	Mn	Mo	Ni	Column
HNO$_3$	311.8	<1.0	33.6	<0.8	11.7	yes
HNO$_3$	<1.0	4.4	<0.3	<1.5	<1.5	no
EDA-TA	9.5	<1.5	<0.3	<1.6	<1.6	yes
EDA-TA	1.9	<1.3	0.4	<1.5	<1.5	no
EDA-CA	20.0	<1.4	<0.3	<1.7	<1.7	yes
EDA-CA	3.8	<2.6	<0.6	<3.2	<3.2	no
KHP	<1.2	3.2	<0.3	<1.7	<1.7	yes

a. <u>Key to eluent identities.</u> HNO$_3$: 10 mM nitric acid. EDA-TA: 0.5 mM ethylenediamine and 1.3 mM tartaric acid. EDA-CA: 3.5 mM ethylenediamine and 10 mM citric acid. KHP: 1 mM potassium hydrogen phthalate.

Perhaps the most commonly encountered example of column poisoning by the sample is the binding on the column of organic sample components. This problem can be easily circumvented by pretreatment of the sample with a cartridge column, use of an ion-exchange guard column, and periodical flushing of the analytical column with a water-methanol mobile phase when the column packing is sufficiently resistant to the use of organic solvents. A more insidious problem is the effect of metal ions on both anion and cation separations. In the former case, metal ions such as calcium and magnesium may form complexes with mobile phase components (e.g.. gluconate), leading to system peaks or an unstable baseline in the final chromatogram (ref. 59). Silica-based anion-exchangers can also show retention of metal ions, with iron (III), aluminium (III) and mercury (II) being strongly retained, whilst copper(II), lead(II) and zinc(II) may elute at retention times similar to those observed for anions on the same column (ref.60). It has recently been shown (ref. 61) that the latter group of metal ions form complexes with phthalate eluents which are retained to varying degrees on silica and polymeric anion-exchangers. It is therefore evident that polyvalent metal ions should be removed from samples on which anion determinations are to be performed, and this may be achieved conveniently by passage of the sample through a cation-exchange cartridge column. Fig. 11 shows the effect of treating a drinking water sample in this manner and it is clear that the removal of calcium and magnesium resulted in an improved baseline.

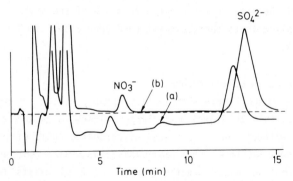

Fig. 11 Effect of calcium and magnesium on the baseline produced in the
determination of anions in drinking water. Chromatogram (a) is
for a sample contaminated with calcium and magnesium and chromatogram (b) is the same sample after treatment by passage through
a cation-exchange cartridge column. Column: Waters IC PAK A.
Eluent: 1.3 mM $Na_2B_4O_7$ - 5.8 mM_3BO_3 - 1.4 mM K-gluconate in
water-acetonitrile (88:12).

Polyvalent cations can exert a detrimental effect on the determination
of monovalent cations by cation-exchange. The relatively weak eluents
used for the monovalent cations are unable to elute polyvalent cations,
with the result that these species remain strongly bound to the column.
The net result of this is that chromatographic performance of the column
for monovalent cations is degraded in terms of decreased retention, loss
of efficiency, poor resolution and reduced peak heights (ref. 62). The
latter three characteristics are directly attributable to the fact that
the solute ions do not compress into a compact band at the head of the
column when ion-exchange sites are occupied by polyvalent metal ions. For
these reasons it is therefore advisable that a cation-exchange guard
column be inserted into the flow-path prior to the analytical column.

5.   SAMPLE HANDLING FOR ULTRA-TRACE ANALYSIS

5.1   INTRODUCTION

The sensitivity of an ion chromatographic method is strongly dependent
on the type of detector used, with amperometric and direct UV absorption
detection being among the most sensitive, and refractive index being
relatively insensitive (ref. 3). Most of the universal detection modes
such as conductivity and indirect UV absorption have comparable practical
detection limits of about 100 ppb for a 100 μl injection. This means that
for reliable quantitation to be achieved, analysis by direct sample
injection can be most conveniently used for solute concentrations of
about 1 ppm or higher. Below this concentration level, either very large
injection volumes must be employed or a sample preconcentration method is

necessary. Fig. 12 shows a schematic representation of the sample concentration ranges applicable to the abovementioned approaches, each of which is discussed below.

5.2   USE OF LARGE INJECTION VOLUMES

Because of the zone compression effect occurring on ion-exchange columns, it is possible to inject very large volumes of sample onto an analytical column without significant loss of chromatographic efficiency. Several authors have utilised this approach (refs. 63 - 65) and the results suggest that an upper practical limit of 2 ml exists for the sample injection volume, otherwise the large solvent peak in the final chromatogram may mask early eluting solutes. Moreover, indirect UV absorption detection has been shown to be superior to conductivity detection because the former method shows a more rapid return to baseline after passage of the large injection peak (refs. 64, 65). Fig. 13 shows a chromatogram obtained using indirect UV absorption detection and a 1 ml sample injection volume.

To increase the volume of sample above the practical limit mentioned above, it is essential that the size of the solvent peak be decreased so that interference with early eluting solutes does not occur. One method to achieve this is to flush the interstitial sample from the column by pumping a measured volume of eluent through the column in the reverse direction to that used for sample elution (ref. 65). The solute ions remain bound to the column during this operation, but the injection peak is reduced because the sample solvent has been displaced from the column. This method permits the use of sample volumes up to 5 ml, but requires the use of two switching valves and therefore seems to offer little advantage over the use of similar hardware with preconcentration columns, as discussed below.

5.3   USE OF PRECONCENTRATION COLUMNS

The most widely used approach to trace enrichment in ion chromatography involves the use of a separate precolumn designed to trap trace levels of solutes from a large volume of sample (ref. 5). The precolumn method is popular because it is simple and convenient to apply, is amenable to automation and offers high enrichment factors.

In precolumn sample enrichment, an accurately known volume of sample is pumped at a precise flow-rate through a small ion-exchange precolumn (or a reversed-phase precolumn coated with an ionic ion-interaction reagent), called the concentrator column. Solute ions contained in the sample are selectively trapped on the concentrator column and are then eluted onto an ion-exchange analytical column for separation and quantitation. This procedure can be effective as an analytical method only if the processes of binding of solute ions on the concentrator column and their subsequent transfer to the analytical column are quantitative.

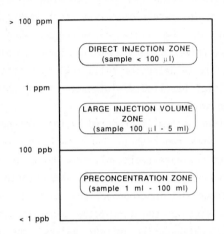

Fig. 12 Working concentration ranges in ion chromatography.

5.3.1   HARDWARE CONSIDERATIONS

The simplest form of sample preconcentration device utilises a single, six-port, high pressure, switching valve which can be actuated either manually or automatically (ref. 66). Fig. 14 shows the plumbing arrangement and operation of this system. In the first step, the concentrator and analytical columns are equilibrated with eluent. The valve is then rotated and the concentrator column is removed from the flow-path whilst eluent continues to be pumped through the analytical column. A measured volume of sample is passed through the concentrator column using either a pump or a large volume syringe, with the effluent being directed to waste. At this stage, the solute ions from the sample are assumed to be retained on the concentrator column and in the sub-sequent step, rotation of the valve permits eluent to be pumped in the reverse direction (ie. the flow direction opposite to that used for sample loading) through the concentrator column. This operation is known as backflushing and is designed to transfer the solute ions to the analytical column in a small volume of eluent. In the final step, the valve is again rotated and eluent then carries the solute ions through the analytical column for separation and detection.

This system has the advantages of simplicity and ease of operation. The backflush volume is generally selected to be high enough to guarantee that all the ions are transfered. It should be noted here that the volume of eluent used to backflush the solute ions to the analytical column will necessarily have lower concentration of eluent ions than that present in the bulk eluent. This is because some eluent ions are required to re-equilibrate the concentrator column which has been depleted of eluent ions during the passage of sample. The result of this is that severe base-line disturbances often occur in the final chromatogram when the detection method employed is sensitive to the background level of eluent ions in the mobile phase. Conductivity detection falls into this category and when this is used, the initial baseline disturbance in the chroma-togram often masks early eluting solutes.

A more flexible preconcentration system is produced by the combination of a single, programmable pump with two high pressure switching valves and a low pressure solvent selection valve (refs. 67, 68). Here, the same pump can be used to deliver eluent and to load the sample onto the concentrator column. Fig. 15 shows the interconnections used for these valves and Fig. 16 illustrates some of the flow-paths achievable with this system. Using a suitable configuration of the valves, the pump tubing and interconnecting lines can be flushed with sample solution or eluent with both the concentrator and analytical columns removed from the flow-path. In Fig. 16a a measured volume of sample is loaded onto the concentrator column at a precise flow-rate, after which a small, accurately known volume of eluent is pumped through the concentrator column in the same flow direction as that used for sample loading (Fig. 16b). This is termed a "wash" step and serves to partially re-equilibrate the concentrator column with eluent ions without loss of bound solute ions. Fig. 16c shows the sample stripping step in which the solute ions are backflushed from the concentrator column onto the analytical column using an accurately known volume of eluent. In the final step of the analysis, the concentrator column is removed from the flow-path and the eluent is pumped directly to the analytical column.

Clearly, this multi-valve system is more complex than the single-valve approach and requires the use of a sophisticated pump. It does however offer the advantages of unlimited and precise control over the volumes of eluent used for the washing and stripping steps, and these may be readily manipulated to adapt to the requirements of a particular sample. In addition, the baseline produced in the final chromatogram is superior to that obtained with a single-valve preconcentration system.

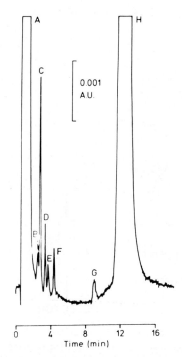

Fig. 13 Use of large injection volumes for anion analysis. Column: Vydac 302 IC 4.6. Eluent: 2.5 mM potassium hydrogen phthalate at pH 4.0. Injection volume: 1 ml. Detection: UV absorption at 285 nm. Solute concentrations: 50 - 200 ppb. Peak identies: A - solvent peak, B - dihydrogenphosphate, C - chloride, D - nitrite, E -bromide, F - nitrate, G - iodide, H - system peak.

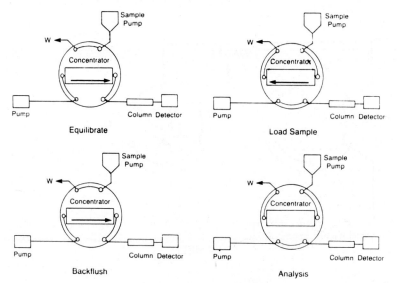

Fig. 14 Sample preconcentration using two pumps, a single high-pressure switching valve and a concentrator column.

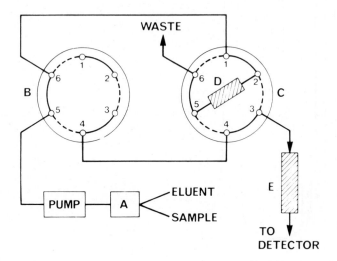

Fig. 15 Apparatus for sample preconcentration using a single pump, a low pressure solvent selection valve (A), two high-pressure switching valves (B and C) and a concentrator column (D). E is the analytical column.

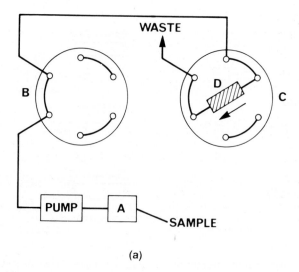

(a)

Fig. 16

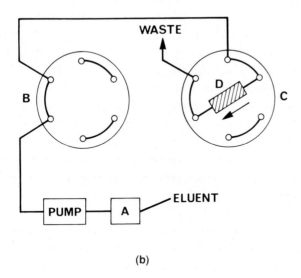

(b)

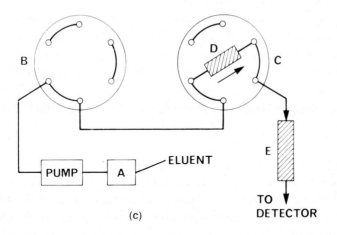

(c)

Fig. 16   Important flow-paths in the preconcentration of a sample using
the apparatus shown in Fig. 15. (a) sample loading, (b) concentrator column washing, (c) sample stripping. Reproduced with permission from ref. 68.

5.3.2 CHOICE OF ELUENT

The most important realisation in selection of an appropriate eluent for a preconcentration method is that eluents which are perfectly suitable for direct injection ion chromatography may be quite inappropriate for use with preconcentration techniques. In the latter case, the eluent must perform three distinct functions: it must permit solute ions to bind onto the concentrator column during the sample loading step, it must transfer quantitatively these solute ions from the concentrator column to the analytical column during the stripping step, and it must provide adequate resolution of the sample components on the analytical column. Clearly these multiple requirements will limit the number of eluents which are suitable for preconcentration methods. The following desirable eluent characteristics may be enumerated for preconcentration using conductivity detection:

(1) Selectivity. The solute ions should elute within a range of capacity factors of 4-30. The lower limit is chosen to minimise interference of early eluting solutes from the relatively large solvent peak which invariably results in preconcentration chromatograms, whilst the upper limit ensures that excessive retention does not preclude reliable quantitation.

(2) Sensitivity. Since the purpose of sample preconcentration is to improve the sensitivity of an ion chromatographic method, the eluent should be chosen to maximise the detectability of the solute ions. For this reason, an eluent anion with low limiting equivalent ionic conductance is prefered. In addition it is desirable that the eluent anion be singly charged in order to provide the greatest detector response to the elution of univalent solute anions.

(3) Eluent pH. Apart from the above considerations, the pH of the eluent exerts two additional important effects on the final chromatogram. In the first place, the presence of neutral, protonated forms of the eluent can result in the appearance of system peaks due to elution of these neutral components under a reversed-phase mechanism. Secondly, bicarbonate ion is present in the majority of samples due to absorption of carbon dioxide from the atmosphere and if quantitation of this species is not required, the resultant large peak can represent a major interference to early eluting solutes. Both of these problems can be circumvented through the use of a fully ionised eluent operated at pH <6. Under these conditions, bicarbonate becomes fully protonated and elutes from the column with the solvent front.

A large number of aromatic carboxylic and sulphonic acids has been evaluated for use as eluents in preconcentration methods (ref. 69). Aromatic monosulphonic acids such as p-toluenesulphonic acid and 2-naphthylamine-1-sulphonic acid have proved to be the most suitable for use with conductivity detection, whilst aliphatic sulphonic acids with methyl-, heptyl- or octyl-side chains are applicable when direct UV absorption detection is employed. In addition the longer chain aliphatic sulphonic acids can also be used with conductivity detection, provided that their surfactant properties are not intrusive.

5.3.3   CONCENTRATOR COLUMN CHARACTERISTICS

It is clear that sample preconcentration is not an open-ended technique and that practical limitations must exist on the amount of sample which can be loaded and recovered quantitatively. It is desirable that large sample volumes can be accommodated and that the sample be loaded at a high flow-rate in order to minimise the time required for the analysis. Studies with fixed-site anion exchange concentrator columns (ref. 70) have shown that the maximum permissible flow-rates and sample volumes are dependent on the nature of the eluent used. As the ion-exchange affinity of the eluent increases, then binding of weakly retained solutes onto the concentrator column reduces markedly with larger sample volumes. However if the eluent conforms to the requirements listed above, sample volumes as high as 100 ml may be loaded at a flow-rate of 8 ml/min with quantitative binding of solute ions being maintained.

It might appear at first sight that the ion-exchange capacity of the concentrator column should be as high as possible in order to provide ample ion-exchange sites for the binding of solute ions. Certainly this situation does encourage quantitative binding, but as the ion-exchange capacity of the concentrator column increases, it becomes more difficult to transfer the bound ions onto the analytical column using a small volume of eluent. Attempts to use a high eluent strength for sample stripping and a lower strength for sample elution have not proved successful (ref. 68), thus the same eluent should be used for both purposes. In this case, the optimal ion-exchange capacity of the concentrator column is approximately 40% of that of the analytical column (ref. 71). Increasing the ion-exchange capacity of the concentrator column beyond this value leads to the requirement for larger strip volumes, causing interference with early eluting solutes and band broadening effects.

The nature of the resin used to support the bonded ion-exchange functionalities can also exert a considerable effect on the preconcentration. Studies have shown (ref. 71) that concentrator columns packed with aminated methacrylate and aminated styrene-divinylbenzene resins of similar ion-exchange capacity gave markedly different performance in breakthrough experiments using a mixture of chloride, nitrate and sulphate. Both chloride and nitrate showed very poor retention on the styrene-divinylbenzene resin in comparison to that obtained using the methacrylate resin.

Finally, it is pertinent to comment on the possibility of replacing the fixed-site ion-exchanger in the concentrator column with a neutral, reversed-phase material which has been coated with a very hydrophobic ion-interaction reagent. In the case of anion preconcentration, this ion-interaction reagent could be cetylpyridinium ion or cetyltrimethyl-ammonium ion. Provided that the ion-interaction reagent remains permanently bound to the stationary phase surface during sample loading, then an ion-exchange column can be produced. The main attractions of this approach are that the nature of the functional group may be varied by recoating the stationary phase with an alternative ion-interaction reagent, and the ion-exchange capacity of the concentrator column can be easily manipulated by altering the conditions under which the concentrator column is coated. It has been shown (ref. 72) that concentrator columns prepared in this way do not show the degree of binding of solute ions expected from consideration of their ion-exchange capacities alone. A comparison of a fixed-site concentrator column with one prepared by the permanent coating method showed that equivalent retention of solute ions was obtained only when the ion-exchange capacity of the latter column was a factor of fifteen times higher than that of the former column.

5.3.4   APPLICATION TO SAMPLES OF LOW IONIC STRENGTH

Most of the samples used for preconcentration methods consist of very dilute aqueous solutions. Examples include rain water, purified water and boiler feed water for power station generators. In each case the sample contains parts-per-billion levels of ionic species and is relatively free from organic impurities. Despite the low levels of ions present, accurate analysis may be essential in order to prevent damage to expensive plant such as steam turbines.

For such samples, a fixed-site ion-exchange concentrator column used with a singly charged aromatic sulphonic acid eluent at pH 6 provides optimal results, expecially when coupled with conductivity detection. A typical chromatogram obtained for the preconcentration of a standard mixture of anions is shown in Fig. 17, which illustrates the excellent separation obtained with this eluent. When applied to the analysis of deionised water containing low ppb levels of ions (Fig. 18), peaks were evident for chloride, nitrate and sulphate. It is noteworthy that the peak widths are similar for Figs. 17 and 18, despite the wide disparity in sample volumes used. An alternative eluent, 2-naphthylamine-1-sulphonic acid, has been applied in Fig. 19 to the determination of anions in water purified by reverse osmosis.

The analysis of samples of the above type is relatively straight-forward and provided that the correct eluent is chosen, the condition of the concentrator column is periodically monitored, and the performance of the system is routinely assessed using recovery experiments, then a high level of confidence can be assigned to the results.

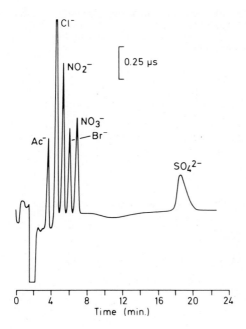

Fig. 17   Preconcentration of a trace solution of anions. Waters IC PAK A analytical column and anion concentrator. Eluent: 3.5 mM toluenesulphonic acid at pH 6.0. Wash volume: 100 μl . Strip volume: 500 μl . Sample: 10 ml of a solution containing 100 ppb of each of the indicated anions, except acetate (Ac⁻) which was present at 500 ppb. Detection: conductivity. Reproduced with permission from ref. 73.

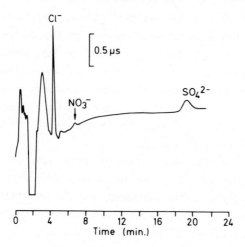

Fig. 18 Preconcentration of anions in deionised water. Sample volume: 100 ml. Solute concentrations: 4 ppb chloride, 0.5 ppb nitrate and 3 ppb sulphate. Other conditions as for Fig. 17. Reproduced with permission from ref. 73.

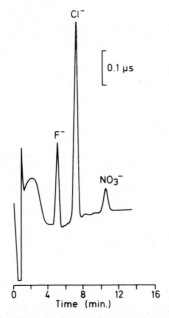

Fig. 19 Preconcentration of anions in water purified by reverse-osmosis. Eluent: 0.4 mM 2-naphthylamine-1-sulphonic acid at pH 6.0. Sample volume: 6 ml. Solute concentrations: 5 ppb fluoride, 20 ppb chloride and 3 ppb nitrate. Other conditions as for Fig. 17. Reproduced with permission from ref. 73.

5.3.5   APPLICATION TO SAMPLES OF HIGH IONIC STRENGTH

It is a challenging prospect to attempt preconcentration analysis of trace components in samples which contain high levels of other ionic species. In such cases it is likely that binding of the trace components to the concentrator column will be adversely affected by mass-action influences of the bulk constituents. Indeed, successful preconcentration can be anticipated only after sample cleanup or when the ions of interest have a much higher affinity for the ion-exchange resin than do the matrix components.

An example of the latter case is the determination of phosphate, sulphate and oxalate in a leaf litter extract taken from coastal vegetation (ref. 73). Preliminary analysis of the extract showed high levels of chloride and nitrate. Two strategies were therefore employed to successfully analyse this sample. First, a high eluent pH was selected in order to convert the phosphate to HPO_4^{2-} and so increase its affinity for the ion-exchange resin in the concentrator column. Second, a styrene-divinylbenzene based ion-exchange material was used in the concentrator column because this resin has been shown to have poor affinity for chloride and nitrate. The final chromatogram obtained using gluconate-borate buffer at pH 8.5 as eluent with a concentrator column packed with aminated polystyrene-divinylbenzene resin is shown in Fig. 20.

In many cases it is beneficial to couple selective detection with sample preconcentration to achieve the desired sensitivity. An example of this approach is the determination of anions using direct UV absorption detection in the wavelength range 200-220 nm. Only a relatively small number of anions show absorbance under these conditions and if a suitable UV-transparent eluent is chosen, then selective analysis is possible.

5.3.6   CONCLUSIONS

Sample preconcentration is a complex procedure and should not be considered to be a simple extension of direct injection ion chromatography. Care must be applied to the selection of the eluent and the concentrator column to ensure quantitative retention on the concentrator column of the solute ions of interest, and their subsequent quantitative transfer onto the analytical column. The ability to optimise the wash and strip volumes of eluent for individual samples provides a high level of flexibility in attaining these goals. The technique can be applied to very dilute aqueous samples of low ionic strength and to more complex samples containing high levels of interfering ionic species.

5.4   USE OF DIALYTIC PRECONCENTRATION METHODS

Trace enrichment using concentrator columns suffers from the disadvantage of being matrix dependent. It is therefore of interest to note that preconcentration can also be performed by Donnan dialysis; moreover, this approach results in the sample ions being transferred to a solution of known composition, regardless of the nature of the sample matrix itself. That is, matrix normalisation occurs.

The principles of Donnan dialysis wherein solute ions are transferred from the sample to a receiving electrolyte solution via an ion-exchange membrane have been discussed in section 3.3.2. When applied to preconcentration, all that needs to be done is to ensure that the volume of the receiver solution is considerably less than that of the sample. Transfer of the solute ions is never quantitative, so the enrichment factors achieved are somewhat less than the volume ratio existing between the sample and receiver solutions. The volume of the receiver solution should therefore be kept as low as possible and typical apparatus (refs. 74, 75) consists of a membrane-covered tube inserted into the stirred sample (as shown schematically in Fig. 21).

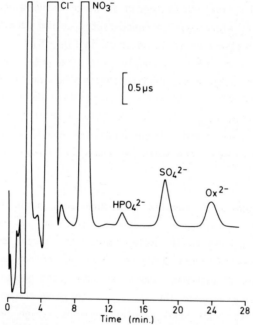

Fig. 20   Preconcentration of an aqueous extract of coastal vegetation leaf litter. Concentrator column packed with aminated styrene-divinylbenzene resin. Eluent: 1.0 mM tetraborate, 4.2 mM boric acid and 1.0 mM gluconic acid. Wash volume: 200 μl . Strip volume: 650 μl . Solute concentrations: 50 ppb phosphate, 150 ppb sulphate and 200 ppb oxalate. Analytical column and detection as for Fig. 17. Reproduced with permission from ref. 73.

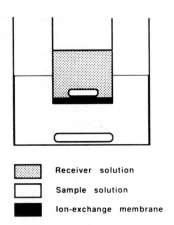

　　□ Receiver solution

　　□ Sample solution

　　■ Ion-exchange membrane

Fig. 21 Schematic representation of apparatus for Donnan dialysis preconcentration.

Whilst sample preconcentration is relatively straightforward to
accomplish, some important steps must be taken before the receiver
solution can be injected onto an ion chromatograph. The reason for this
is the high ionic strength of the receiver, which is essential for
effective Donnan dialysis to occur. In the case of cations, the optimal
receiver solution consists of 0.1 M $Al_2(SO_4)_3$ adjusted to pH 0.5 with
sulphuric acid (ref. 74). When injected into an eluent comprising 0.01 M
tartrate buffer at pH 5, the buffer capacity of the eluent greatly
reduces the hydrogen ion concentration and the aluminium(III) forms a
strong, anionic complex with tartrate and so is effectively eliminated
from interacting with the cation-exchange column. Anions may be dialysed
into a receiver solution comprised of 0.04 M Na_2CO_3 and 0.16 M $NaHCO_3$
(ref. 75) and after dialysis, the receiver solution is subjected to dual
ion-exchange treatment (see section 3.3.2) with a cation-exchange resin
in the hydrogen form. This process converts the receiver components to
water and carbon dioxide, which is removed by application of a vacuum.
The final solution therefore consisted of the solute anions in water.

The enrichment factors achieved in the above examples were approximately 80 for cations after 1 hour of dialysis, and approximately 15
for anions after 30 min of dialysis. Donnan dialysis therefore offers
useful preconcentration but is clearly very limited in its application
because of constraints on the nature of the receiver solution and the
eluent to be used for the final ion chromatographic analysis.

6.   MATRIX ELIMINATION METHODS

6.1   INTRODUCTION

To conclude this chapter, the technique of matrix elimination will be briefly discussed. Matrix elimination can be defined as any <u>instrumental</u> method whereby matrix components are removed from the sample. It is therefore quite distinct from the removal of matrix components using such sample treatment procedures as the use of cartridge columns. Matrix elimination can be applied on-column, usually with the aid of switching valves, or post-column, and has the chief advantage that it can be easily automated. These approaches are described below.

6.2   ON-COLUMN MATRIX ELIMINATION

One simple way of eliminating the detector signal produced by interfering matrix components is to divert the column effluent to waste during elution of these matrix components. This approach has been described for the determination of ppm levels of sulphate in the presence of 0.05% chloride (ref. 76). Here, the sample was injected onto an analytical column and a six-port valve was used to direct the column effluent either to waste or to a second analytical column. The eluent fraction corresponding to chloride was vented to waste, whilst that corresponding to sulphate was passed to the second column and thence to the detector.

Matrix elimination can be combined with sample preconcentration using a concentrator column, provided the matrix components do not interfere with the concentration process. The method is therefore especially suited to samples in which the analytes show very strong retention on the concentrator column. An example of such an application is the determination of aurocyanide in tailings solutions produced from a cyanidation process for the extraction of gold from its ores. These tailings solutions contain very low levels of aurocyanide (10-20 ppb) in the presence of much higher levels of other metal cyano complexes, and require sample preconcentration for reliable analysis. Preconcentration of the tailings solution on a C18 concentrator column conditioned with tetrabutylammonium ion as the ion-interaction reagent yields quantitative binding of gold for sample volumes less than 2 ml (ref. 77). This occurs because of the very strong affinity of aurocyanide for the ion-exchange sites on the column (ref. 78), however other metal cyanides are also retained to some extent and produce severe interference in the final chromatogram (Fig. 22a). Matrix elimination on the concentrator column can be achieved using similar apparatus to that shown in Fig. 15, but adapted to permit the use of two eluents. By careful control of the strength of the eluent and the volume delivered to the concentrator column, it is possible to largely eliminate the interferences without loss of aurocyanide (Fig. 22b). This process can be readily applied to a variety of other sample types.

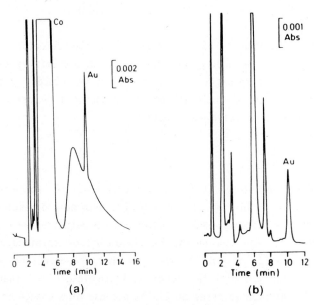

Fig. 22 Matrix elimination in the analysis of gold(I) cyanide in mine
         process liquors. Chromatogram(a) shows the interference of 5 ppm
         of hexacyanocobalt (III) on the preconcentration determination
         of 2 ml of 50 ppb gold(I) cyanide. Chromatogram (b) shows a
         process liquor containing 25 ppb gold and large excesses of
         other metal cyano complexes preconcentrated using the two valve
         system shown in Fig. 15. Separator column: Waters Nova Pak C18.
         Concentrator column: Waters C18 Guard Pak. Eluent: 30:70 (v/v)
         acetonitrile-water containing 5 mM Waters Low UV PIC A. Wash
         volume: 800 μl . Strip volume: 1600 μl . Detection: UV ab-
         sorption at 214 nm. Reproduced with permission from ref. 77.

6.3.   POST-COLUMN MATRIX ELIMINATION

In some cases, it is possible to eliminate a matrix peak after it has eluted from the analytical column. For example, Brown et al. (ref. 79) have observed that the interference of dopamine on the determination of anions in a pharmaceutical product using a phthalate eluent and indirect UV absorption detection could be eliminated by passing the column effluent through a hollow-fibre suppressor device. Under the conditions used, the dopamine became protonated and diffused through the cation-exchange membrane comprising the hollow fibre suppressor, and no interfering peak was monitored by the UV detector.

When the interfering matrix component is UV absorbing and the indirect UV absorption detection mode is employed, it may be possible to eliminate the matrix peak by a judicious choice of detection wavelength. The

78

detector response is given by ref. 3.

$$A = (\epsilon_S - \epsilon_E) \cdot C_S \cdot l \qquad \ldots\ldots\ldots(8)$$

where A is the absorbance change monitored by the detector during passage of the sample, ϵ_S and ϵ_E are the molar absorptivities of the solute and eluent ions, respectively, C_S is the molar concentration of the solute, and l is the path length of the detector flow cell. If a detection wavelength is selected such that:

$$\epsilon_S = \epsilon_E \qquad \ldots\ldots\ldots(9)$$

then no peak should be observed for that particular solute. For example, the peak for nitrate ion in samples prepared by nitric acid digestion can be eliminated using a benzenesulphonate eluent and a detection wavelength of 239 mn (ref. 80). Fig. 23 shows the chromatograms obtained for a mixture of anions at 239 nm and 225 nm. Note that the detection sensitivities for chloride and nitrite change with wavelength and that the peak direction for nitrite reverses at 239 nm because of a change in sign for eqn. 8 at this wavelength.

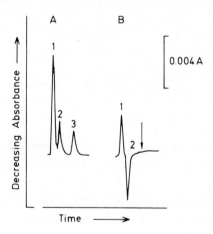

Fig. 23   Elimination of the peak of an absorbing anion (nitrate). Column: TSK-GEL IC Anion PW. Eluent: 1 mM benzenesulphonate. Detection: UV absorption at 225 nm (a0 and 239 nm (b). Peak identities: (1)chloride, (2) nitrite, (3) nitrate. The arrow in (b) shows the elution position of nitrate. Reproduced with permission from ref. 80.

7. CONCLUSION

Sample handling in ion chromatography incorporates many of the same principles governing conventional HPLC methods. Nevertheless, some unique procedures apply to ion chromatography in terms of the sample dissolution methods employed, use of cartridge columns for sample cleanup, contamination effects from cleanup devices and from chromatographic hardware, and sample preconcentration using ion-exchange concentrator columns or dialytic methods. Coupling of these sample handling methods with the powerful separation ability of ion chromatography should ensure that the technique can be applied to a very wide range of sample types.

8. ACKNOWLEDGEMENTS

The author wishes to thank Dr. Allan Heckenberg of the Millipore Corporation and Dr. Art Fitchett of the Dionex Corporation for the provision of unpublished documents relating to sample handling.

REFERENCES

1. D.T. Gjerde and J.S. Fritz, Ion Chromatography, Huethig, Heidelberg, 1987.
2. J.G. Tarter (Ed.), Ion Chromatography, Dekker, New York, 1987.
3. P.R. Haddad and P. Jandik, in J.G. Tarter (Ed.) Ion chromatography, Dekker, New York, 1987, Ch. 4, p. 87.
4. P.E. Jackson, P.R. Haddad and S. Dilli, J. Chromatogr., 295 (1984) 471.
5. W.S. Kim, C.L. Geraci Jr., and R.E. Kupel, Am. Ind. Hyg. Assoc., J., 41 (1980) 334-339.
6. P. Buttini, V. Di Palo and M. Possanzini, Sci. Tot. Envir., 61 (1987) 59-72.
7. D. Grosjean and J.D. Nies, Anal. Lett., 17(A2) (1984) 89-96.
8. P.F. Reigler, N.J. Smith and V.T. Turkelson, Anal. Chem., 54 (1982) 84-87.
9. H. Saitoh, K. Oikawa, T. Takano and K. Kamimura, J. Chromatogr., 281 (1983) 397-402.
10. W.T. Kennedy, W.B. Hubbard and J.G. Tarter, Anal., Lett., 16(A15) (1983) 1133-1148.
11. J. Dolezal, P. Povondra and Z. Sulcek, Decompositon Techniques in Inorganic Analysis, Elsevier, New York 1968.
12. T. Okada and T. Kuwamoto, Anal. Chem., 55 (1983) 1001.
13. S.A. Wilson and C.A. Gent, Anal. Lett., 15(A10) (1982) 851-864.
14. S.A. Wilson and E.A. Gent, Anal. Chim. Acta, 148 (1983) 299-303.
15. C. Mc Crory-Joy, Anal. Chim. Acta, 181 (1986) 277-282.
16. L.M. Busman, R.P. Dick and M.A. Tabatabi, Soli Sci. Soc. Am. J., 47 (1983) 1167-1170.
17. H. Saitoh and K. Oikawa, Bunseki Kagaku, 31 (1982) E375-E380.
18. W.J. Hurst, K.P. Snyder and R.A. Martin Jr., J. Liq. Chromatogr., 6 (1983) 2067-2077.
19. K.L. Evans and C.B. Moore, Anal. Chem., 52 (1980) 1908-1912.
20. K.L. Evans, J.G. Tarter and C.B. Moore, Anal. Chem., 53 (1981) 925-928.
21. V. Abraham and J.M. de Man, J. Assoc. Off. Anal. Chem., 64 (1987) 384-387.

22. J.R. Kreling, F. Block, G.T. Louthan and J. DeZwaan, Microchem. J., 34 (1986) 158-165.
23. P. Viswanadham, D. Smick, J. Pisney and D. Dilworth, Anal. Chem., 54 (1982) 2431-2433.
24. S. Matsushita, Anal. Chim. Acta, 172 (1985) 249-255.
25. H. Small and T. E. Miller Jr., Anal. Chem., 54 (1982) 462.
26. P.E. Jackson and P.R. Haddad, J. Chromatogr., 346 (1985) 125.
27. F.R. Nordmeyer and L.D. Hansen, Anal. Chem., 54 (1982) 2605-2607.
28. H.F. Hamil, USA Environmental Protection Agency Project Summary EPA/600/S2-85/080, 1985.
29. J.A. Cox and N. Tanaka, Anal. Chem., 57 (1985) 385-387.
30. J.A. Cox and N. Tanaka, Anal. Chem. 57 (1985) 383-385.
31. J.A. Cox and E. Dabek-Zlotorzynska, Anal. Chem., 59 (1987) 534-536.
32. B. Kagedal, M. Kallberg, J. Martensson and B. Sorbo, J. Chromatogr., 274 (1983)95-102.
33. P.R. Haddad and M.Y. Croft, Chromatographia, 21 (1986) 648-650.
34. N.M. Ferguson, S.E. Lindberg and J.D. Vargo, Int. J. Environ. Anal. Chem., 11 (1982) 61-65.
35. A.R. Wellburn, New. Phytol., 100 (1985) 329-339.
36. J. Osterloh and D. Goldfield, J. Liq. Chromatogr., 7 (1984) 753-763.
37. D.D. Siemer, Anal. Chem., 52 (1980) 1874-1877.
38. R.A. Hill, J. High Res. Chrom. & Chrom. Comm., 6 (1983) 275-277.
39. L.K. Tan and J.E. Dutrizac, Anal. Chem., 57 (1985) 2615-2620.
40. T. Kamiura, Y. Mori and M. Tanaka, Anal. Chim. Acta, 154 (1983) 319-322.
41. Yu. A. Zolotov, G.I. Malofeeva, O.M. Petrukhin and A.R. Timerbaev, Pure & Appl. Chem., 59 (1987) 497-504.
42. G. Marko-Varga, I. Csiky and J.A. Jonsson, Anal. Chem., 56 (1984) 2066-2069.
43. R.A. Slingsby, Sample Pretreatment with Dionex OnGuard Cartridges, Dionex Corporation Internal Report, 1987.
44. R.W. Frei and J.F. Lawrence (Eds.), Chemical Derivatisation in Analytical Chemistry, Vol. 1, Plenum, New York, 1981.
45. W.G. Robertson and D.S. Scurr, Clin. Chim. Acta, 140 (1984) 97 - 99.
46. P.R. Haddad and A.L.Heckenberg, unpublished results.
47. T. Okada and T. Kuwamoto, Anal. Chem., 57 (1985) 258-262.
48. D.L. DuVal, J.S. Fritz and D.T. Gjerde, Anal. Chem., 54 (1982) 830-832.
49. G.J. Sevenich and J.S. Fritz, Anal. Chem., 55 (1983) 12-16.
50. P.R. Haddad, unpublished results.
51. A.L. Heckenberg, Millipore Corporation Internal Report, 1987.
52. D. Bunk, Millipore Corporation Internal Report, 1987.
53. R. Bagchi and P.R. Haddad, J. Chromatogr., 351 (1986) 541-547.
54. P.R. Haddad, unpublished results.
55. R.A. Mowery, Jr., J. Chromatogr. Sci., 23 (1985) 22.
56. C.N. Trumbore, R.D. Trembley, J.T. Penrose, M. Mercer and F.M. Kelleher, J. Chromatogr., 280 (1983) 43.
57. S.R. Hutchins, P.R. Haddad and S. Dilli, J. Chromatogr., 252 (1982) 185.
58. P.R. Haddad and R.C.L. Foley, J. Chromatogr., 407 (1987) 133.
59. C. Erkelens, H.A.H. Billiet, L. de Galan and E.W.B. de Leer, J. Chromatogr., 404 (1987) 67.
60. D.R. Jenke and G.K. Pagenkopf, Anal. Chem., 55 (1983) 1168.
61. A. Siriraks, J.E. Girard and P.E. Buell, Anal. Chem., in press.
62. D.R. Jenke, J. Chromatogr., 370 (1986) 419-426.
63. A.E. Bucholz, C.I. Verplough and J.L. Smith, J. Chromatogr. Sci., 20 (1982)499.
64. A.L. Heckenberg and P.R. Haddad, J. Chromatogr., 299 (1984) 301-305.

65. T. Okada and T. Kuwamoto, J. Chromatogr., 350 (1985) 317-323.
66. R.A. Wetzel, C.L. Anderson, H. Schleicher and G.D. Crook, Anal. Chem., 51 (1979) 1532-1535.
67. P.R. Haddad and A.L. Heckenberg, J. Chromatogr., 318 (1985) 279.
68. A.L. Heckenberg and P.R. Haddad, J. Chromatogr., 330 (1985) 95.
69. P.E. Jackson and P.R. Haddad, J. Chromatogr., 355 (1986) 87.
70. P.R. Haddad and P.E. Jackson, J. Chromatogr., 367 (1986) 301.
71. P.E. Jackson and P.R. Haddad, J. Chromatogr., 389 (1987) 65.
72. P.R. Haddad and P.E. Jackson, J. Chromatogr., 407 (1987) 121.
73. P.E. Jackson and P.R. Haddad, J. Chromatogr., submitted for publication.
74. J.E. DiNunzio and M. Jubara, Anal. Chem., 55 (1983) 1013-1016.
75. J.A. Cox and N. Tanaka, Anal. Chem., 57 (1985) 2370.
76. P.J. Naish, Analyst (London), 109 (1984) 809-812.
77. P.R. Haddad and N.E. Rochester, J. Chromatogr., submitted for publication.
78. D.F. Hilton and P.R. Haddad, J. Chromatogr., 361 (1986) 141 - 150.
79. D. Brown, R. Playton and D. Jenke, Anal. Chem., 57 (1985) 2264-2267.
80. T. Okada and T. Kuwamoto, J. Chromatogr., 325 (1985) 327.

CHAPTER III

WHOLE BLOOD SAMPLE CLEAN-UP FOR CHROMATOGRAPHIC ANALYSIS

U. CHRISTIANS AND K.-FR. SEWING

1.   INTRODUCTION

From all biological fluids, blood is of the greatest analytical interest, since it is the most important transport medium in the human body and blood levels of most therapeutic and diagnostic substances correlate with their function. This chapter is divided into two main parts: in the first part, general strategies for blood sample handling for various drugs are presented: in the second part, as an example measurement of cyclosporine and its metabolites is discussed in detail. Cyclosporine proved to be a good example since blood level measurement is of high clinical impact and the whole spectrum of sample preparation strategies has been applied for cyclosporine determination.

Developing an adequate sample preparation method several factors must be considered:

1. the chemical properties of the constituents in question.
2. the biological matrix.

The substance of interest must be extracted from its biological matrix prior to chromatographic analysis. Proteins and other macromolecules may interfere with detection and columns may get plugged or rapidly in-activated. A good extraction procedure should be reproducible with little loss of the material of analytical interest. It should be rapid and allow several samples to be analyzed in a short period of time and it should be inexpensive. Because of better handling the use of plasma or serum as biological matrix is preferable over blood. Plasma and serum is produced by removing the cellular components of blood by centrifugation or natural clotting. This step must be regarded as a pre-analysis purification. However, there are some conditions under which drugs require measurement in blood rather than plasma or serum:

1. The partition between corpuscular components and plasma depends on conditions which cannot be easily controlled and/or when the drug is preferentially bound to blood cells (e.g. cyclosporine).

2. The sample volume is small (e.g. in pediatrics, experimental animals or in vitro systems).

3. Blood samples are dehiscent and decomposed so that the production of plasma or serum is impossible (e.g. in forensic medicine).

4. The drug develops its effects in the blood cells (e.g. chloroquine).

5. Blood levels reflect the therapeutic and toxic effects better than plasma or serum levels (e.g. some drugs acting on the central nervous system) (ref. 1).

For development of an extraction procedure pK_a, partitition coefficients in organic solvents and binding to blood components should be available. The distribution in the blood components influences the choice of an adequate matrix. Basic drugs often have a large volume of distribution and are detectable in blood only in low concentrations, especially when administered at low doses (<1 mg/kg body weight) (ref. 2). Most acidic and amphoteric drugs can be quantitatively determined in serum, plasma or urine. They usually remain in the intravasal compartment and have a high affinity to plasma proteins.

Serum is produced by natural clotting of fibrinogen acquainted with a removal of hemo- and lipoproteins, which bears the risk of loosing drugs into the clot. On the other hand plasma, which must be anticoagulated, is rich in lipids and lipoproteins, since these compounds cannot be removed by centrifugation. Freshly drawn blood with its corpuscular components, lipids, lipoproteins and proteins requires a purification step before further extraction, usually hemolysis and protein precipitation.

The following steps of the extraction procedure are similar to those used for extraction from plasma or serum. To remove remaining material interfering with the analysis usually more extended purification steps are required.

The extent of sample purity required depends on the analytical system used and how much impurities the detection system tolerates.

For blood sample preparation various strategies have been applied: liquid-liquid extraction and solid-liquid extraction including column-switch techniques. In most cases anticoagulants are added to blood samples. Anticoagulants like ethylenediamine tetraacetate (EDTA) and acid-citrate dextrose (ACD) contain ultraviolet-absorbing impurities, which can interfere with the material to be detected (ref. 3). The anticoagulant can also influence the accuracy of the measurement (ref. 4), since with EDTA anticoagulated blood can be pipetted more reproducibly than heparin anticoagulated blood particularly after storage for several days.

The effectiveness of the sample preparation procedure depends on the completeness of protein removal. Thus a step of hemolysis and protein precipitation is essential for all extraction procedures from blood, plasma and serum. However, although it is almost mandatory to remove all inert proteins, the components of analytical interest should be recovered. Hemolysis without protein precipitation is achieved by

1. freezing (<-20°C) and subsequent thawing of the sample,
2. ultra-sound
3. osmotic shock.

Other techniques like mechanical membrane disrupture and enzymes are seldom in blood analysis.

The following deproteineation techniques are used (ref. 5):

1. Change in pH by adding a strong acid to the sample (e.g. trichloroacetic acid, perchloric acid, hydrochloric acid).
2. Change in the ionic strength by addition of salts (e.g. ammonium sulfate).
3. Change in temperature by heating the sample and denaturating the proteins.
4. Change in the dielectric constant by addition of organic solvents (e.g. acetonitrile, methanol, ethanol).
5. Filtration and ultra-filtration.

During or after protein precipitation the pH required for the extraction procedure has to be adjusted.

2.   EXTRACTION PROCEDURES FOR BLOOD SAMPLES
2.1   LIQUID-LIQUID EXTRACTION

Blood sample preparation procedures using liquid-liquid extraction can be divided into four main steps:

1. hemolysis and protein precipitation,
2. extraction of the components of interest,
3. purification and removal of interfering materials,
4. volume reduction and reconstitution for chromatographic analysis.

Generally deproteination and hemolysis are associated with pH adjustment for the following extraction step into an organic solvent. Thus the method chosen for deproteination depends on the pK_a of the material to be analysed. Acidic drugs are extractable at pH < 5.5 and basic drugs at a pH > 5.5. In blood one fraction of drugs is bound to plasma proteins and the other blood components and the other fraction is free. By deproteination and extraction the protein bonds must be broken or the recovery may be decreased. A decrease in recovery can also occur if the compounds of interest are co-precipitated or physically entrapped in the protein precipitate. Basic drugs can be extracted from blood without prior procedures by the use of appropriate buffer solutions with a pH ranging from 6 to 14. Usually the pH to be chosen is 3 units above the pK_a because then more than 99% of the basic drug is in its unionized form and can be extracted into an organic solvent. Due to their ionic strengths these buffer solutions cause protein denaturation with minimal loss of the drug (refs. 2,6,7). Acidic drugs can be extracted after adjusting the pH < 5.5. The low pH causes protein precipitation with the risk of co-precipitating the compounds of interest. For liquid-liquid purification of blood samples four strategies are used:

1. The drug is converted into its ionized form by changing the pH and can be extracted into an aqueous phase. The organic layer is removed and discarded. In a second step the lipophilic, unionized form of the drug is re-constituted by changing the pH into the opposite direction and the drug can be back-extracted into an organic solvent (Fig. 2, side columns).

2. The drug is dissolved in an aqueous/organic solvent, e.g. water/ acetonitrile, and the interfering compounds are removed by washing the sample with a lipophilic solvent, that is not miscible with the aqueous layer e.g. hexane. Compounds of interest dissolved in a lipophilic solvent can also be purified by washing with an aqueous solution (Fig. 1, middle column).

Liquid-liquid extraction can be combined with solid-liquid purification steps:

1. The extracted sample is spread on a TLC plate. After development the circle of silica adsorbing the compounds of interest is scraped off the plate and the silica gel is extracted (ref. 8).
2. Interfering materials are removed by adsorption on a solid phase or after liquid-liquid extraction the compounds of interest are adsorbed on a solid phase and thus separated from interfering materials.

Fig. 1 is a schematic flow-diagram of steps used for liquid-liquid sample clean-up of basic and acidic drugs. The diagram demonstrates the way of blood sample preparation for GC and HPLC analysis. For reversed phase chromatography the aqueous back-extracts can be directly injected into the HPLC system The side columns of the diagram show purification steps by basic or acidic back-extraction, the middle column the removal of interfering materials with an organic solvent, inmiscible with the aqueous layer containing the compounds of interest, in which these materials are insoluble.

Table 1 shows an overview of blood sample extraction strategies with subsequent chromatographic analysis. The extraction procedures listed are used for sample preparation prior to chromatographic analysis (HPLC, GC, TLC). Usually the authors describe the extraction from several different kinds of biological matrices but only procedures and values for blood sample preparation and analysis are listed here. The procedures are divided into the four main steps as discussed above. In addition de Silva (refs. 9 and 10) describes schematically the extraction of various 1,4-benzodiazepines and Foerster et al. (ref. 11) the extraction of multiple acidic and neutral drugs from blood. Basic back-extraction describes the following procedure: the sample is acidified and the drugs are extracted into an aqueous phase. Then the pH is raised and the drugs are re-extracted into an organic solvent. If the solvents used for extraction and back-extraction are identical the solvent used for back-extraction is not mentioned.

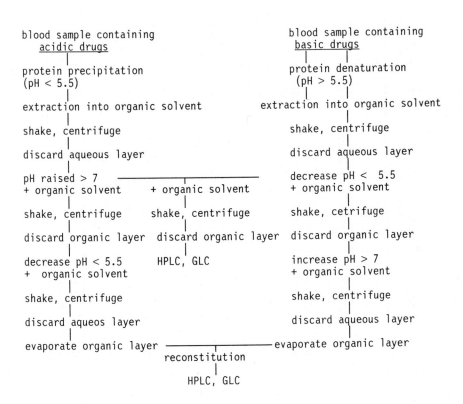

Fig. 1   Liquid-liquid clean-up procedures for blood sample analysis.

TABLE I   Liquid-liquid extraction of blood samples

Substance (s)	pK_a	Ref.	Deproteinization/ pH adjustment	Extraction	Purification/ Derivatization	rec. %	Analysis/ Detection	det. $\mu g/l$
Acebutolol	9.4	12	distilled water, 2 N NaOH	ethylacetate	acidic back-extraction	n.r.	HPLC / UV (240 nm)	10
1-α acetyl-methadol		13	pH 9.2	n-butylchloride	basic back-extraction with $CHCl_3$	>90	GC / MS	5
Acetazolamide	7.2 9.0	14	acetate buffer pH 5	CH_2Cl_2/diethylether /2-propanol (6/4/2)	methylation with trimethylphenyl-ammonium hydroxide	80-90	GC/^{63}NiECD	25
Antipyrine		15	filtration, ethanol	CH_2Cl_2/n-pentane (50/50)		90g± 9.9	HPLC / UV (254nm)	6
Atenolol	9.6	16	0.1 N ammonium-hydroxide	n-butanol/cyclo-hexane (70/30)	basic back-extraction	55	GC/^{63}NiECD	10
Bacmecillinam	6.8	17		CH_3Cl/hexane (1/9)		96-104	HPLC/UV (230 nm)	0.6
Barbiturates	4.0	18		CH_2Cl_2	acidic back-extrac-tion, methylation	83-113	GC/FID	100
Barbiturates	4.0	19		acetone/ether (50/50)	derivatization with methyl iodide and K_2CO_3	70	GC/FID	16 ng/l
Benzodiazepine derivative		20	phosphate buffer pH 9	diethyl ether/ CH_3Cl (70/30)		92± 5.4	HPLC/UV (230 nm)	50

TABLE I   (continued)

Compound	pH	No.	Aqueous phase	Organic solvent	Procedure	Recovery	Method	Detection limit
Benzodiazepine derivatives		21	distilled water, ammonium hydroxide	diethyl ether		70-100	HPLC / UV (240 nm)	n.r.
Butaperazine		1	distilled water, sodium carbonate	n-pentane/iso-propanol (97/3)		82-95	HPLC/UV	<40
Carbamazepine		22		CH_2Cl_2	hexane wash, methylation	94±12	GC/FID	n.r.
Carprofen		23	acetate buffer pH 5	diethyl ether	TLC, derivatization	42±83	HPLC/UV (254 nm)	270
Chloroquine	8.7	24	1.8% barium hydroxide	CH_3Cl		94-105	GC/FID	10
Chloroquine	8.4 10.8	25	buffer pH 10	diethyl ether	basic back-extraction from heptane	n.r.	GC/FID	25
Chloroquine	8.4 10.8	26	distilled water NaOH (pH >11)	hexane/1-pentanol (90/10)	basic back-extraction in $CHCl_3$	85-105	GC/NSD	10 nMol/l
Chloroquine	8.4 10.8	27	distilled water, dipotassium hydrogen phosphate pH 9.5	CH_2Cl_2	acidic extraction into the aqueous phase	73-97	HPLC/UV (254 nm)	3-4
Chloroquine	8.4 10.8	28	deionized water, 0.001 N HCl	hexane	basic back-extraction	95-103	GC/NSD	5-15
Cimetidine	6.8	29	freeze and thaw, 1 N NaOH pH 9.0	1-octanol	basic back-extraction in ethanol	98-106	HPLC/UV (228 nm)	50

TABLE I   (continued)

	pKa	No.	Medium	Extraction solvent	Procedure	Recovery (%)	Method	Detection
Clonazepam	1.5 10.5	30	borate buffer pH 10	isoamyl alcohol/ hexane (10/90)	hydrolysis	35-50	GC/ECD	5
Debrisoquine	11.9	31	distilled water	diethyl ether	basic back-extraction in cyclohexane	n.r.	GC/FID or NSD	3
Diazepam	3.3	32	1 M phosphate buffer pH 7.0	diethyl ether	basic back-extraction	91-116	HPLC/UV (240 nm)	20-30
Diazepam	3.3	33	phosphate buffer pH 7.0	n-heptane	basic back-extraction	90-95	GC/^{63}NiECD	50
Dipyridamole	6.4	34	sodium hydroxide	diethyl ether		93.9	HPLC/FD(285/ 430 nm)	1
Enflurane		35		n-heptane		98-100	GC/TCD	4.1 μMol /l
Flestolol	<8.0	36	acetonitrile	acetonitrile/ CH_2Cl_2 (2/5)	acidic extraction into aqueous phase	38	HPLC/UV (229 nm)	10
Tetrahydro-cannabinol	10.6	37	2 N HCl (pH 4)	hexane/ iso-amyl alcohol (98/2)	acidic back-extrac-tion into hexane	n.r.	TLC/FD	0.4
II$_\delta$-9-tetrahydro-cannabinol	10.6 10.4	38		$CHCl_3$	filtration, TLC	98-100	GC/MS	0.5
Hydroxychloroquine		39	distilled water, ammonia (pH 13)	diethyl ether	extraction of the organic layer into acetonitrile and phosphoric acid	85	HPLC/FD (337/370 nm)	1

TABLE I   (continued)

Compound	pKa	No.	Reagent	Solvent	Procedure	Recovery	Method	Limit
Imipramine	9.5	40	ammonium hydroxide	butanol/hexane (20/80)		92.5	HPLC/FD (240/370nm)	25
Ketamine	7.5	41	ammonium hydroxide pH 10.1	$CHCl_3$: isopropanol (75:25)		n.r.	GC/FI	100
Mefloquine		42	freeze and thaw	isopropyl acetate		86$\pm$9	GC/MS	3
Mefloquine		43	phosphate buffer pH 7.4	ethyl acetate		100$\pm$9.9	HPLC/UV	50
Mefloquine		44	0.2 N H_2SO_4	diethyl ether	wash acidified sample with ether, derivatization	93$\pm$9.7	GC/^{63}Ni ECD and FID	1
Methadone	8.3	45	4 M Na_2CO_3	1-chlorobutane	basic back-extraction into $CHCl_3$	93$\pm$2	GC/FID	5
Morphine	8.0 9.9	46	phosphate buffer pH 8.7 - 9.0	ethyl acetate	aluminium oxide, derivatization	83	GC/^{63}Ni ECD	1
Morphine-3-glucuronide		47	acetate buffer pH 5, ß-glucuronidase	ethylacetate/isopropanol (90/10)	basic back-extraction	81	HPLC/ECD	0.5
Naloxone	7.9	48	1 M carbonate buffer pH 10	diethyl ether	acidic extraction in aqueous phase	78$\pm$3.2	HPLC/UV (214 nm)	1
Pentacaine		49	distilled water, 5M HCl	1,2 dichloroethane	heptane, Na_2CO_3 filtration	75-92	GC/MS	5
Phenobarbital, Phenytoin, Primidone	7.4 8.3	50	glacial acetic acid	$CHCl_3$	basic back-extraction	90-110	HPLC/UV (254 nm)	100 200 300

TABLE I   (continued)

	pH	No.	Buffer/reagent	Solvent	Extraction	Recovery (%)	Detection	Sensitivity
Phentolamine	7.7	51	1 M ammonium hydroxide	diethyl ether	acidic extraction in aqueous phase	83	HPLC/UV (280 nm)	15
Promethazine	9.1	52	borax buffer pH 10	n-heptane/iso-pentanol (99/1)	basic back-extraction	97-99	GC/NSD	5
Propranolol	9.5	53	5 N NaOH	isoamyl-alcohol: n-heptane (1.5:98.5)		n.r.	HPLC/FD	5
Pyramidobenzazepine		54	4.8 M KCl pH 6.1	benzene	derivarization	98 ± 8.9	GC/ECD	2
Theophylline	<1 8.1	55	glacial acetic acid	$CHCl_3$		90-110	GC/FID	1000
Thiopental		56	phosphate buffer pH 5.5	CH_2Cl_2		n.r.	HPLC/UV (290 nm)	200
Tocainide	7.8	57	1 N NaOH, destilled water	CH_3Cl		67-90	HPLC/UV (225 nm)	0.03 ppm
Warfarin	5.0	58	5 N HCl	$CHCl_3$	filtration, wash with aqueous sodium pyrophosphate		HPLC/UV (270 nm)	

ECD: electron capture detector, FD: fluorescence detector, FID: flame ionization detector, GC: gas chromatography, HPLC: high performance liquid chromatography, MS: mass spectrometry, NSD: nitrogen selective detector, TCD: thermal conductivity detector, TLC: thin layer chromatography, UV: ultraviolet absorbance detector, n.r.: not reported.

Another form of liquid-liquid extraction is the use of silica material like Extrelut[R] (refs. 59-62). Though the extraction columns contain solid phase material, the basic principle is a liquid-liquid extraction. Extraction with diatomaceous earth obeys the same basic mechanism (refs. 63-65). Silica gels are porous carrier materials. Water molecules distribute on the surface of the silica gel and become the stationary phase. Compounds are dissolved in the water phase and are eluted from the columns by organic solvents, unmiscible with water. Such columns can be used at a pH range from 1-13. After protein precipitation by acid or buffer the aqueous blood sample is pulled by vacuum through the column (refs. 61,62). Silica gel can also be used for sample purification (refs. 46,59) by absorbing interfering materials from blood without absorbing the components to be eluted.

2.2 SOLID-LIQUID EXTRACTION

Several methods for the extraction of compounds from blood have been reported using solid sorbents as an alternative to liquid-liquid extraction. For the extraction of blood samples the use of solid phase material has the following advantages (refs. 66,67):

1. The formation of emulsions disturbing extraction is avoided.
2. Little volume of solvents are necessary.
3. Acidic drugs can be extracted with high recovery.
4. Fatty acids, their esters and cholesterol are not co-extracted.

The solid-liquid extraction procedures of blood samples can be divided into 4 main steps:

1. hemolysis, deproteination and pH-adjustment,
2. adsorption of the compounds of interest on the solid phase material,
3. purification by washing the adsorbent with lipophilic or hydrophilic solvents,
4. elution of the drugs from the adsorbent,
5. volume reduction and if necessary derivatization.

Liquid-liquid extraction of acidic drugs is sometimes complicated by the co-extraction of lipids and lipoproteins. Co-extraction of these compounds is less in solid-liquid extraction and thus further clean-up steps like back-extractions often reducing recovery are usually not required. For solid-liquid extraction the following solid sorbents are used:

1. Bonded phase silica gels, also available as pre-packed disposable columns (refs. 68-70),
2. anion and cation exchange resins (ref. 71),
3. non-ionic resins like activated charcoal (ref. 72) and Amberlite XAD-2 (refs. 73-75).

Blood samples must be prepared for extraction on bonded phase silica gel columns by hemolysis and protein precipitation before sucked by vacuum through the extraction columns. The columns are previously primed with the same solvent, as used for elution of the drugs from the columns, in order to remove interfering substances. With a polar solvent (usually water) the conditions used for extraction are established by loading the columns with polar groups. The drugs are retained by the column and can further be purified by washing the adsorbed materials with lipophilic or hydrophilic solvents in which they have a small partition coefficient. The columns are also suitable for an extraction by ion-pair chromatography (ref. 68) and can be cleaned and reused. The life span of extraction columns used for extracts from blood is shorter than for those from serum or plasma, since large amounts of blood components like lipoids and lipoproteins are co-extracted and can plug the columns. The columns should not be confronted with solvents with a pH > 9. The extraction procedure with disposable solid phase extraction columns can be automated using extraction systems such as an advanced automated sample processing unit (AASP, Varian, Walnut Creek, CA, USA). The pre-extracted blood samples are automatically loaded on the extraction columns, purified and injected into the HPLC-system (refs. 76,77).

Ionic and non-ionic resins can be added to the blood sample in extraction columns (refs. 78-80), in capsules (ref. 69), in nylon fabric bags (ref. 78) or as resin slurry (refs. 74-75). Anion exchange resins are suitable for the extraction of acidic drugs such as barbiturates, salicylates and phenylbutazone, cation exchange resins for the extraction of basic drugs such as quinidine, chlorpromazine, strychnine, and morphine. Charcoal is ineffective in binding most basic drugs except strychnine and proved valuable in binding non-ionic organic compounds like gluthetimide, meprobamate and carbromal (ref. 71). Ion exchange resins are also used to remove ionic impurities from blood samples (ref. 81) . The anionic resin Amberlite XAD-2, introduced into pharmacological and toxicological analysis by Fujimoto et al. (ref. 82) is a nonpolar styrene-divinylbenzene copolymer with a particle size of 50-100 μm (ref. 66). It allows the extraction of drugs from blood without preceeding deproteination (ref. 80). The resin slurry is prepared by washing the resin subsequently with water, methanol and acetone. The resin is stored in water or a buffer solution (refs. 66,73,75). After adsorption of the compounds of interest the XAD-2 particles are filtered and extracted with an organic solvent.

Schlicht et al. (ref. 66) and Ibrahim et al. (ref. 79) described the extraction of several drugs from blood samples using XAD-2 resins, Ford et al. the extraction of acidic drugs from blood using C_{18} bonded silica columns (ref. 67) and Missen et al. (ref. 75) compared the extraction of benzodiazepines with various resins. The extraction of drugs from blood using solid phase materials is acquainted with some disadvantages that must be taken into account.

1. The extraction may give variable recoveries due to the pH and nature of the eluting solvent and the sorbent.
2. The resins and column materials loose their adsorption efficiency the more often they are reused.
3. The frits and the column material can be plugged by not sufficiently deproteinized samples or if the columns are reused too often.

Since it is often not possible to perform 'digital chromatography' on the extraction columns, an internal standard may help to correct recovery of a compound and to make extraction more reliable and reproducible. A good internal standard should

1. structurally be as similar to the compound of interest as possible,
2. have the same distribution coefficients in organic solvents,
3. have the same binding characteristics to the blood compounds as the compound of interest,
4. have a retention time in chromatographic analysis close to the compound of interest,
5. be clearly separated from the compound of interest during analysis,
6. have the same properties concerning the detection system used.

The internal standard is added in a known amount to the sample prior to sample preparation and analysis. A good internal standard is able to eliminate the bias caused by losses and compensates random errors during extraction or analysis (ref. 83).

Fig. 2 shows a flow-chart of solid-liquid extraction procedures. If XAD-2 material for the extraction of neutral and basic drugs is used it can be renounced at the deproteination step (ref. 80). The purification step can also be performed after elution of the drugs from the solid sorbent using liquid-liquid extraction.

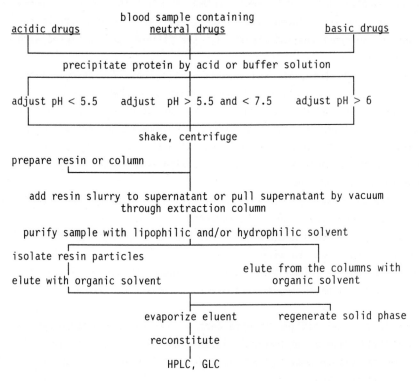

Fig. 2   Extraction of blood samples by liquid-solid extraction

2.3   COLUMN-SWITCHING

On-line sample preparation using column-switching has been applied to plasma, serum and urine samples and is discussed in detail in Volume I. Blood sample analysis requires a preceeding purification step and is basically equal to analysis in plasma, serum or urine. Column-switching techniques for cyclosporine blood samples are described in part 3.2.3 of this chapter.

3.   BLOOD SAMPLE PREPARATION AND HPLC ANALYSIS OF Sandimmun[R] (CYLOSPORINE)
3.1   INTRODUCTION

Sandimmun[R] (Cyclosporine A, cyclosporine, Sandoz OL 27-400 N) is an immunosuppressive agent and its application after organ transplantation has proved to be of great value (refs. 84-85). Due to its narrow thera-peutic range and its pharmacokinetic properties, blood level monitoring is mandatory. (ref. 86).

Simultaneous measurement of the parent compound and the cyclosporine metabolites in blood by HPLC is of great clinical relevance, since the commercially available and commonly used monoclonal radioimmuno assay (RIA) kits (Sandoz) (ref. 86) measure the parent compound or all metabolites to an mostly unknown extent. With HPLC it is possible to determine the metabolites and to quantify each of the metabolites separately. This will be of special value if one or more of the metabolites prove to be responsible for the cyclosporine adverse effects especially nephrotoxicity.

Cyclosporine is a neutral, lipophilic and cyclic undecapeptide with a molecular weight of 1202.6. All its amino acids are S-configurated except D-alanine in position 8 (Fig. 1). Amino acids in positions 1, 3, 4, 6, 9, 10, and 11 are N-methylated. The amino acid in position 1 is a ß-hydroxilated, N-methylated and unsaturated C9-amino acid. The tertiary structure of cyclosporine is an antiparallel ß-pleated sheet conformation. Its partition coefficient octanol/water is 120/1. The cyclosporine molecule lacks of chromophoric substituents, making UV-detection more unspecific and demanding more extensive extraction procedures. The molar absorption coefficient at the wave-length maximum (195 nm) is 66 000 1/mol x cm. It shows good solubility in alcohols, ether, acetone and chlorinated hydrocarbons and poor solubility in water and saturated hydrocarbons (refs. 88-91). Cyclosporine is metabolized by microsomal cytochrome P450 (ref. 92) in the liver to more than 30 metabolites (ref. 93). The structures of the metabolites 1, 8, 9, 10, 13, 16, 17, 18, 21, 25, 26 (refs. 94 and 95), 203-218 (ref. 96) and two aldehyde metabolites (ref. 97) have been elucidated. All metabolites retain the cyclic undecapeptide structure and prove to be more hydrophilic than the parent compound. The reactions involved in cyclosporine degradation are demethylation, hydroxilation, oxidation and cyclization (Table II).

Choice of the biological matrix (ref. 98)

For routine drug monitoring cyclosporine is usually measured in blood. However, the question of the biological matrix is still under discussion. 58% of cyclosporine are bound to the erythrocytes in blood, 10 to 20% to the lymphocytes. In plasma cyclosporine is bound to lipoproteins, preferentially those of high and low density (refs. 99-103). The free fraction is 1-1.5% at 37°C (ref. 104). The distribution between blood and plasma is temperature dependent and is lowered from 37°C to room temperature (refs. 99, 105-108). Binding of cyclosporine to the lipoproteins is also temperature dependent being highest at body temperature and decreasing linearly with lower temperature. The cyclosporine metabolites 1 and 17 are associated with the erythrocytes

(>90%) (refs. 109-111). The relative distribution in blood is constant until cyclosporine concentrations > 1000 $\mu g/l$. Furthermore the relation between concentration in blood and plasma varies significantly with the hematocrit (refs. 112 and 113). The reasons for choosing blood as the biological matrix are:

1. There are no technical problems because of the temperature dependent distribution between erythrocytes and plasma.
2. Measurement is independent of the hematocrit.
3. Plasma levels are considerably increased in hemolysed blood samples.

The choice of the anticoagulant used for cyclosporine blood samples proved to be of importance. In routine heparinized specimens stored > 1 days contain small blood clots. Since cyclosporine is bound to a great percentage to the corpuscular blood components clotting causes a decrease in the concentration measured and the clots plug the extraction columns in solid-phase extraction procedures (refs. 4, 98, 114-116).

The methods developed for the quantitative determination of cyclosporine and its metabolites in blood cover almost the whole spectrum of blood sample preparation strategies.

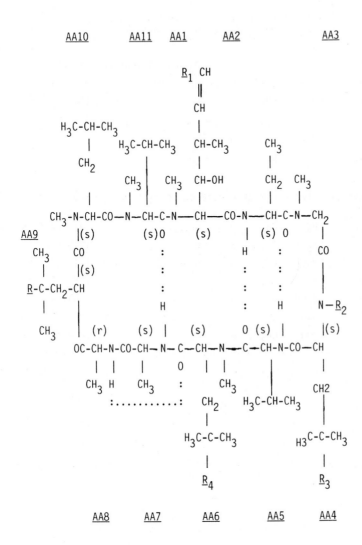

Fig. 3   Structural formula of cyclosporine.

TABLE II Structures of the cyclosporine (Cs) metabolites, hitherto characterized (refs. 94-96) with their molecular weights.

Metabolite	R	R_1	R_2	R_3	R_4	modifications	weight
Cs	H	CH_3	CH_3	H	H		1202.64
1	OH	CH_3	CH_3	H	H		1218.64
8	OH	CH_2OH	CH_3	H	H		1234.64
9	OH	CH_3	H	H	OH		1220.62
10	OH	CH_3	CH_3	OH	H		1234.64
13	hydroxylated and N-demethylated derivative of Cs						1204.62
16	OH	CH_3	CH_3	H	OH		1234.64
17	H	CH_2OH	CH_3	H	H		1218.64
18	H	CH_2OH	CH_3	H	H	AA1:cyclization	1218.64
21	H	CH_3	H	H	H		1188.62
25	H	CH_2OH	H	H	H		1204.64
26	OH	CH_2OH	CH_3	H	H	AA1:cyclization	1234.62
203-218	H	COOH	CH_3	H	H		1232.62

TABLE III Characteristics of various HPLC procedures for quantitative determination of cyclosporine

		Extraction			HPLC			CV	
Ref.	matrix	preparation	extraction	clean-up	recovery	column	elution	det.-limit (μg/l)	%	comments
117	plasma urine	+ water	diethyl ether	-	76±5% 104±5%	RP8	gradient	20	n.r.	
118	plasma		extraction identical with ref. 117		n.r.	RP8	isocratic	5	4.4	derivatization of cyclosporine with 2-naphthylene-selenylchloride
119	blood plasma	+ distilled water	diethyl ether	hexane	74% 49%	RP18	isocratic	25	9.2	
120	serum	methanol	C18 cartridge (Sep-Pak, Waters)	water/ methanol	90±10%	RP18	isocratic	n.r.	n.r.	
121	plasma blood	heat (55°C), freeze + thaw	column-switching		91.9±0.9%	RP8, RP18	step gradient	25	3	
122	serum		CN cartridge (Baker)	water/ methanol	50-70%	RP18	gradient	50	9.3-14.1	
123	plasma		Clin Elut cartridge (Fisher)		77.3±6%	TMS	isocratic	100	n.r.	
124	plasma blood	acidification (HCl)	diethyl ether	NaOH	83-99%	RP8 ultra-sphere	isocratic	31	3.6-6.0	

TABLE III (continued)

125	blood serum	Tris-buffer pH 9.8	diethyl ether/ CN cartridge (Baker)	aceton-nitrile/water, hexane	34.7%	RP8	isocratic	25	21	automated sample preparation
126/ 127	plasma blood	acetonitrile/water, column-switching		hexane	100%	RP8, RP18	column-switching	5 15	0.5-11.1	
81	blood	freeze + thaw acetonitrile		hexane, Dowex ion exchange resin	90%±5%	RP18	isocratic	25	0.3-8.0	
128	blood plasma	freeze+thaw buffer pH 10	diethyl ether	acidifi-cation, hexane	74% 85%	RP18	isocratic	25	7.0	
129	serum	phosphoric acid in acetonitrile	colum-switching	n.r.	n.r.	RP8, RP18	column-switching	n.r.	2.5-12.5	modification of refs. 126, 127
130	blood	freeze+thaw acetonitrile	charcoal slurry ethyl acetate		80%	RP18	isocratic	50	8.6	
131	blood, plasma	-	diethyl ether	alkalized acidified methanol	96±6%	CN	isocratic	100	6.04	
132	blood	10% iso-propanol in acetonitrile	C18 cartridge (Baker Bond)	70% methanol	86±108%	CN	isocratic	50	n.r.	

TABLE III (continued)

	Sample				Recovery	Column	Elution	Time (min)	pH	Remarks
110	blood serum, plasma	acetonitrile/ water (30/70, v/v)	CN cartridge (Bond Elut^R)	acetonitrile/ acetic acid	90% 98%	CN	isocratic	15	2.6-6.5	determination of a cell-bound metabolite
133	blood	acetonitrile/ dimethyl sulfoxide	C18 cartridge (Bond Elut^R)	acetonitrile water, hexane	75-80%	CN ultrahere	isocratic	<10	6.4-6.6	
134	blood	--	diethyl ether	acidification, hexane	n.r.	RP18	isocratic	n.r.	n.r.	modification of ref. 119
135	blood	acetonitrile	column-switching	acetonitrile/ water	75±3%	Ultrapore RPSC (Altex) RP18, 3 μm	column-switching	2	0.1-6.2	
136	blood	HCl	diethyl ether	NAOH	n.r.	RP8	isocratic	20	1.8-4.5	
137	blood	HCl	diethyl ether	heptane	>90%	RP8, 3 μm	isocratic	10	3.9-5.7	ion-pair chromatography (ammonium sulfate)
138	blood	acetonitrile	column-switching		98.4-100.2%	CN, TMS	column-switching	5	0.7-1.8	
139	blood	freeze+thaw	diethyl ether	hexane	108%	RP6, RP18	column-switching	25	5.7-6.3	extraction equal to ref. 119, microbore analytical column

TABLE III   (continued)

Ref, sample									
140, 141 blood	acetonitrile/methanol(9/1)	C18 cartridge (Bond Elut)	hexane, silica gel cartridge	n.r.	RP1	isocratic	25	4.9	
142 blood	HCl	methyl-t-butyl-ether	NaOH, heptane	88-90%	RP8 microbore	isocratic	1.5	4.8-5.9	small sample volume (0.2-0.5 ml)
76 blood	acetonitrile/dimethylsulfoxide	C8 cartridge (Bond Elut)	hexane, acetonitrile/water	98-104%	RP18, 3 μm	isocratic	20	4.5-7.8	extraction with advanced automatic sample unit AASP (Varian)
143 blood		C18 cartridge (Bond Elut)		95-108%	CN	isocratic	15	6.6-6.9	modification of ref. 135
144 plasma, bile, urine	HCl	diethyl ether	NaOH	75-83.9 $\pm$7.7%	RP18	isocratic	50	18.4-5.6	
77 serum	acetonitrile	C8 cartridge (Analytichem)	hexane, acetonitrile/water	87%	CN, 3 μm	isocratic	20	3.8-12.5	extraction with advanced automatic sample unit AASP (Varian), normal phase chromatography
145 blood, plasma, bile	HCl	diethyl ether	NaOH	>96%	RP18, 3 μm	gradient	22	5.6	determination of metabolite 1, 17, 18, 21

TABLE III   (continued)

No.	Sample	Extraction	Clean-up	Mobile phase	%	Column	Mode			Remarks
146	blood, plasma	diethyl ether	silica cartridge (Sep Pak, Waters), acetyl-acetate/hexane		45±2%	RP18	isocratic	10–20	6	
147	blood	HCl, diethyl ether	heptane, NaOH, hexane		70%	RP8, 3 μm	isocratic	25	5.3–11.5	
148	blood	acetonitrile/water (30/70)	CN cartridge (Bond Elut R)	acetonitrile/water	47–95%	CN isocratic alternatively RP 8, silica gel semi-preparative isolation of metabolites		15–25	7.1–9.6	determination of metabolite 1, 8, 13, 17, 18, 21, 25, 26, 203–218 and 1 yet un-identified metabolite
149 150 151	blood, bile, urine	acetonitrile/ water(30/70)	C8 extraction columns	acetonitrile/ water, hexane	73–85%	RP8	gradient	25	5.6–12.6	determination of metabolite 1, 8, 9, 10, 13, 16, 17, 18, 21, 25, 26, 203–218 and 7 yet un-identified metabolites

n.r.: not reported, RP: reversed phase.

3.2   BLOOD SAMPLE PREPARATION FOR SANDIMMUN[R] (CYCLOSPORINE) MEASUREMENT

3.2.1 LIQUID-LIQUID EXTRACTION PROCEDURES

All methods published until now use cyclosporine C or D as internal standard. The extraction procedures consist of four steps:

1. hemolysis and deproteination,
2. extraction of cyclosporine,
3. sample purification,
4. volume reduction and transfer into the mobile phase

Hemolysis was achieved by rapid thawing and freezing (ref. 129) or by adding distilled water (ref. 119), acetonitrile (ref. 81) and hydro-chloric acid (137).

In routine analysis cyclosporine was extracted from blood by diethyl ether (refs. 117,119,124,127,128,136-138), methyl-t-butyl ether (ref. 142) and acetonitrile (ref. 81). The advantage of methyl-t-butyl ether over diethyl ether are its resistance to peroxide formation and cleaner extracts than obtained by diethyl ether extraction (ref. 142). The use of acetonitrile combines its properties as an effective extraction solvent and its protein precipitating potency (ref. 81). Since the extracts contain interfering lipophilic material and acidic, basic and ionic contamination (ref. 81) which may cause damage to the column, purification steps are required. Purification was achieved by washing the sample with hexane or heptane (refs. 119,128,134,137,139,142) with acidic and basic solutions (refs. 124,128,131,134,142) or by adding ion exchange resins (ref. 81). After evaporation of the purified layer and resuspension in the mobile phase, some methods use a second purification step by extracting interfering substances with a final hexane or heptane wash (refs. 137, 142). Back extraction of cyclosporine from an aqueous phase by changing pH is not possible because of its chemical properties. Thus the organic layer containing cyclosporine is washed by basic and acidic solutions and the aqueous layer has to be discarded in either case. Gfeller et. al. (ref. 118) used a derivatization of cyclosporine with 2-naphthylselenylchloride to improve the detection limit.

The method of Sawchuck and Cartier (ref. 119) introduced a hexane washing step into cyclosporine analysis and many later published liquid-liquid extraction methods used a modification of this extraction pro-cedure (refs. 128,137,139,142). Blood, distilled water and the internal-standard Cyclosporine D were given into a centrifuge tube. Diethyl ether was added and the sample shaken and centrifuged. The aqueous phase was discarded and the organic layer was evaporated. The sample was taken up in methanol and was washed with hexane twice. The hexane layers were removed, the aqueous layer was basified with NaOH and cyclosporine was extracted by diethyl ether. The diethyl ether phase was evaporated and the remaining materials were reconstituted with the mobile phase.

Most of these extraction procedures use an internal standard: Cyclosporine D or C. Cyclosporine D is cyclosporine with valin and cyclosporine C with threonine as amino acid 2 (Fig. 3). These cyclosporine derivatives represent only a small modification of the whole molecule. They have distribution coefficients in organic solvents equal to cyclosporine and almost the same UV-absorbing properties. The use of these internal standards for the quantification of cyclosporine metabolites is critical (ref. 152). The behavior during extraction is considerably different from the metabolites as shwon in bile in ref. 151 and Table V.

3.2.2 SOLID-LIQUID EXTRACTION PROCEDURES

Until now all column extraction methods described for cyclosporine included 5 steps:

1. Hemolysis of the corpuscular blood ingredients and deproteination,
2. sample loading on the extraction column,
3. sample purification,
4. elution of cyclosporine and its metabolites from the extraction column,
5. volume reduction for HPLC analysis.

Yee et al. (ref. 122) used a protein precipitation step with acetonitrile containing the internal standard Cyclosporine D. The sample was then pulled by vacuum through a prepacked disposable cyanopropyl column, being washed with acetonitrile and water. The column was washed with methanol/water 40/60 (v/v) and cyclosporine was eluted using methanol.

Kates et al. (ref. 125) combined a diethyl ether extraction with purification on prepacked disposable cyano columns. Blood samples were adjusted at pH 9.8 and extracted with diethyl ether. Diethyl ether was evaporated, the sample dissolved in methanol/water was diluted with water and drawn through the column with water. The sample was subsequently cleaned by acetonitrile/water 25/75 (v/v) and hexane was then eluted from the column with methanol.

The method developed in our laboratory (refs. 149,150,151, Fig. 4) uses 3 ml glas extraction columns filled with 25-40 μm RP 8 material (LiChroprep[R], Merck, Darmstadt, FRG). The internal standard Cyclosporine D was dissolved in acetonitrile/water (pH 3.0) 50/50 (v/v) at a concentration of 10 μg/ml. 25 μl of the internal standard solution were pipetted into a 10 ml centrifuge tube. Subsequently 1 ml blood and 2.1 ml acetonitrile/water (pH 3.0) (30/70 v/v) were added. Each sample was vortexed for 20 s and centrifuged for 5 min at 2 500 rpm. The supernatant was pulled by vacuum through the extraction columns. The extraction columns were previously primed with 3 ml acetonitrile and 3 ml water. The samples were washed with 3.2 ml acetonitrile /water (pH 3.0) (20/80 v/v) and with 0.5 ml hexane. The column was dried by sucking air through it for 1 min. To elute cyclosporine and its metabolites the extraction column was set into a diethyl ether cleaned 10 ml centrifuge tube and 2 ml dichloromethane was centrifuged through the extraction columns (700 rpm, 5 min). Dichloromethane was evaporated and the remaining materials were taken up in 300 μl acetonitrile/ water (pH 3.0) (50/50 v/v). 500 μl hexane were added and the sample was vortexed for 10 s. Phases were separated and 100 ml of the aqueous phase were injected into the HPLC system.

This extraction procedure is a modification of the method published by Lensmeyer and Fields (ref. 110). The first step of the extraction procedure consists of adding a mixture of acidified water (pH 3.0) /acetonitrile (30/70 v/v) resulting in a final acetonitrile concentration of 20% in the sample. According to ref. 110 gross protein precipitation occurs at a final acetonitrile concentration of more than 21%. At the acetonitrile concentrations reached in the sample blood cells are hemolysed and some protein blood components precipitate. The recovery is considerably lower at a higher pH of the dilution mixture. The recovery drops to about 20% when gross protein precipitation occurs due to plugged extraction columns. Critical conditions for gross protein precipitation are high temperatures over 25°C as reached when centrifuging the sample in a warm centrifuge. Another reason for a decreased recovery is the extraction of deep frozen or samples stored at +4°C for more than 1 week. The first step also adjusts the sample to the conditions required for column extraction. After centrifugation the supernatant which has a clear red color is given onto the extraction columns. We chose no commercially available disposable prepacked columns but refillable glas columns with removable teflon frits for the following reasons:

1. To reduce costs of external column extraction procedures the extraction columns are reused. The more often they are reused the more reproducibility and recovery decrease and the chance of loosing a

sample because of a plugged column increases. After analysis the solid phase of the glass columns is removed and the frits reusable for at least three times are cleaned by ultra-sound in acetonitrile.

2. One of the main problems of cyclosporine analysis are interfering materials like plastic softeners which have similar chromatographic and spectral properties like cyclosporine itself. It has been reported that interfering material can be leached from the extraction columns. Glass is an inert material.

A 'digital chromatography' of cyclosporine on the extraction columns is not possible resulting in variable recoveries of 70-95% in our system. In our method this variability can be compensated by using an internal standard (Table IV). The columns are filled with 100 mg RP 8 solid phase material. Variation of the packing volume up to 50% does not influence recovery. pH adjustment of the sample to an acid pH increases retention on the columns especially of the carboxylated metabolite 203-218. After loading cyclosporine and its metabolites onto the extraction columns, the sample is washed with an acetonitrile/water (pH 3.0) (80/20 v/v) mixture, decreasing the amount of potentially interfering material. During this step the recovery is not reduced when the water is acidified. The next step, washing the column with hexane is to remove lipohilic impurities. cyclosporine and its metabolites are almost insoluble in hexane. Up to this step the solvents are sucked through the column by vacuum. To elute the compounds of interest from the extraction column the columns are set in centrifuge tubes. The eluent dichloromethane is centrifuged through the extraction columns and the eluate containing cyclosporine and its metabolites is collected at the bottom of the centrifuge tube. In contrast to other eluents like methanol or acetonitrile dichloromethane can faster be evaporated. The amount of coeluted interfering material is equal to an elution by other solvents. For evaporation of dichloromethane an apparatus equipped with glass tubes for nitrogen insufflation should be used, since plastic tubes are a potential source of pollution of the sample with plastizisers (Fig. 10). The final hexane wash used in our method removed interfering material, stemming from the laboratory equipment. This step was not essential but it made extraction more reliable.

With slight modifications the method could be adapted to analyse urine and bile samples. 1 ml urine was pipetted into a 10 ml centrifuge tube and 300 μl acetonitrile were added. After centrifugation the supernatant was passed through the extraction columns. 1 ml of a bile sample and 2 ml acetonitrile/water (pH 3.0) (30/70 v/v) were washed with 3 ml hexane. The acetonitrile/water (pH 3.0) phase was separated and the hexane layer discarded. After loading cyclosporine and its metabolites on the extraction columns the extraction procedure was continued as described for blood samples.

```
                    1 ml blood
                  + 25 μl internal standard (containing 250 ng
                                               cyclosporine D)
                  + 2.1 ml acetonitrile/water (pH 3.0) 30/70 (v/v)
                            |
                            |
    extraction column           shake 20 s (vortex-mixer)
  + 3.2 ml acetonitrile         centrifuge 5 min, 2500 rpm
  + 3.2 ml water (pH 3.0)       |
         └──────────────────────┤
                            |
       draw supernatant through the extraction column
                            |
       + 3.2 ml acetonitrile/water (pH 3.0) (20/80 v/v)
       + 0.5 ml hexane
       dry column by air stream
                            |
    centrifuge 2.0 ml dichloromethane through the column
                        ├──────────────┐
                        | remove column material
                        | clean teflon frits
                        |
    evaporate eluate at 50°C under a stream of nitrogen
                            |
       + 0.3 ml acetonitrile/water (pH 3.0) (50/50 v/v)
       + 0.5 ml hexane
                            |
             shake 20 s (vortex-mixer)
             centrifuge 2 min, 2500 rpm
                            |
            inject 75 μl into HPLC-system
```

Fig. 4   Extraction of cyclosporine and its metabolites from blood by solid liquid extraction (refs. 149, 150, 151)

Modifications for the extraction of urine samples

```
        1 ml urine
    +   25 µl internal standard
    +  300 µl acetonitrile
                        |
       vortex mix (40s)
       centrifuge 2 min 2500 x g
                        |


    (extraction continued like blood)
```

Modifications for the extraction of bile samples

```
        1 ml bile
    +   25 µl internal standard
    +   2.1 ml acetonitrile/water (pH 3.0) (30/70 v/v)
    +   2   ml hexane
                        |
       vortex mix (1 min)
       centrifuge 2 min 2500 x g
                        |                   discard
                        |_____
                                            |
       suck acetonitrile/water phase        hexane layer
       through extraction columns
                        |


    (extraction continued like blood)
```

Reproducibility, linearity and detection limit of the method used in our laboratory are listed in Table IV.

TABLE IV   Calibration curve, detection limit and CV of the method described above

	Calibration curve range checked	r	detection limit	CV
blood	0-3 mg/l	1.0	25 µg/l	6.3%
bile	0-6 mg/l	0.989	50 µg/l	7.2%
urine	0-30 mg/l	0.996	50 µg/l	12.3%

The CV includes the variation of the cyclosporine metabolites. The recovery in blood ranged from 72-85% with an average of 79.2%. The recoveries of the metabolites 8, 26, 17 and the internal standard Cyclosporine D are shown in table V. The cyclosporine metabolites, the parent compound and the internal standard differ in their lipophilic properties. Thus, the possibility must be taken into account that the recoveries of these compounds are not identical during the extraction procedure. In the table it is shown that the recovery of the internal standard is significantly (p<0.001) different from the cyclosporine metabolites. Cyclosporine D cannot be used as internal standard for the cyclosporine metabolites in bile. Cyclosporine D ist almost insoluble in hexane. 85% of the internal standard were removed into the hexane layer during the first hexane washing step. The large amounts of lipids and lipoids extracted from bile samples into the hexane phase caused coextraction of Cyclosporine D. The recovery of the parent compound was also affected but not the recoveries of the metabolites. Cyclosporine and its metabolites can be measured concomitantly in bile by adding one internal standard (Cyclosporine D) for cyclosporine to the sample before and a second one (Cyclosporine C) for the cyclosporine metabolites after the first hexane wash (ref. 150). A significant discrepancy between the recoveries of the metabolites in bile, in urine and blood samples could not be found.

TABLE V   Recoveries of the cyclosporine metabolites and the internal standard Cyclosporine D from bile and urine samples

	urine			bile		
Metabolite	Recovery % ± SD		n	Recovery % ± SD		n
Cyclosporine D	78	± 18	8	15.2	± 4.2	7
Metabolite 8	73.4	± 8.9	3	78.7	± 8.2	3
Metabolite 26	78.1		2	71.8	± 5.2	3
Metabolite 17	85.1	± 4.2	3	69.4	± 16.6	3
AV (metabolites)	79.5	± 6.9	8	73.3	± 10.7	9

AV: average.

Kabra et al. (ref. 76) described an automated external column extraction by using an advanced automated sample processing unit (AASP, Varian, Walnut Creek, CA, USA) online with the LC. Blood samples were preextracted with acetonitrile/dimethyl sulfoxide (94/4 v/v) and hexane. C8 extraction columns were found to give a good recovery and precision (refs. 76 and 77). The columns were primed with acetonitrile/water (2/3 v/v) and the aqueous phase + 1 ml water followed by acetonitrile/water (2/3 v/v) were passed through the columns. The columns were loaded on the AASP and purged with acetonitrile/water (12/13 v/v).

Wallemacq et al. (ref. 77) also described a cyclosporine extraction procedure using the AASP, applicated to serum samples. This method combines the standard reversed phase extraction with normal phase chromatography. With a modification it is possible to extract and analyse already structurally elucidated and until now unknown cyclosporine metabolites. The authors tested C18, C8, C2, 2-OH, CN and Si extraction column for their extraction procedure summarized below and found C8 material to give the best results. The recoveries for these columns were 73% (C18), 70% (C2), 64% (CN), 64% (2-OH), 20% (SI) and 87% for C8. For protein precipitation acetonitrile and the internal standard were added to a serum sample. The sample was washed with hexane and the acetonitrile/water phase was separated and diluted with water into the AASP cartridges. Using the AASP the sample are purified with acetonitrile/water (33/67 v/v) and hexane and dried by sucking air through the extraction columns prior to injection into the HPLC system.

To move interfering components of the blood samples Moyer et al. (refs. 140 and 141) developed a double cartridge solid phase extraction procedure using two subsequent extraction columns filled with C18 bonded phase and silica gel. After protein precipitation with acetonitrile/methanol (9:1) the supernatant is rinsed through the C18 extraction column by vacuum and is washed with water/methanol (70/30 v/v) and acetone/hexane (1/99 v/v). The cyclosporins were eluted with iso-propanol and ethyl acetate (1/3 v/v) and the eluate was passed through a silica gel extraction column, which did not retain the cyclosporins but interfering materials. After evaporation and reconstitution in the mobile phase the sample was injected into the HPLC.

Another approach was the extraction of cyclosporine by charcoal resins (ref. 133). After hemolysis by freezing and thawing and extraction with acetonitrile 5 ml aqueous slurry containing 10 mg charcoal were pipetted to the supernatant. After agitating the mixture on a shaker and centrifugation the aqueous portion was decanted and the remaining charcoal extracted with ethyl acetate. The ethyl acetate was evaporated and the residue reconstituted in the mobile phase.

3.2.3 EXTRACTION AND ANALYSIS BY COLUMN-SWITCHING

On-line column-switching techniques have been applied in most cases in the analysis of drugs in plasma or serum and urine (refs. 153 and 154). Blood sample clean-up with this technique has been developed to determine cyclosporine. On-line column-switching techniques have the following advantages compared with liquid-liquid and solid-liquid extraction procedures (ref. 155):

1. minimal loss of the material to be analysed,
2. short total analysis time,
3. highly selective and reproducible analysis because of fully automated sample preparation,
4. facilitated concentration in trace analysis.

Column-switching techniques for purification and analysis of blood samples consist of the following four steps:

1. hemolysis, deproteinization and pH-adjustment,
2. sample purification on a precolumn,
3. separation on the analytical column,
4. column wash procedures and reequilibration of the system.

In contrast to plasma or serum blood contains more lipids and proteins, which cause column overloading and an increase in column back-pressure if the sample is injected directly requiring a short sample preparation prior to injection. Hemolysis and deproteinization is achieved by adding equal amounts of acetonitrile to the blood samples. This mixture is shaken and centrifuged and the supernatant containing the material to be analysed is loaded on the extraction column inside the HPLC-system (refs. 121, 135, 138). On the pre-column the sample is purified by washing with non-eluting organic solvents or organic solvent/water mixtures. By combining front-cut and end-cut procedures (ref. 155) early and late eluted components are eliminated and only the relevant part of the chromatogram is switched to the analytical column (heart-cut) (ref. 155). Most column-switching blood sample clean-up techniques for cyclosporine use a heart-cut procedure (ref. 121, 135, 138). Most column-switching techniques described for cyclosporine analysis use no internal standard but an actual calibration curve for cyclosporine quantification. They show excellent reproducibility, recovery and detection limit (Table III).

The column-switching technique described by Nussbaumer et al. (ref. 121) was a modification of the column-switching technique proposed by Erni et al. (ref. 155). After protein precipitation with acetonitrile the sample was injected onto a 40 x 4.6 mm 30-40 μm RP 8 column kept at room temperature where it was washed. Purification and analysis of the sample was performed by a step gradient using subsequently methanol/water/acetonitrile, water/acetonitrile, tetrahydrofuran and methanol as eluents. After clean-up the cyclosporine containing fraction was switched by forward flushing the pre-column (ref. 156) to the 150 x 4.6 mm 5 μm RP 18 analytical column by heart-cut. The analytical column was kept at 70°C.

The method described by Schran et al. (ref. 127) and Smith et al. (ref. 126) included an automated on-line liquid-liquid purification step. Protein precipitation was achieved by adding acetonitrile/water (97.5/2.5 v/v) to the blood sample. The supernatant was injected into the microprocessor controlled Technicon system (Technicon Instruments, Tarrytown, NY, USA), wherein the sample was washed automatically with hexane and was injected onto a C8 column, which was kept at 75°C. The high temperature of the extraction column improves separation resulting in smaller cut times and less interfering material on the analytical column. Cyclosporine was eluted with acetonitrile/water 55/45 (v/v) and was loaded onto a C18 column using a heart-cut procedure. From the C18 column cyclosporine was eluted isocratically by acetonitrile/water 75/25 (v/v).

The basic strategies of two later published methods (refs. 135,138) were similar to that of Nussbaumer et al. (ref. 121). After protein precipitation with acetonitrile samples were loaded on the extraction columns and purified by acetonitrile/water. During elution from the extraction column the cyclosporine containing fraction was switched to the analytical column by heart-cut. The main difference to the earlier method was the use of a protein separation column, 5 μm, 75 x 4.6 mm and a subsequent 75 x 4.6 mm, 3 μm octadecyl column as analytical column (ref. 135) and 30 x 4 mm cyano-propyl 5 μm precolumn combined with a CLC-TMS, 5 μm column (Shimadzu, Kyoto, Japan) which allow lower oven temperatures (60 °C) for both analytical and extraction column and low flow rates to obtain a sufficient separation. The pre-columns were kept at the same temperature as the analytical columns.

3.3 CHROMATOGRAPHIC ANALYSIS OF SANDIMMUN[R] (CYCLOSPORINE) AND ITS METABOLITES

Chromatography of cyclosporine and its metabolites is discussed in detail in this chapter since chromatography of cyclosporine is acquainted with some interesting analytical problems that must also be considered in chromatographic analysis of other drugs. However, the measurement of cyclosporine metabolites is currently under discussion and will be of great impact when one or more of the metabolites will prove to be biologically active, which seems to be very likely.

One of the chromatographic difficulties in HPLC analysis of cyclosporine is the characteristic peak broadening. It is the result of an interconversion of several cyclosporine conformers exhibiting different chromatographic characteristics. Chromatography at $4^{o}C$ results in a splitting into single conformer peaks. High temperatures accelerate interconversion of the conformers resulting in sharper peaks, representing an average conformer composition (ref. 157). Further analytical conditions to overcome separation of cyclosporine conformers resulting in peak broadening are:

1. low pH at 3.0 - 5.0 of the mobile phase,
2. composition of the mobile phase, e.g. gradient elution,
3. using a more polar stationary phase for example C1, C2, or CN.

Further positive effects of HPLC at elevated temperatures are (ref. 158):

1. High temperatures decrease viscosity of the eluents allowing the use of sorbents with a smaller particle size and/ or pellicular sorbents, resulting in a greater column efficiency.
2. Increasing temperature accelerates sorption by increasing mass transfer and kinetic rates. This effect is of special benefit for chromatography of large or very lipophilic molecules, as which cyclosporine must be regarded.

Adverse effect of increased temperatures are:

1. Faster degradation of the stationary phase.
2. Acceleration of not desirable on-column reactions, as expressed by the dimensionless Damköhler number (ref. 159).

Both effects are of practical value for HPLC analysis of cyclosporine as discussed below. Assessment of cyclosporine analysis by supercritical fluid chromatography (SFC) will be the subject of further investigations.

Some authors (ref. 119, 123) observed a fast degradation of their analytical columns within 100-150 h. Other authors reported considerably longer column life times up to 3000 injected samples (refs. 137, 157). Reversed phase columns can be operated by temperatures up to $100^{\circ}C$ without degradation (personal communication, Merck, Darmstadt, FRG). At the beginning of cyclosporine measurement in our laboratory our analytical columns (C8) had also a very reduced life span. The main reasons turned out to be too fast and frequent heating and cooling of the column and the injection of poorly purified samples. Thus our analytical column was kept at 75 °C with a small flow of 0.1 ml/min during periods without analysis. The analytical column was guarded by a precolumn and after each analysis the column was flushed with acetonitrile/water (pH 3.0) (90/10 v/v) to remove peaks not being eluted during analysis time. We found 2 peaks eluting after injecting blood sample extracts that could not be detected after injection of cyclosporine standard solutions which were eliminated by the acetonitrile flush. Without the column flushing step degradation of the analytical column occurred after a few hundred analyses resulting in a proceeding decrease of the theoretical plates of the column and an increasing back pressure of the analytical system. In our system the rise in pressure was caused by plugged inlet frits of the precolumn and the analytical column. It disappeared after exchanging the metal frits by teflon frits (refs. 149 and 150). Lensmeyer et al. (refs. 110,148) set a silica saturating column between the pumps and the injector valve to minimize dissolution of the silica based sorbent in the analytical columns. Another approach to extend column life time was the use of a cyanopropyl analytical column which allows reduction of the column temperature to 60°C as described above.

Ultraviolet absorption from compounds not derived from cyclosporine can produce peaks in the chromatograms. Especially plastic softeners from laboratory equipment and from rotor seals, valves and frits inside the HPLC-system can disturb cyclosporine analysis. In our extraction procedure only diethylether cleaned centrifuge tubes and injector vials were used after the elution step from the extraction column and the final hexane wash removed remaining contaminations (2.2).

In our laboratory the following chromatographic system was developed and is currently in use (ref. 150):

The column used are two sequential 250 x 4.0 mm C8, 7 μm LiChrosorb[R] filled manu-fix[R] cartridges guarded by a 50 x 4 mm pre-column filled with the same material (all Merck, Darmstadt, FRG). Analysis were performed on an HP 1090 HPLC system (Hewlett Packard, Karlsruhe, FRG). Solvent A was water adjusted to pH 3 with phosphoric acid; solvent B acetonitrile. The eluents were degassed by ultra-sound and vacuum prior to use and during run by helium insufflation. The flow rate was set at 1.4 ml/min, the detector wavelength at 205 nm and the oven temperature at 75°C. For elution three superimposed sequential linear gradients with increasing steepness resulting in an almost concave gradient were used. The gradient began at 43/57 (v/v) acetonitrile/water (pH 3.0) and increasing to acetonitrile/water (pH 3.0) 49/51 (v/v) until 20 min after injection, to acetonitrile/water (pH 3.0) 59/41 (v/v) until 35 min after injection and to acetonitrile/water (pH 3.0) 80/20 (v/v) until 45 min after injection to stay at this level for 5 min. The gradient was followed by a column clean-up step to remove late eluting lipophilic material otherwise occupying the binding sites of the column material resulting in decreasing efficiency of the column. The column was flushed with acetonitrile/water (pH 3.0) for 5 min and then reequilibrated to the start conditions with an eluent composition of acetonitrile/water (pH 3.0) 43/57 (v/v). Cycle time between two injections was 60 min.

The stationary phase used for HPLC analysis ranged from silica gel to RP18 material. For our system C8 sorbent gave the best results. With more polar sorbents like C4, C1 or CN a reasonable resolution especially of the metabolites 1 and 17 was not possible in our system. Wang et al. compared different columns of the same length and diameter filled with CN, phenyl and four different C18 packings under the same conditions analysing bile samples (ref. 160). In the study C18 columns gave the best results in resolving cyclosporine metabolites, especially 1 and 17. C8 columns were not included in the study. Rocher et al. (ref. 161) found C8 columns superior to C18 columns in cyclosporine measurement. The two columns were tested under the same conditions and the concentrations measured with the C18 column were higher than those measured with the C8 column due to an interfering peak that was present in high concentration in patients with renal insufficiency and could be separated from the cyclosporine peak using a C8 column. Moyer et al. (refs. 140 and 141) proposed the use of a more polar phase to minimize cyclosporine band broadening. Unfortunately more polar columns are less effective in separating cyclosporine metabolites (ref. 160).

One of the major problems in HPLC analysis of cyclosporine and its metabolites is the great number of metabolites. Until now more than 30 could be isolated in our laboratory (Fig. 5). To increase the resolving power of our chromatographic system gradient elution and an elongated analytical column were used. The gradient consists of three linear gradients, since the cyclosporine metabolites detectable with our system are eluted in three main groups (Metabolite 8, 9, 26; 13, 25; 203-218, 17, 1, 18, 21). Each of the linear gradients was designed to get the best resolution possible of each metabolite group. Superposition of the three linear gradient resulted in the concave gradient described (Fig. 6).

TABLE VI   Mass spectrometric data of cyclosporine metabolites corresponding to Fig. 5, isolated from human bile or generated by human liver microsomes.

	metabolite	molecular weight	formula
1	H355	1234.4	$Cs + 2\ O$
2	17	1218.4	$Cs +\ \ 0$
3	8	1234.6	$Cs + 2\ O$
4	-	1232.4	$Cs + 2\ O - 2\ H$
5	H410/420	1216.4	$Cs +\ \ 0 - 2\ H$
6	-	1220.4	$Cs + 2\ O\ \ \ \ \ \ - CH_3$
7	-	1217.4	$Cs + 2\ O - 2\ H - CH_3$
8	203-218	1232.4	$Cs + 2\ O - 2\ H$
9	-	1250.4	$Cs + 3\ O$
10	-	1264.4	$Cs + 4\ O - 2\ H$
11	H230	1248.4	$Cs + 3\ O - 2\ H$
12	18	1218.4	$Cs +\ \ 0$
13	H310	1234.6	$Cs + 2\ O$
14	26	1234.6	$Cs + 2\ O$
15	H235	1394.2	$Cs +\ \ 0 + $ glucoronic acid
16	-	1219.4	$Cs + 2\ O\ \ \ \ \ \ - CH_3$
17	-	1250.4	$Cs + 3\ O$
18	1	1218.4	$Cs +\ \ 0$
19	10	1234.4	$Cs + 2\ O$
20	-	1251.4	$Cs + 4\ O\ \ \ \ \ \ - CH_3$
21	-	1248.4	$Cs + 3\ O - 2\ H$
22	16	1234.6	$Cs + 2\ O$
23	21	1188.4	$Cs\ \ \ \ \ \ \ \ \ \ \ - CH_3$
24	-	1203.4	$Cs +\ \ 0\ \ \ \ \ \ - CH_3$
25	13	1203.4	$Cs +\ \ 0\ \ \ \ \ \ - CH_3$
26	-	1219.4	$Cs + 2\ O\ \ \ \ \ \ - CH_3$
27	9	1219.4	$Cs + 2\ O\ \ \ \ \ \ - CH_3$
28	-	1235.4	$Cs + 3\ O\ \ \ \ \ \ - CH_3$
29	25	1204.4	$CS +\ \ 0\ \ \ \ \ \ - CH_3$

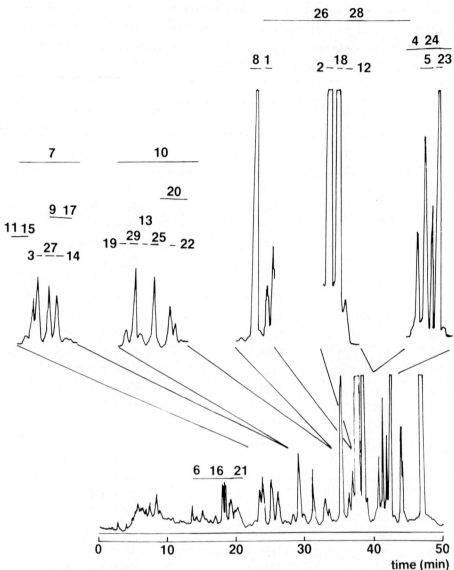

Fig. 5   Shows a chromatogram of a sample containing human microsomes at a concentration of 3 mg protein/ml and cyclosporine. Cyclosporine and the microsomes were incubated for 2 h prior to extraction and analysis as described in the text (2.2, 3.3) and refs. 149-151. Cyclosporine D was used as internal standard. The corresponding fractions were isolated by HPLC and further analysed by FAB-MS. The mass spectrometric data are listed in Table VI. The numbers of the metabolites in the Table correspond to the fraction numbers in the chromatogram. The nomenclature as proposed by refs. 94, 95 and 151 is used in the metabolite column of Table VI. HPLC parameters Figure 5: Detector wavelength: 205 nm, oven temperature: 75°C, plot attenuation: 2^4, flow: 1.4 ml/min. Abbreviation Cs: cyclosporine

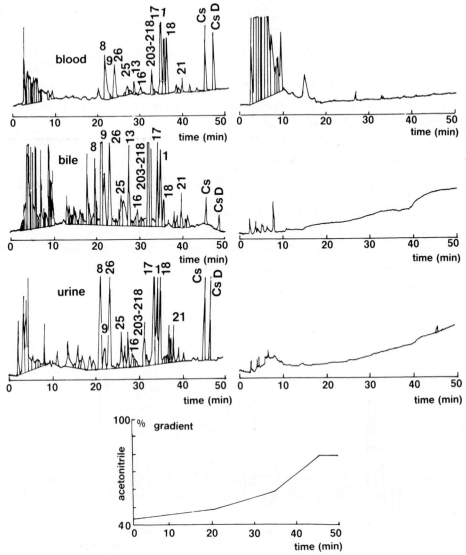

Fig. 6   6 Chromatograms, one blood (top), bile (middle) and urine (below) sample and the corresponding blank samples. The cyclosporine containing samples stem from a liver grafted patient after his first intra-venous cyclosporine dose of 14 mg/kg body weight. The samples were taken 12 hours after application of cyclosporine. The samples were extracted and analysed as described in 2.2 and 3.3 and refs. 149-151. The detector wavelength was set at 205 nm, the attenuation at 2^4, the flow at 1.4 ml/min and the oven temperature at 75^0C. The diagram (bottom) shows the acetonitrile/water (pH 3) gradient used for analysis. Analysis were performed on a HPLC system HP 1090 (Hewlet Packard) and chromatograms plotted on a 7470A (Hewlett Packard). Numbers in the chromatograms indicate the cyclosporine metabolites using the nomenclature of Maurer et al. (refs. 94-95). Cs: cyclosporine, Cs D: cyclosporine D. Arrow indicates injection.

   With the HPLC procedure described, 11 structurally elucidated as well
as 7 not yet structurally known cyclosporine metabolites could be
quantified and isolated from urine and bile of grafted patients (Fig. 7).
For peak identification the chromatogram was fractionated and the
corresponding material was isolated by semi-preparative HPLC.
Constituents of an isolated fraction were regarded to be identical with
authentic cyclosporine metabolites if:

1. no peak with an equal retention time was found in blank samples of not
   cyclosporine treated patients,
2. the unspecific cyclosporine radioimmunoassay cross-reacted with the
   fraction in question,
3. the material in the fraction showed mass spectrometric properties
   implying the structure of a cyclosporine derivative,
4. the isolated fraction and the authentic metabolite have the same
   retention time under the chromatographic conditions described above
   (Fig. 8)

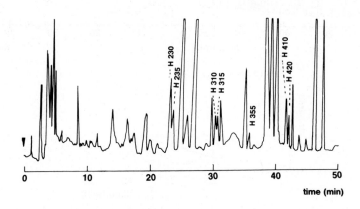

Fig. 7   Chromatogram of a urine extract of a liver grafted patient. The
         sample was extracted and analysed according to 2.2, 3.3, refs.
         149-151. The HPLC system and parameters were the same as des-
         cribed in Fig. 6. The numbers indicate cyclosporine metabolites
         not yet structurally identfied using the nomenclature developed
         in our laboratory (ref. 151). These metabolites could be iso-
         lated by semi-preparative HPLC and their biological activity was
         evaluated. They can be quantitatively determined from the chroma-
         tograms using the internal standard cyclosporine D (Cs D). The
         corresponding FAB-MS data are listed in Table 6.
         Cs: cyclosporine. Arrow indicates injection.

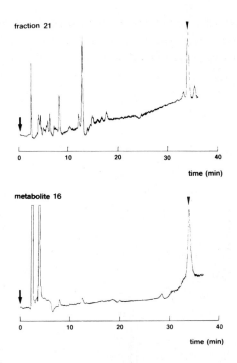

Fig. 8   Chromatograms of fraction 21 (top) isolated from a urine extract
          of a liver grafted and cyclosporine treated patient and the
          authentic metabolite 16 (bottom), isolated and identified as
          described in refs. 94 and 95. The chromatographic conditions and
          HPLC parameters were the same as described in 3.3 and ref. 150.
          The analysis were stopped after 40 min. Arrow indicates
          injection. Both peaks (↓) are eluted from the column with a
          retention time of 33.7 min.

In HPLC analysis of cyclosporine the following problems must be
considered:

1. The large number of cyclosporine metabolites and their only small
   structural differences make a reasonable resolution almost impossible.
   Figure 5 shows the metabolites that could be isolated by semi-
   preparative HPLC from a suspension of human liver microsomes (3 mg/ml
   protein) after incubation with 10 μg cyclosporine for 2 h. The numbers
   indicate the peaks or regions in the chromatogram where the corresponding metabolite could be isolated and are assigned to the
   structural data as elucidated by fast atom bombardment mass
   spectrometry (FAB-MS). These data indicate that several peaks in the
   chromatogram represent a conglomerate of several metabolites. As

example the isolated fraction containing metabolite 9 is shown in Fig. 9.
Although the peaks seem symmetric and pure after injecting the isolated
fraction into the HPLC, FAB-MS detects three peaks representing different
cyclosporine derivatives.

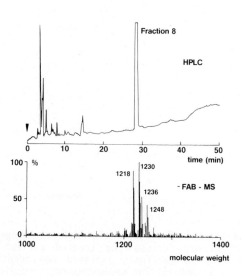

Fig. 9   The chromatogram (top) shows fraction 8 containing metabolite 9
          (nomenclature ref. 94), isolated by semi-preparative HPLC from
          urine of a liver grafted patient and reinjected into the HPLC
          system. The chromatographic conditions and HPLC parameters are
          the same as described in 3.3, Fig. 6, refs. 150 and 151.
          Although the peak is symmetric and seems to be caused by a mono-
          substance, the corresponding FAB-MS (negative mode) (bottom)
          analysis showed at least 3 different cyclosporine derivatives
          (ref. 151). Arrow indicates injection.

2. It is very likely that the chromatographic conditions (pH 3.0, 75°C)
   change the cyclosporine metabolite pattern detected. Henricsson et al.
   (ref. 162) reported the identification of a sulfated cyclosporine derivative, isolated by ion-exchange chromatography. Unter HPLC conditions
   described above the binding of sulfated or glucuronated derivatives
   can be hydrolyzed.

3. Plastic material leached from laboratory tools have similar spectral
   properties as the cyclosporins and similar retention times. Fig. 10

shows a chromatogram of a blood sample of a liver grafted patient without cyclosporine therapy. The sample had contact with a polypropylen tube of the evaporation apparatus. Several peaks in the range of the cyclosporine metabolites were detected. The chromatogram below in the same Fig. shows a chromatogram of the same re-extracted sample without interfering plastic material. Interference of other drugs with the HPLC assay has been reported for sulfamethoxazole (ref. 163).

4. The quantification of cyclosporine metabolites is influenced by the chromatograpic system used. A peak separation down to the baseline is not possible for most of the cyclosporine metabolites. The peak area therefore depends on the integration software and integration method used. As outlined above most peaks do not represent a monosubstance. Using an HPLC system with a greater resolving power may result in two or more smaller peaks (ref. 161). Therefore the reproducibility is good for one HPLC system but comparison of the metabolite concentrations measured with other laboratories is difficult.

Lensmeyer et al. (ref. 148) compared three chromatographic systems using a 80 x 4 mm 5 μm octyl column, a 160x4 mm 5 μm silica gel column and a 250 x 4.6 mm cyano-propyl column. The cyano propyl-column was preferred and run with water/ acetonitrile/tetrahydrofurane/acetic acid/n-butylamine (600/390/20/0.16/ 0.1 v/v/v/v/v/) gave the best results in the systems tested. The chromatograms show an incomplete separation of the metabolites 1 and 18.

Bowers et al. (ref. 145) simultaneously analysed cyclosporine and its metabolites on a 50x4.6 mm 3 μm octadecyl column using an isocratic gradient with water/methanol/acetonitrile (34/20/46 v/v/v/) at an oven temperature of 70°C. Incomplete separation of the hydrophilic group of cyclosporine metabolites was achieved and the metabolites were not quantitively determined in bile.

Burckart et al. (ref. 164) extracted cyclosporine and its metabolites from pooled bile using a modified extraction procedure according to Sawchuk et al. (ref. 111). cyclosporine and its metabolites were eluted from the analytical column by an acetonitrile/water gradient.

Rosano et al. (ref. 109) extracted cyclosporine and its metabolites from blood using a liquid-liquid extraction procedure with diethyl ether. Cyclosporine and its metabolites 1, 17 and 21 were seperated isocraticly.

Wallemacq et al. (refs. 77 and 165) used a normal phase HPLC system for analysis of cyclosporine and its metabolites. A 3 μm particle size CN column and hexane: isopropanol (85/15 v/v) as mobile phase were used

126

(refs. 77 and 165). Column temperature was 50 °C. 28 metabolites could be separated by semi-preparative HPLC and their mass spectrometric properties were evaluated.

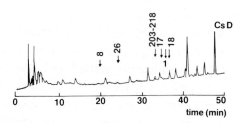

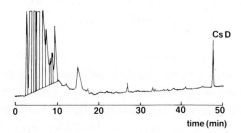

Fig. 10   The chromatograms show the HPLC analysis of the same blood
          sample received from a liver grafted patient without cyclosporine
          therapy for 10 days. The extraction procedure, the chroma-
          tographic conditions and HPLC parameters were the same as
          described before (2.2, 3.3, Fig. 6, refs. 149-151). During
          extraction, the sample showed in the chromatogram above had come
          into contact with a polypropylen tube of the evaporation
          apparatus resulting in contamination with plastic material. The
          arrows with the numbers according to Maurer's nomenclature
          (refs. 94 and 95) indicate the retention times of the cyclo-
          sporine metabolites. The chromatogram shows the analysis of the
          same blood sample without contaminating materials. Both samples
          contain the internal standard cyclosporine D (CsD). Arrow in-
          dicates injection.

4.   TROUBLE SHOOTING IN DEVELOPMENT OF BLOOD SAMPLE CLEAN-UP PROCEDURES
      Unexpected losses of the compounds of interest during blood sample
clean-up can be caused by the following factors (ref. 2):

1. Chemical breakdown or adsorption of the compounds of interest because
   of a too long storing period or inadequate storing conditions,
2. adsorption of the drugs on metal, glass or plastic surfaces of
   laboratory equipment,

3. co-precipitation during deproteinization,
4. inadequate pH of the buffers and solvents used,
5. low partition coefficient in the organic extraction solvent,
6. insufficient adsorption on the solid phase material used,
7. chemical breakdown because of chemical, biological, photochemical and/or thermal instability,
8. chelation e.g. with heavy metals or intercalation,
9. loss during the evaporation step because of volatility of the drug,
10. insoluble residue after evaporation,
11. insufficient derivatization because of an incomplete reaction, by-products and removal of excess reagents

Contamination of the sample during extraction and analysis can be caused by:
1. coextraction of interfering material such as lipoproteins and lipids/lipoids,
2. use of contaminated solvents or solvents of a minor purity,
3. interfering residues in glass-ware and laboratory tools,
4. leached plastic softeners from laboratory tools and extraction columns,
5. unadequately primed solid phase column material,
6. contaminated injector needle of the injection system,
7. solubilised plastic material from the vials and the vial caps of the injection system,
8. solubilised plastic softeners from the valves, rotorseals and capillaries of the HPLC-system.

REFERENCES

1. D.L. Garver, H. Dekirmenjian, J.M. Davis, R. Casper, and S. Ericksen, Am. J. Psychiatry, 134 (1977) 304-307.
2. J.A.F de Silva, in: Trace organic sample handling, E. Reid (ed.), Ellis Horwood, Chichester, 1981, pp 192-204.
3. M. Zakaria and P.R. Brown, Anal. Biochem., 25 (1982) 120.
4. R. Prasad, M.S. Maddux, M.F. Mozes, N.S. Bishop and A. Maturen, Transplantation, 39 (1985) 667-669.
5. P.R. Brown, Chromatogr. Sci., 28 (1984) 31-48.
6. J.A.F. de Silva J. Chromatogr., 273 (1983) 19-42.
7. W.O. Pierce, T.C. Lamoreaux, F.M. Urry, L. Kopjak and B.S. Finkle, J. Anal. Toxicol., 2 (1978) 26-31.
8. D. Sohn, J. Simon, M.A. Hanna and G. Ghali, J. Chromatogr. Sci., 10 (1972) 294-296.
9. J.A.F. de Silva, J. Chromatogr., 340 (1985) 3-30.
10. J.A.F. de Silva, Anorexigenics, antipsychotics, and antianxiety drugs, in K. Tsuji (ed.), GLC and HPLC determination of therapeutic agents, vol. 2, Chromatogr. Sci., Marcel Dekker, 1978, pp. 581-636.
11. E.H. Foerster, J. Dempsey and J.C. Garriott, J. Anal. Toxicol., 3 (1979) 87-91.

128

12. T.W. Guentert, G.M. Wientjes, R.A. Upton, D.L. Combs and S. Riegelman, J. Chromatogr., 163 (1979) 373-382.
13. T.A. Jennison, B.S. Finkle, D.M. Chinn and D.J. Crouch, J. Chromatogr. Sci., 17 (1979) 64-74.
14. S.H. Wallace, V.P. Shah and S. Riegelman, J. Pharm. Sci., 66 (1977) 527-530.
15. S. Loche, A.B. Rifkind, E. Stoner, A. Faedda, M.C. Garabedian and M.I. New, Ther. Drug Monit., 8 (1986) 214-218.
16. B. Scales and P.B. Copsey, J. Pharm. Pharmacol., 27 (1975) 430-433.
17. D. Westerlund, B. Petterson and J. Carlqvist, J. Pharm. Sci., 71 (1982) 1148-1151.
18. R.D. Budd and D.F. Mathis, J. Anal. Toxicol., 6 (1982) 317-320.
19. W. Duenges, E. Bergheim-Irps, H. Straubs and R.E. Kaiser, J. Chromatogr., 145 (1978) 265-274.
20. C.V. Puglisi and J.A.F. de Silva, J. Chromatogr., 226 (1981) 135-146.
21. H.M. Stevens, J. Forensic Sci. Soc., 25 (1985) 67-80.
22. K. Worm, Z. Rechtsmed., 77 (1975) 41-45.
23. J.K. Stoltenborg, C.V. Puglisi, F. Rubio and F.M. Vane, J. Pharm. Sci., 70 (1981) 1207-1212.
24. J.E. O'Brien, V. Abbey, G. Hinsvark, J. Perel and M. Finster, J. Pharm. Sci., 68 (1979) 75-78.
25. A. Viala, J.P. Cano and A. Durand, J. Chromatogr., 111 (1975) 299-303.
26. Y. Bergqvist and S. Eckerbom, J. Chromatogr., 306 (1984) 147-153.
27. E. Pussard, F. Verdier and M.C. Blayo, J. Chromatogr., 374 (1986) 111-118.
28. F.C. Churchill, D.L. Mount and I.K. Schwartz, J. Chromatogr., 274 (1983) 111-120.
29. W.C. Randolph, V.L. Osborne, S.S. Walkenstein and A.P. Intoccia, J. Pharm. Sci., 66 (1977) 1148-1150.
30. D. Shapcott and B. Lemieux, Clin. Biochem., 8 (1975) 283-287.
31. M.S. Lennard, J.H. Silas, A.J. Smith and G.T. Tucker, J. Chromatogr., 133 (1977) 165-166.
32. P.M. Kabra, G.L. Steven and L.J. Marton, J. Chromatogr., 150 (1978) 355-360.
33. N. Scagliola, R. Julien and C. Dreux, Ann. Biol. Clin. (Paris), 34 (1976) 27-32.
34. K.M. Wolfram and T.D. Bjornsson, J. Chromatogr., 183 (1980) 57-64.
35. M.S. Miller and A.J. Gandolfi, Anaesthesiology, 51 (1979) 542-544.
36. P. Moore, K. Mai and C.M. Lai, J. Pharm. Sci., 75 (1986) 424-426.
37. J. A. Vinson, D.D. Patel and A.H. Patel, Anal. Chem., 49 (1977) 163-175.
38. J.N. Pirl, V.M. Pappa and J.J. Spikes, J. Anal. Toxicol., 3 (1979) 129-132.
39. S.E. Tett, D.J. Cutler and K.F. Brown, J. Chromatogr., 344 (1983) 241-248.
40. S.A. Stout and C.L. De Vane, Psychopharmacology (Berlin), 84 (1984) 39-41.
41. M.M. Kochar, L.T. Bavda and R.S. Bushan, Res. Commun. Chem. Pathol. Pharmacol., 14 (1976) 367-376.
42. D.E. Schwartz and U.B. Ranalder, Biomed. Mass Spectrom., 8 (1981) 589-592.
43. J.M. Grindel, P.F. Tilton and R.D. Shaffer, J. Pharm. Sci., 68 (1979) 718-721.
44. T. Nakagawa, T. Higuchi, J.L. Haslam and R.D. Shaffer, J. Pharm. Sci., 68 (1979) 718-721.
45. R.K. Lynn, R.H. Leger, W.P. Gordon, G.D. Olsen and N. Gerber, J. Chromatogr., 131 (1977) 329-340.
46. S. Felby, Forensic Sci. Int., 13 (1979) 145-150.
47. M.W. White, J. Chromatogr., 178 (1979) 229-240.

48. L.A. Asali, R.L. Nation and K.F. Brown, J. Chromatogr., 278 (1983) 326-335.
49. M. Stefek, L. Benes and V. Kovacik, Biomed. Mass Spectrom., 12 (1985) 388-392.
50. P.M. Kabra, G. Gotelli, R. Stanfill and L.J. Marton, Clin. Chem., 22 (1976) 824-827.
51. F. de Bros and E.M. Wolshin, Anal. Chem., 50 (1978) 521-525.
52. C.J. Reddrop, W. Riess and T.F. Slater, J. Chromatogr. 192 (1980) 375-386.
53. A.J. Wood, K. Carr, R.E Vestal, S. Belcher, G.R. Wilkinson and D.G. Shand, Br. J. Clin. Pharmacol., 6 (1978) 345-350.
54. C.V. Puglisi, F.J. Ferrara Jr., J.A.F. de Silva, J. Chromatogr., 275 (1983) 319-333.
55. H. Sheehan and P. Haythorn, J. Chromatogr., 117 (1976) 392-398.
56. B. Levine, R. Blanke and J. Valentour, J. Anal. Toxicol., 7 (1983) 207-208.
57. E.M. Wolshin, M.H. Cavanough, C.V. Manion, M.B. Meyer, E. Milano, C.R. Reardon and S.M. Wolshin, J. Pharm. Sci., 67 (1978) 1962-1695.
58. D.E. Mundy, M.P. Quick and A.F. Machin, J. Chromatogr., 121 (1976) 335-342.
59. H.S. Chun Hong, R.J. Steltenkamp and N.L. Smith, J. Pharm. Sci., 64 (1975) 2007-2008.
60. P.O. Edlund, J. Chromatogr., 206 (1981) 109-116.
61. J. Breiter, R. Helger and H. Lang, Forensic Sci., 7 (1976) 131-140.
62. J. Breiter, Arzneim. Forsch., 28 (1978) 1941-1944.
63. A.W. Missen, Clin. Chem., 22 (1976) 927-928.
64. H.H. McCurdy, J. Anal. Toxicol., 4 (1980) 82-85.
65. H.H. McCurdy, L.J. Lewellen , J.C. Cagle and E.T. Solomons, J. Anal. Toxicol., 5 (1981) 253-257.
66. H.J. Schlicht and H.P. Gelbke, Z. Rechtsmed, 81 (1978) 25-30.
67. B. Ford, J. Vine and T.R. Watson, J. Anal. Toxicol., 7 (1983) 116-118.
68. J.A. Apffel, U.A.T. Brinkman and R.W. Frei, J. Liq. Chromatogr., 5 (1982) 2413-2422.
69. T.R. Mac Gregor, P.R. Farina, M. Hagopian, N. Hay, H.J. Esber and J.J. Keirns, Ther. Drug Monit., 6 (1984) 83-90.
70. S.N. Rao, A.K. Dhar, K. Kutt and M. Okamoto, J. Chromatogr., 231 (1982) 341-348.
71. K.D.G. Edwards and M. McCredie, Med. J. Australia, 54 (1967) 534-539.
72. N. Elahi, J. Anal. Toxicol., 3 (1979) 35-38.
73. J.M.F. Douse, J. Chromatogr., 301 (1984) 137-154.
74. G. Cimbura and E. Koves, J. Anal. Toxicol., 5 (1981) 296-299.
75. A.W. Missen and J.F. Lewin, Clin. Chim. Acta, 53 (1974) 389-390.
76. P.M. Kabra and J.H. Wall, J. Chromatogr., 385 (1987) 305-310.
77. P.E. Wallemacq and M. Lesne, J. Chromatogr., 413 (1987) 131-140.
78. M. Bogusz, J. Gierz and J. Bialka, Forensic Sci., 12 (1978) 73-82.
79. G. Ibrahim, S. Andryauskas and M.L. Bastos, J. Chromatogr., 108 (1975) 107-116.
80. G. Machata and W. Vycudilik, Arch. Toxicol., 33 (1975) 115-122.
81. N.E. Hoffman, A.M. Rustum, E.J. Quebbeman, A.A.R. Hamid and R.K. Ausman, J. Liq. Chromatogr., 8 (1985) 2511-2520.
82. J. Fujimoto and R. Wang, Toxicol. Appl. Pharmacol., 16 (1970) 186-193.
83. H. Ko and E.N. Petzold, Isolation of samples prior to chromatography, in K. Tsuji and W. Morozowich (ed.), GLC and HPLC determination of therapeutic agents, vol. 1, Chromatogr. Sci., Marcel Dekker, (1978), pp 278-299.
84. Canadian Multicentre Trial Group, N. Eng. J. Med., 309 (1983) 809.
85. European Multicentre Trial Group, Lancet, ii (1983) 986.

130

86. K. Uchida, T. Morimoto, N. Nakanisi, N. Yamada, Y. Tominaga, A. Orihara, T. Kondo, T. Morimoto, K. Morozumi, T. Kano and H. Takagi, Dev. Toxicol. Envir. Sci., 14 (1986) 163-166.
87. P. Donatsch, E. Abisch, M. Homberger, R. Traber, M. Trapp and R. Voges, J. Immunoassay, 2 (1981) 19-32.
88. A. Rüegger, M. Kuhn, H. R. Lichti, R. Huguenin, C. Quiquerez and A. von Wartburg, Helv. Chim. Acta, 59 (1976) 1075-1092.
89. H.R. Loosli, H. Kessler, H. Oschkinat, H.P. Weber, T.J. Petcher and A. Widmer, Helv. Chim. Acta, 68 (1985) 682-703.
90. T.J. Petcher, H.P. Weber and A. Rüegger, Helv. Chim. Acta, 59 (1976) 1480-1488.
91. R.M. Wenger, Helv. Chim. Acta, 67 (1984) 502-525.
92. T. Kronbach, V. Fischer and U.A. Meyer, Clin. Pharmacol. Ther., 43 (1988) 630-635.
93. R.J. Ptachcinski, R. Ventkataramanan and G. J. Burckart, Clin. Pharmacokinet., 11 (1986) 107-132.
94. G. Maurer, H.R. Loosli, E. Schreier and B. Keller, Drug. Metab. Dispos., 12 (1984) 120-126.
95. G. Maurer and M. Lemaire, Transplant. Proc., 18 (1986) 25-34.
96. N.R. Hartman, L.A. Trimble, J.C. Vederas and I. Jardine, Biochem. Biophys. Res. Commun., 133 (1985) 964-971.
97. H. Hashem. R. Ventkataramanan, G.J. Burckart, L. Makowka, T.E. Starzl, E. Fu and L.K. Wong, Transplant. Proc., 20 (1988) 176-178.
98. Members of the National Academy of Clinical Biochemistry/American Association for Clinical Chemistry Task Force on Cyclosporine, Clin. Chem., 33 (1987) 1269-1288.
99. M. Lemaire and J.T. Tillement, J. Pharm. Pharmacol., 34 (1982) 715-718.
100. W. Mraz, R.A. Zink and A. Graf, Transplant. Proc., 15 (1983) 2426-2429.
101. B.S. Zaghloul, R.J. Ptachcinski, G.J. Burckart and R. Ventkataramanan, J. Clin. Pharmacol., 27 (1987) 240-2.
102. J. Gurecke, V. Warty and A. Sanghvi, Transplant. Proc., 17 (1985) 1997-2002.
103. W.M. Awni and R. J. Sawchuck, Drug. Metab. Disp., 13 (1985) 127-132.
104. S. Henricsson, J. Pharm. Pharmacol., 39 (1987) 384-385.
105. S. Robson, J. Neuberger, G. Alexander and R. Williams, Br. med. J., 288 (1984) 1317-1318.
106. H. Dieperink, Lancet, i (1983) 416.
107. M. Wenk, F. Follath and E. Abisch, Clin. Chem., 29 (1983) 1865.
108. P. Suit, F. van Lente and C. E. Pippenger, Clin. Chem., 34 (1988) 597.
109. T.G. Rosano, B.M. Freed, J. Cerilli and Lempert N., Transplantation, 42 (1986) 262-266.
110. G.L. Lensmeyer and B.L. Fields, Clin. Chem., 31 (1985) 196-201.
111. D.J. Freeman, A. Laupacis, P.A. Keown, C.R. Stiller and S.G. Carruthers, Brit. J. Clin. Pharmacol., 18 (1984) 887-893.
112. T.G. Rosano, Clin. Chem., 31 (1985) 410-412.
113. R.W. Yatscoff, D.N. Rush and J.R. Jeffrey, Clin. Chem., 30 (1984) 1812-1814.
114. R.P. Agarwal, R.A. McPherson and G.A. Threatle, Ther. Drug Monit., 7 (1985) 61-65.
115. J.M. Potter and H. Self, Ther. Drug Monit., 8 (1986) 122-123.
116. A. Johnston, J.T. Marsden and D.W. Holt, Ther. Drug. Monit., 8 (1986) 200-204.
117. W. Niederberger, P. Schaub and T. Beveridge, J. Chromatogr., 182 (1989) 454-457.
118. J. C. Gfeller, A.K. Beck and D. Seebach, Helv. Chim. Acta, 63 (1980) 728-732.
119. R.J. Sawchuk and L.L. Cartier, Clin. Chem., 27 (1981) 1368-1371.
120. R. Lawrence and M.C. Allwood, J. Pharm. Pharmacol., 32 (1980) 100.

121. K. Nussbaumer, W. Niederberger and H.P. Keller, J. High Resol. Chromatogr., Commun., 5 (1982) 424-427.
122. G.C. Yee, D.J. Gmur and M.S. Kennedy, Clin. Chem. 28 (1982) 2269-2271.
123. B. Leyland-Jones, A. Clark, W. Kreis, R. Dinsmore, R. O'Reilly and C.W. Young, Res. Commun. Chem. Pathol. Pharmacol., 37 (1982) 431-444.
124. S.G. Carruthers, D.J. Freeman, J.C. Koogler, W. Howson, P.A. Keown, A. Laupacis and C. R. Stiller, Clin. Chem. 29 (1983) 180-183.
125. R.L. Kates and P. Latini, J. Chromatogr., 309 (1984) 441-447.
126. H.P. Smith and W.T. Robinson, J. Chromatogr., 305 (1984) 353-362.
127. H.F. Schran, W.T. Robinson and E. Abisch, Bioanalytical considerations, in J.F. Borel (ed.), Ciclosporin, Progress in Allergy, Karger, Basle, vol. 38, (1986) pp 73-92.
128. R. Garaffo and P. Lapalus, J. Chromatogr., 337 (1985) 416-422.
129. D.J. Gmur, G.C. Yee and M.S. Kennedy, J. Chromatogr., 344 (1985) 422-427.
130. M.K. Aravind, J.N. Miceli and R.E. Kauffman, J. Chromatogr., 344 (1985) 428-432.
131. K. Takada, N. Shibata, H. Yoshimura, H. Yoshikawa and S. Muraniski, Res. Commun. Chem. Pathol. Pharmacol., 48 (1985) 369-380.
132. A.K. Shibabi, J. Scaro and R.M. David, J. Liq. Chromatogr., 8 (1985) 2641-2648.
133. P.M. Kabra, J.H. Wall and N. Blanckeart, Clin. Chem., 31 (1985) 1717-1720.
134. A. Sanghvi, H. Seltman and W. Diven, Arch. Pathol. Lab. Med., 109 (1985) 1072-1075.
135. G. Hamilton, E. Roth, E. Wallisch and F. Tichy, J. Chromatogr., 341 (1985) 411-419.
136. V. Toulet, M.C. Saux, C. Rouquette, F. Penouil and A. Brachet-Liermain, Ann. Biol. Chem. (Paris), 44 (1986) 517-521.
137. G.C. Kahn, L.M. Shaw and M.D. Kane, J. Anal. Toxicol., 10 (1986) 28-34.
138. H. Hosotsubo, J. Takezawa, N. Taenaka, K. Hosotsubo, and I. Yoshiya, J. Chromatogr., 383 (1986) 349-355.
139. G. Schumann, M. Oellerich, K. Wonigeit and M. Wrenger, Fresenius Z. Anal. Chem., 324 (1986) 328-329.
140. T.P. Moyer, J.R. Charlson and L.E. Ebnet, Ther. Drug Monit., 8 (1986) 466-468.
141. T.P. Moyer, P. Johnson, S.M. Faynor and S. Sterioff, Clin. Biochem. (Ottawa), 19 (1986) 83-89.
142. T. Annesley, K. Matz, T. Balogh, L. Clayton and D. Giacherio, Clin. Chem., 32 (1986) 1407-1409.
143. J. Klima, R. Petrasek and V. Kocandrle, J. Chromatogr., 385 (1987) 357-361.
144. R.G. Buice, F.B. Stentz and B.J. Gurley, J. Liq. Chromatogr., 10 (1987) 421-438.
145. L.D. Bowers and J. Singh, J. Liq. Chromatogr., 10 (1987) 411-420.
146. S. Maguire, F. Kyne and U.D. Conaill, Ann. Clin. Biochem. (Warwick), 24 (1987) 161-166.
147. L. Sangalli and M. Bonati, Ther. Drug. Monit., 9 (1987) 353-357.
148. G.L. Lensmeyer, D.A. Wiebe and I.M. Carlson, Clin. Chem., 33 (1987) 1841-1850.
149. U. Christians, K.O. Zimmer, K. Wonigeit and K.F. Sewing, J. Chromatogr., 413 (1987) 121-129.
150. U. Christians, K.O. Zimmer, G. Maurer, K. Wonigeit, and K.F. Sewing, Clin. Chem., 34 (1988) 34-39.
151. U. Christians, H.J. Schlitt, J.S. Bleck, H.M. Schiebel, R. Kownatzki, G. Maurer, S.S. Strohmeyer, R. Schottmann, K. Wonigeit, R. Pichlmayr and K.F. Sewing, Transplant. Proc., 20 (1988) 609-613.

132

152. G.J. Burckart, C.P. Wang, R. Venkataramanan and R.J. Ptachcinski, Transplantation, 43 (1987) 932.
153. J.A. Apffel, T.V. Alfredson and R.E. Majors, J. Chromatogr., 206 (1981) 43-57.
154. P. Schauwecker, R.W. Frei and F. Erni, J. Chromatogr., 136 (1977) 63-72.
155. F. Erni, H.P. Keller, C. Morin and M. Schmitt, J. Chromatogr., 204 (1981) 65-76.
156. J.C. Gfeller and M. Stockmeyer, J. Chromatogr., 198 (1980) 162-168.
157. L.D. Bowers and S.E. Mathews, J. Chromatogr., 333 (1985) 231-238.
158. F.D. Antia and C. Horvath, J. Chromatogr., 435 (1988) 1-15.
159. W.R. Melander, H.J. Lin, J. Jacobson and C. Horvath, J. Phys. Chem., 88 (1984) 4527-4536.
160. C.P. Wang, G.J. Burckart and R. Ventkataramanan, Ther. Drug Monit., 10 (1988) 1988.
161. L.L. Rocher, D. Giacherio and T. Annesley, Transplant. Proc., 18 (1986) 652-654.
162. S. Henricsson and A. Lindholm, Transplant. Proc., in press.
163. P.L. Kimmel, T.M. Philips, N.C. Kramer and A. M. Thompson, Kidney Int., 27 (1985) 343.
164. G.J. Burckart, T.E. Starzl, R. Venkataramanan, H. Hashim, L. Wong, P. Wang, L. Makowka, A. Zeevi, R.J. Ptachcinski, J.E. Knapp, S. Iwatsuki, C. Esquivel, A. Sanghvi and D.H. van Thiel, Transplant. Proc., 6 (1986) 46-49.
165. P.E. Wallemacq, G. Lhoest, D. Latinne and M. Bruyere, Transplant. Proc., in press.

CHAPTER IV

RADIO-COLUMN LIQUID CHROMATOGRAPHY

A.C. VELTKAMP

List of Abbreviations and Symbols

CLC	:	Column Liquid Chromatography
RP-HPLC	:	Reversed-Phase High-Performance Liquid Chromatography
ß	:	Beta particle (negatron)
E_b	:	End-point energy of ß-spectrum
γ	:	Gamma (electromagnetic radiation > 30 keV)
SC	:	Scintillation Counting

LSC	:	Liquid Scintillation Counting
HSC	:	Heterogeneous (solid) Scintillation Counting
DPM	:	Desintegration per minute
CPM	:	Counts per minute
Bq	:	Becquerel = desintergration per second (DPS)
Ci	:	Curie; 1 Ci = 37 GBq
CPM(b)	:	Background count rate
NCA	:	'No Carrier Added'
Spec. Act.:		Bq/mmol (or DPM/mol)
DPM(s)	:	Absolute activity sample
E	:	Counting efficiency; $E = CPM/DPM(s)$
FM	:	Figure of Merit; $FM = E^2/CPM(b)$
T_d	:	Counting time (min)
T_b	:	Counting time of background
C_b	:	Background counts accumulated in T_b
C_s	:	Net peak area in radiograms (in counts)
$RSD(C_s)$	:	Relative standard deviation in C_s
F_t	:	Total flow rate through radioactivity detector (ml/min)
F_e	:	Flow rate eluent (ml/min)
F_s	:	Flow rate liquid scintillator (ml/min)
S	:	Split ratio of column eluate
V_d	:	Flow cell volume radioactivity detector (ml)
d_p	:	Particle size
T_w	:	Peak-base width (min)
Δw	:	Time between neighbouring peaks (min)
σ_w	:	Standard deviation peak
h	:	Reduced plate height
N	:	Plate number
V_o	:	Column void volume (ml)
	:	Solvent viscosity (kg/ (m x s))
K_o	:	Permeability constant
L	:	Cell length (in m)
P	:	Pressure drop

1. INTRODUCTION

1.1 THE USE OF RADIOISOTOPES IN CHEMICAL ANALYSIS

Since in the late fourties cyclotron- and reactor-produced radio-isotopes became available on a commercial scale, they have proven to be an indispensable tool for solving a wide variety of analytical problems (refs. 1, 2). The main advantages reside in the sensitivity (Table I) and selectivity of radioactivity determinations. Basically, applications in radioanalysis can be categorized as activation analysis, tracer and assay techniques.

TABLE I   Approximate detection limits for various analytical techniques[1,2]

Analytical technique	Detection limit (g)	
Radioactivity detection	10^{-12}	- 10^{-14}
Mass spectrometry	10^{-10}	- 10^{-12}
Fluorimetry		10^{-12}
Electrochemistry		10^{-12}
Ultraviolet spectrometry		10^{-9}
Infrared spectrometry	10^{-6}	- 10^{-9}
Refractive index		10^{-6}
NMR spectrometry	10^{-4}	- 10^{-6}

[1] For an ideal compound under ideal conditions
[2] Adapted from refs. 3, 4

In activation analysis, stable elements are activated to form Y-emitting radioisotopes, from which the elemental compositon of biochemical or geological samples can be established by γ-spectrometry (Fig. 1). Radiotracers provide a means to selectively study, for instance, the distribution and metabolic pathways of organic and inorganic analytes. Finally, radiolabeled compounds are extensively used in assay techniques for localization, identification or quantification of biologically active analytes. Examples in this category include immuno assay, receptor assay and enzyme assay.

Many radioanalytical procedures rely upon the separation of radioelements or radiolabeled compounds in one or more stages of the experiment. For example, in tracer or assay experiments, it is normally mandatory to use a radiochemically pure component, which necessitates isolation of the desired compound from radiolabeled impurities formed during preparation or storage of the compound. Radioenzyme assays require separation of substrates and products. In the elucidation of metabolic pathways, metabolites have to be separated from parent compounds.

Although the activity of the radioanalytes can be quite high, corresponding amounts are relatively low, normally less than 1 mg, depending on the specific activities involved. As a consequence, chromatographic techniques play a major role in the isolation and separation of

136

radiolabeled analytes. Applications in these areas, and the role of column liquid chromatography in radioanalysis are considered in more detail in section 4.

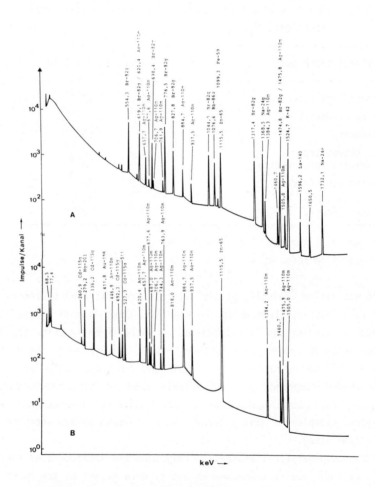

Fig. 1   γ-spectrum of mushroom sample after neutron activation, either before (A) or after (B) radiochemical group separation on Dowex 1 X 8. (Reprinted by permission from Fresenius Z. Anal. Chem., Springer-Verlag, ref. 5)

1.2   CHARACTERISTICS OF RADIO-COLUMN LIQUID CHROMATOGRAPHY

Radio-column liquid chromatography, by definition, involves the separation of radioelements or radiolabeled compounds by ion exchange, size exclusion or HPLC, and does not necessarily imply radioactivity determination of the column eluate. In preparative radio-chromatography, for instance, quantitative radioactivity determination may be hampered by limitations in the dynamic range of the radioactivity detector used due to the high activities involved, while UV, fluorescence, or electro- chemical detectors may still respond linearly to increasing amounts of the radioactive compound. Most applications, however, depend on the combination of a chromatographic separation and radioactivity counting, for instance by scintillation counting (SC) techniques.

Since this chapter deals with radio-column liquid chromatography, it is of interest to briefly summarize its main characteristics (Table II), some of which are more thoroughly dealt with in following sections.

TABLE II   Characteristics of radio-column liquid chromatography

Selectivity,	optimal as with other isotope techniques.
Sensitivity,	strongly determined by the counting time and efficiency of the detector system.
Quantitation,	normally based on determination of single radioisotope only (one calibration curve); for short-lived isotopes, correction factors have to be used to account for decay during chromatography.
Chromatography,	occasionally, isotope effects on retention behaviour may occur (refs. 164, 165, 172); differences between elution of 'no carrier added' and 'carrier added' may also be observed (refs. 8-10)
Hazards,	normally negligible with low to medium energy ß-emitters (^{3}H, ^{14}C); proper lead shielding of apparatus required for higher energy ß-emitters (^{32}P) and γ-emitters.

Selectivity

Under optimal conditions, activity eluting from the column will be detected if it is in excess over a stable background signal, with no interference from non-radioactive endogenous material or gradient elution (Fig. 2).

Apparently, selectivity in detection is the major advantage of radiotracers, especially for pharmaceutical and biochemical studies. In contrast to most other detection principles, analyte preconcentration becomes very effective because the corresponding preconcentration of

138

(unlabeled) components from the matrix normally does not interfere. In
most applications an additional advantage is that only one type of radioisotope has to be determined; therefore a single calibration curve
suffices for the quantitation of all compounds labeled with that isotope.
It thus avoids the need for standards.

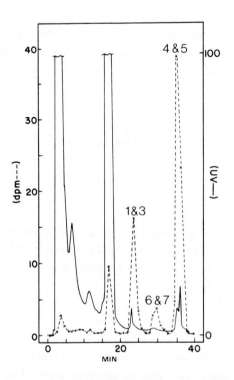

Fig. 2   Gradient HPLC-UV and ^{14}C-determination of nonconjugated fecal
metabolites of ^{14}C-doxylamine succinate in rats. Column, Supelco
LC-CN (250 x 4.6 mm); gradient elution with methanol/0.01 M
phosphate buffer (containing 0.01-0.02 M triethylamine, pH 7.8);
detection, UV (254 nm) and off-line liquid scintillation
counting of 1 ml fractions, respectively. (Reprinted by
permission from the J. Anal. Toxicol., Preston Publications,
Inc., ref. 6).

Sensitivity

The sensitivity of radioactivity counting is defined as the product of
counting efficiency, E, and the counting time, T_d, of the sample. E
represents the efficiency by which the radioactive desintegrations are
collected in the counting device ($0 < E < 1$). The number of counts, C_{s+b},
of a radioactive sample accumulated in T_d minutes is given by

$$C_{s+b} = E \times T_d \times DPM(s) + C_b \tag{1}$$

where DPM(s) is the absolute activity of the sample (in desintegrations per minute) and C_b the number of background counts. The column eluate may be fractionated for subsequent off-line counting, the sensitivity of which can be increased at will via T_d.

In flow-through ('on-line') detection, T_d equals the mean residence time in the flow cell of the radioactivity detector (also called cell-transit or cell-clearance time) and is calculated from

$$T_d = V_d/F_t \tag{2}$$

in which V_d is the flow cell volume (in ml) and F_t the total flow rate through the flow cell (in ml/min). It thus follows that one has to compromise between the sensitivity of flow-through counting and chromatographic criteria such as the maximum allowable band broadening from the relatively large cell volume and the total run time. Fig. 3 shows the impact of T_d on the sensitivity of flow-through radioactivity determinations. After their separation by RP-HPLC, the ^{14}C-labeled pesticides carbaryl and parathion were transported through the radio-activity monitor at varying flow rates F_t. Corresponding counting times T_d were 0.026 (3A), 0.51 (3B) or 6.60 (3C) min (see also sections 2.4.1 and 3.5).

Other characteristics

In working with isotopes with short half-life times as compared to the time needed to record the chromatogram, correction factors have to be used to account for the decay during chromatography. This is the case, for instance, in the determination of rare-earth fission products or radiopharmaceuticals labeled with ^{11}C, ^{13}N, ^{15}O or ^{18}F. These radio-pharmaceuticals are normally prepared under 'No Carrier Added' (NCA) conditions. This in effect means that all molecules of interest are labeled and the specific activity obtained(in DPM/mol or Bq/mmol) approaches the theoretical maximum. The corresponding masses are low, often less than 0.1 μg. Under such circumstances the only expedient to characterize the reaction products is by chromatography, in particular HPLC, for example by comparing the capacity ratios obtained from the ra-diogram with those of 'cold' (unlabeled) standards. Care has to be taken in this comparison, because the retention behaviour of isotopically labeled compounds may differ from their unlabeled analogues (refs. 164, 165, 172), especially under NCA conditions (refs. 8-10).

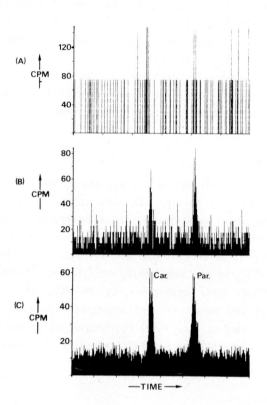

Fig. 3   Flow-through radiograms of ^{14}C-labeled pesticides, recorded at different counting times T_d. Column, Brownlee Cyano Spheri-5 (100 x 4.6 mm); eluent, acetonitrile/water (30/70, v/v), 1.0 ml/min); sample, ^{14}C-carbaryl (5 Bq) and ^{14}C-parathion (4 Bq) (for structural formuleas see Fig. 19); detection after post-column extraction and segmentation of the eluate with water-immiscible liquid scintillator. During the separation, the segmented stream is stored in a capillary storage loop. After the separation has finished, the contents of the loop is introduced into the ß-detector (V_d = 0.065 ml) at lower flow rate to increase T_d. In these examples, T_d is 0.026 (A), 0.51 (B) or 6.60 (C) min. (Reproduced from ref. 7; see also refs. 89-93).

Finally, it should be noted that in general some sort of written permit from the authorities is needed in order to work with certain types and amounts of radioactivity. Such permission will depend on the level of organization and facilities of the institute. Stringent laboratory planning, the use of dedicated apparatus (preferable automated) and an

experienced staff is required for most radioanalytical procedures. Automation and shielding of apparatus used in highly radioactive environments
are discussed in ref. 11.

1.3   OBJECTIVES AND OUTLINE

Radio-chromatography (electrophoresis, paper, thin-layer, column liquid and gas chromatography) of ß-labeled compounds has been reviewed in
1978 by Roberts (ref. 12). Since that time, a number of flow-through
radioactivity detectors have become commercially available and there has
been an increase of publications on procedures and applications of
flow-through radioactivity detection. More recent discussions of radio-
column liquid chromatography can be found in refs. 13-16.

Analytical and chromatographic techniques in radiopharmaceutical
chemistry are treated in ref. 17. This book gives updated and expanded
contributions from a symposium, organized in 1984 in Los Angeles, CA,
USA, and is concerned mainly with short-lived, γ-emitting isotopes for
nuclear medicine. It thus appears that there is a need for a literature
survey of recent developments in radio-column liquid chromatography,
covering its broad range of procedures and applications.

In section 2, the outline and mathematical aspects of scintillation
counting (SC) in radio-column liquid chromatography are considered. Some
parameters for optimization of flow-through ß-counting are discussed in
section 3, with emphasis on methods for decoupling the chromatographic
and counting steps or, in other words, possibilities to overcome the
ultimate problem of coupling a fast separation with slow counting steps.
In section 4, applications are reviewed. In section 5, some conclusions
are drawn from the discussions in the preceding sections.

2.   PRINCIPLES OF SCINTILLATION COUNTING IN COLUMN LIQUID
   CHROMATOGRAPHY

2.1   INTRODUCTORY COMMENTS ON NUCLEAR RADIATION AND DETECTION

Radioisotopes are defined as isotopes having limited life times,
caused by excess of energy in the nucleus. By spontaneous decay, this
energy is released as radiation and, eventually, stable isotopes are
formed. The actual life time of a single nucleus is unpredictable and
virtually independent of its physical or chemical state.

The radiation appears as electromagnetic radiation or as particles.
Examples from the first category are X-ray and γ-radiation (photons with
energies less than 30 keV and over 30 keV, respectively). Examples of the
last kind include positrons and negatrons (particles of mass 0.911 x

10^{-30} kg, with single positive and negative electrical charge, respectively). Negatrons are normally indicated as betas (represented by ß).

In many cases decay is via the simultaneous emission of electromagnetic radiation and particles. In that case, determination of γ-radiation is normally more convenient. Furthermore, positron emitters can be detected via annihilation photons of 511 keV that are produced after recombination of the emitted positrons with electrons. Commonly used positron emitters are ^{11}C, ^{13}N, ^{15}O, or ^{18}F. Radioisotopes are characterized by their half-life time, T1/2 - that is the time-lapse during which a population of radioisotopes decreases by a factor of two - the maximum theoretical specific activity (expressed as Bq/mmol) and the type(s) and energies of radiation. The main properties of commonly used radioisotopes have been summarized in table III.

TABLE III   Main properties of commonly used radioisotopes

	E_γ(MeV)	E_b(keV)	T 1/2	Spec. act.[1]	Range (μm)[2]
^{3}H		19	12.4y	1 TBq/mmol	4
^{11}C	0.511		20.4m		
^{14}C		155	5760y	2 GBq/mmol	264
^{13}N	0.511		9.96m		
^{15}O	0.511	2.03m			
^{18}F	0.511	109.7m			
^{32}P		1710	14.3d	200 TBq/mmol	7870
^{35}S		167	87.4d	30 TBq/mmol	302
^{36}Cl		700	3×10^5y	3 kBq/mmol	2707
^{45}Ca		258	164d	45 GBq/mmol	636
^{59}Fe	1.1;1.2	500;1600	44.6d	40 GBq/mmol	
^{63}Ni		67	100y	20 GBq/mmol	54
^{67}Ga	0.18;0.30		78h	25 TBq/mmol	
^{99m}Tc	0.14		6h		
^{125}I	0.035(X-ray)		60d	75 TBq/mmol	
^{131}I	0.36	610	8d	550 TBq/mmol	
^{140}La	0.49;1.6	1380	40h	5 GBq/mmol	

[1] Specific activities given are about the maximum values specified for commercially available radiolabeled compounds or elements.
[2] The maximum range given is calculated for pure ß-emitters in water as described in ref. 18.

The detectors used in radio-column liquid chromatography can be classified either as charge collectors or scintillation counters. With charge collectors, the electric charge produced on the interaction of nuclear radiation with the detector medium is collected and converted into electrical pulses. The pulse height and pulse rate can be related to the incident nuclear energy and activity. Examples of charge collectors are gas ionization chambers (Geiger-Müller counters) and semiconductors (Ge(Li) counters).

In scintillation counting, nuclear radiation is absorbed in liquid or solid scintillator material and converted into photons with wavelenghts from between 360 to 600 nm that are detected by photomultiplier tubes.

Although most detectors are capable of detecting both γ's and β 's (e.g. E > 0), they are normally referred to as either γ-or β-detectors, mainly because of large differences in E for the two types of radiation. The choice of detector further depends on the state of the sample (gaseous, liquid or solid) to be counted, the nuclear energies of the radioisotopes involved and, with mixtures of γ-emitting isotopes, the ability to discriminate between different γ-energies. For the latter, Ge(Li) spectroscopy is normally preferred above NaI(Tl) because of its superior energy-resolving power (Fig. 1).

Determination of β-radioactivity in liquids is relatively inconvenient because of (i) the small range of β particles (see Table III) which for efficient detection necessitates close contact between sample and detector medium, and (ii) the continueous energy distribution (Fig. 4A, B, C). The β spectra are specified by the maximum end point energy $E_b(max)$, and radioisotopes are correspondingly classified as 'low' (^3H, ^{63}Ni), 'medium' (^{14}C, ^{35}S) or 'high' (^{32}P, ^{36}Cl) energy emitters. As a consequence, individual quantification in mixtures of pure β-emitters is possible only for isotopes with large differences in E_b, as is the case in the dual-isotope determination of, for example, $^3H/^{14}C$ or $^3H/^{32}P$.

β-scintillation counting

Low to medium energy β's are normally determined by liquid or heterogeneous (solid) scintillation counting (LSC and HSC, respectively). In LSC (refs. 18, 19), the sample is thoroughly mixed with a liquid scintillator cocktail. The cocktail is usually based on toluene, pseudocumene or dioxane to which fluorofors such as 1.4-diphenyloxazole (PPO) and/or 1.4-bis-2(5-phenyloxazolyl)benzene(POPOP) and solubilizers have been added, their nature depending on the sample type to be incorporated; alcohols or non-ionic detergents such as Triton X-100 can be used as solubilizers for aqueous samples. The sovent molecules absorb the incident nuclear energy to form an electronically excited state. This energy is transferred to the fluorofor (the scintillator) which returns to its electronic ground-state by emission of photons (scheme 1). Within several nanoseconds, 20-30 (depending on the incident ß-energy) photons are produced per desintegration which are isotropically emitted. This enables coincidence counting techniques to be used (Fig. 10): The sample is placed between two opposite PM tubes, the signals of which are fed to an electronic circuit and checked for coincidence within a pre-set coincidence time. Coincident pulses are counted only. To some extent it is thus possible to distinguish between scintillation pulses and one-photon events, such as chemiluminescence. The overaleffect of the coincidence circuit is to lower the background count rate, CPM(b). The coincidence time may be chosen from several tens to hundreds of ns, depending on the type of scintillator and sample. Too short times, however, may adversely affect the counting efficiencies, and a compromise between optimum E and CPM(b) must be found experientally. Energy pulse height analysis is another means to improve the E/CPM(b) ratio, by setting lower and upper energy discriminators of the window. The lower discriminator is used for reducing low-energy back-ground pulses. Pulse height analysis is also used in dual-isotope detection with two-energy windows chosen, in which counts are collected (Fig. 4A; see also refs. 20, 127). The activity from both isotopes is determined after correction for counts, collected from the high-energy isotope in the low-energy window (cross-over correction). The same detection device can be used for HSC.

In HSC, the sample is in contact with solid scintillator, for example granular anthracene, inorganic glasses such as yttrium silicate or europium-doped calcium fluoride or plastic scintillator, doped with organic fluorofors. The ß-spectra from the glasses (see Fig. 4C) are distinctly different form spectra recorded in LSC, with the maxima shifted to higher energy. This facilitates discriminating radioactivity signals and low-energy pulses arising from PM darkcurrent or chemiluminescence.

Cerenkov counting

High energy ß's with E_b(max) of over about 0.3 MeV can be detected by scintillation and also at satisfactory efficiencies ($0.2 < E < 0.4$) via Cerenkov radiation, produced on the interaction of these ß's (^{24}Na, ^{32}P, ^{36}Cl, ^{40}K) with solvents of high dielectric constant such as, e.g., water (refs. 18, 19). The wavelength range of this type of radiation is from the ultraviolet into the visible, but with largest intensity in the short-wavelength region. Therefore, the PM tubes should preferably be equipped with UV-transparent quartz windows. Alternatively, fluorofors can be added to the sample, the function of which is to accept the Cerenkov radiation and shift the emission wavelength to higher values in order to match the maximal sensitivity of the PM tubes. Cerenkov counting has the obvious advantage that no scintillator cocktail is needed and the sample can be recovered unmodified.

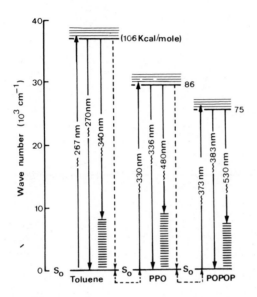

Scheme 1   Coupled energy transfer and energy level diagram for the components in a ternary liquid scintillation system (Reproduced with permission from ref. 19, Amersham, England, 1977).

2.2   PRINCIPLES OF FLOW-THROUGH γ-COUNTING

As will be illustrated below in section 4.1 it follows that γ-column liquid chromatography is used primarily for purification and identification purposes in the synthesis of radiotracers. Due to the high activities loaded on the column, flow-through γ-determination becomes straightforward, allowing the use of flow cell volumes in the order of 10 μl. Thereby, the chromatographic integrity is preserved while maintaining sufficient precision in counting results. As only one radioisotope has to be detected, energy resolution is of less concern. In these situations, scintillation counting (SC) is adequate.

146

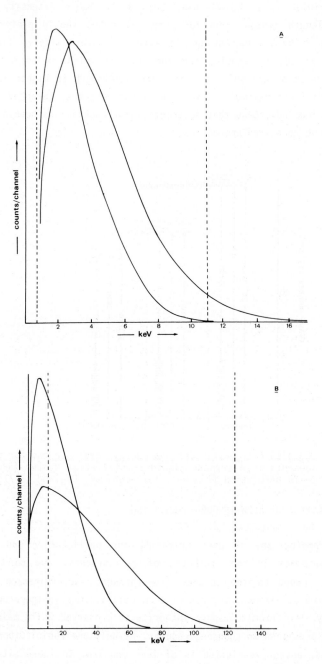

Fig. 4A, B Liquid scintillation spectra of ^{3}H (A) and ^{14}C (B), recorded for unquenched and quenched samples on a Philips PW 4700 scintillation counter.

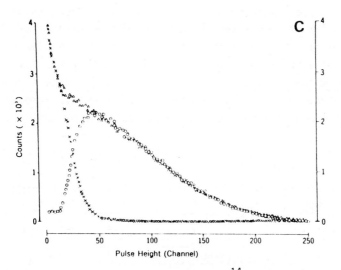

Fig. 4C CaF₂(Eu) scintillation spectrum for ^{14}C, recorded without coincidence (⊿), through coincidence (0) and anti-coincidence (X).
(Reprinted from the J. Chromatogr., Elsevier, from ref. 21).

   Thallium-activated sodium iodide (NaI(T1)) is by far the most popular
scintillator (ref. 22). It is relatively inexpensive, readily available
in different geometries and is applicable to a broad range of γ-energies
at satisfactory counting efficiencies. The penetration power of the
γ-rays in the crystal increases with increasing energy, and high-energy
γ's may escape from the material without generation of scintillating pulses. Large crystal sizes, however, are inconvenient because of the extensive shielding necessary to reduce the background count rate, CPM(b). A
cylindrical 7.6 x 7.6 cm NaI(T1) crystal can yield efficiencies with
values of 0.10-0.30 for energies of 90-600 keV, with an optimum
performance at 90-160 keV.

   Flow cells can be constructed from PTFE with low γ-energies (up to 100
keV) or stainless-steel capillaries with higher γ-energies. They should
be fixed preferably in the central depletion (well) of the crystal for
high and reproducible counting geometry ('4π' counting). Because of the
high penetration power of γ-radiation, the crystal should be lead-
shielded from the environment. Improper shielding may result in
artificial band broadening originating from activity in connecting
capillaries.

   An illustrative block diagram of the apparatus is given in Fig. 5. After injection, the injected plug is first transported through a

148

stainless-steel flow cell which measures the total radioactivity of the injected sample. Subsequently, separation takes place and the column eluate is led through a second flow cell to monitor the radiogram. By comparison of the signals from the first and second flow cell, the recovery of activity from the column can be established. The principle can also be used to indicate detector overload resulting from deadtime losses in the radioactivity detector. It is therefore recommended that the count rate in the flow cell should not exceed 10,000 CPS.

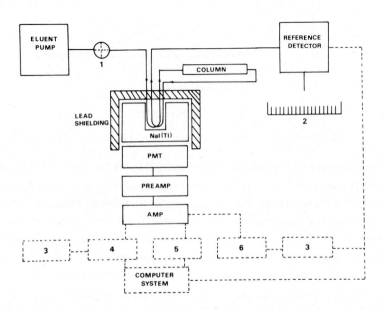

Fig. 5   Apparatus for flow-through NaI(Tl)-scintillation counting with pre- and post-column γ-counting. 1 = injector, optional: 2 = fraction collector, 3 = strip-chart recorder, 4 = ratemeter, 5 = multi-or single-channel analyzer, 6 = frequency/voltage converter.
(Pinciple adapted from ref. 22).

In Fig. 6, an HPLC chromatogram of rare-earth radionuclides is given with flow-through NaI(Tl) counting. The system was used in the investigation of the short-lived fission products of ^{252}Cf.

Some alternatives for flow-through NaI(Tl) detection have been published. Simonnet et al. (ref. 24) adapted a commercially available flow-through radioactivity detector for the determination of ^{125}I-labeled proteins. It was made of a 0.76 mm i.d. polyethylene capillary, positioned in a 20 ml sample vial filled with liquid scintillator in front of two coincident PM tubes (external SC). A counting efficiency of 0.42 was found at CPM(b) = 465. It is anticipated that this principle can be applied for detection of other low-energy - or X-ray emitters.

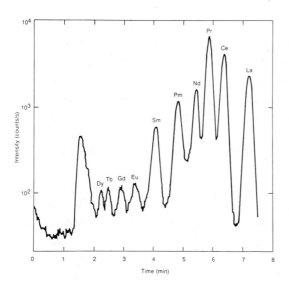

Fig. 6   Flow-through NaI(T1) radiogram of short-lived rare earth fission products of ^{252}Cf. Column, Aminex A-9 (150 x 3.2 mm, at 95 $^{\circ}$C); pH- and concentration-gradient elution with α-hydroxyisobutyric acid from 0.65 M (pH 3.6) to 0.95 M (pH 4.8), 1.0 ml/min. (Reprinted with permission from the J. Radioanal. Chem., Elsevier, ref. 23).

Nowadays most commercial flow-through detectors have provision for external SC, which are normally specified for ^{125}I and ^{99m}Tc only. The flow cells are made of PTFE capillary, encapsuled in liquid, plastic or NaI(T1) scintillator. Few papers on their use have been published. A counting efficiency of 0.05 only has been calculated for the 511 keV annihilation radiation of ^{13}N with a plastic scintillator flow cell (ref. 25) at CPM(b) = 60. This principle was also used in the determination of homogeneous [mono-^{125}I-Tyr^{10}]- and [mono-$^{12\_5}$I-Tyr^{13}-]-glucagon (after HPLC separation) (Fig. 25).

Langström and Lundqvist (ref. 27) constructed a flow cell from stainless steel foil (thickness, 0.07 mm) surrounded by 1.5 mm thick plastic scintillator for the determination of high-energy positron emitters, which for ^{11}C resulted in E = 0.40 at CPM(b) < 180.

Flow-through γ-determination based on semiconductor detectors is rather rare (refs. 28, 29). Needham and Delaney (ref. 28) utilize a small-volume CdTe semiconductor (2.6 x 2.6 x 2.0 mm housed in a 5 mm i.d. x 12 mm long aluminium cylinder). The compactness of this detector device even allowed positioning against the heat exchanger coil of an RI detector. Unfortunately, the counting efficiencies are rather low and decrease rapidly with increasing energy; E values of 0.10, 0.02 and 0.013 are found for the 93, 185 and 300 keV photopeaks of 67Ga, respectively, and of 0.24 for the 140 keV photopeak of ^{99m}Tc. For this isotope, the relative quantitative sensitivity RSQ (as defined in ref. 30) was 30-fold worse as compared to that in NaI(Tl) detection. This was attributed to the better counting geometry of the flow cell in the NaI(Tl)-well.

2.3 PRINCIPLES OF ß-COUNTING IN COLUMN LIQUID CHROMATOGRAPHY
2.3.1 INTRODUCTION
Except for ^{125}I, the most important radioisotopes for biomedical and pharmaceutical research, notably ^{3}H, ^{14}C, ^{35}S and ^{32}P, are pure (100%) ß-emitters. Their relatively long half-lifes facilitate the radio-synthesis of a wide variety of compounds with no time restrain on subsequent experiments, which makes them excellent radiotracers in recovery and metabolic profiling studies of organic compounds.

Developments in ß-column liquid chromatography up to 1976 have been reviewed by Roberts (ref. 12). Until then, most workers used off-line LSC. Flow-through detectors, when available, were assembled largely from readily available electronic components, such as high-voltage supply, PM tubes, amplifiers, pulse height channel analyzers and coincidence circuits. The remaining equipment (flow cells, light-tight housings) were home-made.

In the following sections, basic principles and apparatus for ß-determination in column liquid chromatography are given. A more detailed treatment of performances and alternative methods is given in section 3.

2.3.2 OFF-LINE LIQUID SCINTILLATION COUNTING (LSC)
In off-line LSC, successive eluate fractions are collected in sample vials on a regular time basis. After thorough mixing of the fractions or of aliquots of the fractions with LC, the vials are placed on a tray and counted separately. Plotting of the counts/fraction or the count rate/fraction versus the fraction number gives the radioactivity distribution (radiogram) in the column eluate.

Despite its relatively simple operation and good counting performance, the total procedure of collection, counting and data analysis is rather laborious and difficult to automate. Moreover, the preservation of the chromatographic resolution strongly depends on the fraction volume chosen, as is illustrated in Fig. 7 and refs. 25, 32, 33, 34. Small fraction columes are beneficial in this respect, but increased counting times/fraction are required to maintain precision in the counting results, and the total analysis time may become immoderate. For these reasons, it may be concluded that, in general, off-line LSC is not applicable to high-efficiency HPLC, especially when high sample throughputs are required.

It has further been shown in reversed-phase HPLC, that collection of small (< 0.2 ml) aqueous fractions becomes more capricious if increasing percentages of water are present in the eluent (ref. 35). This has been explained by fluctuations in surface tension of the droplets, which led to significant differences in the collected volume/fraction. Some of these difficulties can at least partially be avoided by collecting total peak volumes. For this, the output of any type of flow-through detector may be triggered to recognize the beginning and end of an eluting peak, which controls an automatic fraction collector. However, the number and positions of peaks normally identified by liquid chromatography detectors are by no means representative for the actual radioactivity distribution in the chromatogram, as can be seen from Figs. 2, 9 and 11. Of course, no problems in selectivity exist when using flow-through radioactivity counting for this purpose, and most commercial ß-detectors have a provision for controlling the fractionation of column eluates.

2.3.3 FLOW-THROUGH LSC

The apparatus needed for flow-through LSC is shown in Fig. 8. A reference detector is used for the determination of mass (if specific activities have to be determined) and facilitates the development of separation conditions. A splitter is included if part of the eluate must be recovered unmodified.

In most applications, reversed-phase HPLC is used (see section 4). Efficient mixing of the aqueous eluate with the organic scintillator cocktail then is of crucial importance for reproducible and efficient counting. Pulse-free pumping of the cocktail to the eluate is accomplished by using a packed 'dummy' column or pressure resistor in the scintillator solvent line. In some examples, mixing is accomplished in a low-volume mixing chamber with magnetic stirrer (refs. 36-38), but it has

152

now been accepted that small-bore (0.25 mm i.d.) T-pieces work equally well, giving efficient mixing with virtually no extra column peak broadening under normal conditions. Under more demanding circumstances, such as ^{3}H counting in eluents of high ionic strength or with high water contents, a specially designed high-efficiency mixing unit may be helpful (ref. 39). Depending on the type of eluent, the mixing ratio scintillator /eluate and the mixing efficiency, typical E values are 0.20-0.35 for ^{3}H and 0.50-0.90 for ^{14}C. Typical background count rates vary from 10 to 60 CPM.

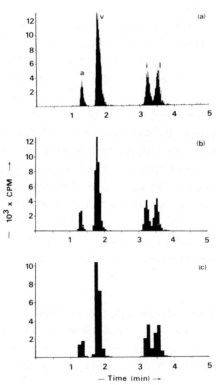

Fig. 7a-c   (a) Flow-through LSC of ^{14}C-labeled amino acids. Column, Brownlee RP-18 Spheri-5 (100 x 4.6 mm); eluent, methanol/water (5/95, v/v) pH 4.2 with 0.02 M acetate buffer containing 0.03 M sodium hexylsulphonate, 1.0 ml/min; liquid scintillator, 1.0 ml/min; flow cell volume, 63 μl; sample frequency, 0.6 s^{-1}. (b, c) represent corresponding simulations for off-line detection with sample frequencies of 3.0 (b) and 6.0 (c) s^{-1}, respectively; at eluent flow rate of 1.0 ml/min, this corresponds to fraction volumes of 0.05 and 0.1 ml, respectively. (Reproduced by permission from Eur. Chromatogr. News, John Wiley & Sons, ref. 31).

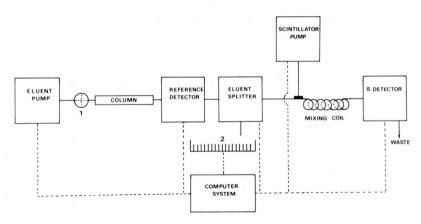

Fig. 8   Apparatus for flow-through liquid scintillation counting. 1 = injector, 2 = fraction collector. For detailed diagram of ß-detector, see Fig. 10

The flow cell in the ß-detector consists of a spiral made out of wound light-transparent material (normally PTFE), which is closely held between two PM tubes for coincidence counting. Dilution of eluate with scintillator allows cell columes V_d to be somewhat larger as compared to cell columes commonly used in, e.g., UV or RI detectors, while maintaining most of the chromatographic integrity. This is illustrated in Fig. 9, in which V_d = 1.0 ml at F_e = 1.0 ml/min and F_s = 4.0 ml/min.

2.3.4 FLOW-THROUGH HETEROGENEOUS SCINTILLATION COUNTING (HSC)

Compared to LSC, the set-up for flow-through HSC is considerably simpler because no LS-pump or eluate splitting is needed (Fig. 10). Three approaches may be distinguished: (1) the flow cell is constructed of UV-transparant glass or PTFE capillary, packed with finely divided, solid scintillator; (2) it is made of plastic scintillator capillary; or (3) it is constructed from a PTFE capillary encapsuled in a liquid or plastic scintillator for external SC.

Little information on ß counting with the latter two approaches is available. However, due to the small range of ß particles in liquids as compared to the dimensions of these cells, counting efficiencies are relatively low. For instance, with a 0.7 mm i.d. NE102A plastic scintillator cell, an efficiency of 0.057 has been claimed for ^{14}C-labeled solvents (ref. 41). The corresponding efficiency for ^{14}C-labeled gaseous carbon dioxide was 0.58. Plastic scintillators are not inert to oxidising acids and most organic solvents, such as acetonitrile and tetrahydrofuran. Their use is therefore restricted to qualitative SC with aqueous eluates. The performance for higher energy ß's from ^{32}P is significantly better; values of over 0.70 may then be obtained (ref. 41).

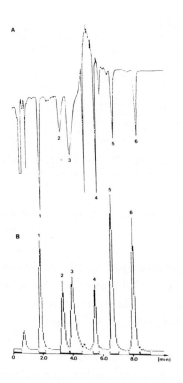

Fig. 9   Representative UV (A) and radiogram (B) of tyrosine containing
metabolites of enkephalin. (B) was obtained after incubation of
an homogenate of rat-astrocytes with added carboxypeptidase A
and [^{3}H-Tyr]enkephalin. Column, Spherisorb ODS II (60 x 4.1 mm);
gradient elution with mobile phases consisting of methanol and
citric acid/phosphate buffer to which 0.3 mM sodium n-octa-
nesulphonate was added; detection, UV (254 nm) and flow-through
LSC, respectively; flow cell volume of the radioactivity
detector was 1 ml; peaks: 1 = Tyr, 2 = Tyr-Gly-Gly, 3 = Tyr-Gly,
4 = Tyr-Gly-Gly-Phe, 5 = Met-enkephalin and 6 = Leu-enkephalin.
(Reprinted with permission from the J. Chromatogr., Elsevier,
ref. 40).

Of the heterogeneous modes, the packed cell (1) guarantees the most intense contact between eluate and scintillator and, thus, optimum geometric counting efficiencies. This approach was first used in amino acid analyzers, with anthracene- (ref. 42) or POPOP-packed cells (ref. 43). The main disadvantages of these types of scintillators are their solubility in organic solvents and the relatively low (< 0.02) counting ef-ficiencies for ^{3}H. These materials have now been replaced by inorganic scintillator glasses, such as cerium-activated lithium silicate or yttri-um silicate, or salts such as europum-doped calcium fluoride. The new types of scintillators are virtually inert to most solvents used in column liquid chromatography. It is, however, recommended to avoid pH values of less than 2 or greater than 8. Furthermore, the calcium fluori-de material is normally not compatible with solvents containing ammonia. Depending on the particle size and the packing density of the cell, counting efficiencies are around 0.05 (^{3}H), 0.70 (^{14}C) and 0.90 (^{32}P).

Fig. 11 shows that a relatively large cell volume can be used without deterioration of the chromatographic resolution in the radiogram as obtained from the reference detector. In this case, the HSC cell has a volume of 0.6 ml, and is packed with yttrium silicate particles.

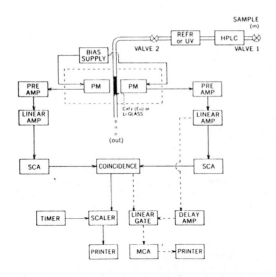

Fig. 10   Schematic diagram for flow-through heterogeneous scintillation counting. AMP = Amplifier, SCA = Single-Channel Analyser, MCA = Multi-Channel Analyser. (Reprinted from the J. Chromatogr., Elsevier, ref. 21).

2.3.5 FLOW-THROUGH CERENKOV COUNTING

To the author's knowledge, only one paper dealing with flow-through Cerenkov counting has been published (ref. 45). This is rather surprising in view of some distinct advantages of this technique in the determination of high-energy β's (see section 2.1). In the paper, inorganic pyrophosphate (PP_i) produced in a cell culture medium was determined by incubation with radioactive orthophosphate ($^{32}P_i$). Intra- and extracellular $^{32}PP_i$ was measured using a weak anion exchange HPLC separation of PP_i from P_i and other phosphor-containing compounds, with flow-through Cerenkov counting at E > 0.99 (Fig. 12).

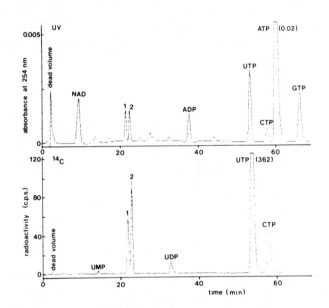

Fig. 11 HPLC-UV and HPLC-radioactivity profiles of a perchloric acid extract of human lymphoblastic cell line MOLT-3 after incubation with ^{14}C-uridine for 2 h. The extracted nucleotides were separated on Partisil-10 SAX radially compressed modules (100 x 8 mm) at 2.0 ml/min; detection was by UV (254 nm) and flow-through HSC, respectively. For the latter, the flow cell was packed with modified yttrium silicate granules, resulting in an empty volume of 0.6 ml. Under these conditions, a detection limit of 5 Bq was found (at E = 0.43); peaks: 1 = UDP-N-acetylglucosamine and/or UDP-N-acetylgalactosamine, 2 = UDP-glucose and/or UDP-galactose. (Reprinted with permission from the J. Chromatogr., Elsevier, ref. 44).

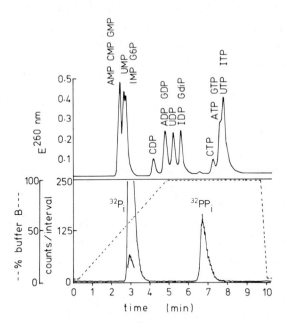

Fig. 12 Chromatographic separation of phosphate-containing compounds. Column, radial-pack Bondapak-NH$_2$ (100 x 8 mm); mobile phase, 0.1 M citric acid (buffer A) and 0.2 M citric acid containing 0.1 M potassium sulphate plus 0.02 M magnesium sulphate (B) (gradient elution); detection by UV (260 nm; upper trace) and flow-through Cerenkov counting of ^{32}P (lower trace). (Reprinted by permission from Anal. Biochem., Academic Press ref. 45)

2.4. DATA ANALYSIS

2.4.1 DEFINITIONS OF DETECTION LIMITS

Relatively few papers have been concerned with definitions of detection limits in flow-through counting in terms of chromatographic parameters and the properties of the detector device, such as the counting efficiency, counting time and background count rate (refs. 30, 32, 46-49). Algorithms derived by Sieswerda et al. (ref. 46) and Klein and Hunt (ref. 47) are briefly summarized in this section. For convenience, symbols and abbreviations as used by these authors have been changed to conformity. A more fundamental treatment on detection limits in radio-activity determination can be found in refs. 50 and 51. The net peak area C_s (in counts) is given by:

$$C_s = C_{s+b} - (T_w/T_b) \times C_b \qquad (3)$$

in which C_{s+b} represents the total counts collected in the peak and C_b the total background counts collected in time T_b (in min). T_w is the total peak counting time (in min), which in flow-through counting equals the base peak width (in min).

If it is assumed that the standard deviation in C_s, SD (C_s), is determined by the statistics of the radioactive decay alone, $SD(C_s)$ can be calculated from poisson statistics, and follows from:

$$SD(C_s) = [C_{s+b} + (T_w/T_b)^2 \times C_b]^{0.5} \tag{4}$$

This assumption is valid for low levels of activity only. At high activity levels, contributions from other sources, such as variations in injection volume, counting efficiency or counting time should be considered as well.

Sieswerda (ref. 46) then defines the precision as the signal-to-noise ratio in C_s, which is the reciprocal of the relative standard deviation in C_s, $RSD(C_s)$. Combining eq. (3) and (4) yields:

$$RSD(C_s) = \frac{SD(C_s)}{C_s} = [\frac{1}{C_s} \times (1 + \frac{C_b \times T_w}{C_s \times T_b} \times (1 + \frac{T_w}{T_b}))]^{0.5} \tag{5}$$

Since $C_s = E \times DPM(s) \times T_d$ and $C_b = CPM(b) \times T_b$, from eq. (5) $RSD(C_s)$ becomes:

$$RSD(C_s) = [\frac{1}{E \times T_d \times DPM(s)} \times (1 + \frac{CPM(b) \times T_w}{E \times T_d \times DPM(s)} \times (1 + \frac{T_w}{T_b}))]^{0.5} \tag{6}$$

Assuming $T_b \gg T_w$, eq. (6) simplifies to:

$$RSD(C_s) = [\frac{1}{E \times T_d \times DPM(s)} \times (1 + \frac{CPM(b) \times T_w}{E \times T_d \times DPM(s)})]^{0.5} \tag{7}$$

From eq (7) detection limits have been calculated as function of $E \times T_d$ and $RSD(C_s)$, they have been plotted in Fig. 13, assuming typical values for T_w and $CPM(b)$.

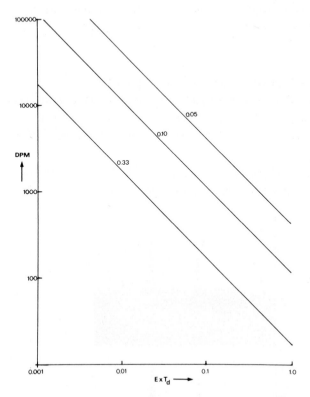

Fig. 13 Detection limits (in DPM) vs. ExT_d, calculated from eq. (7) at $RSD(C_s)$ = 0.33, 0.10 and 0.05, with $CPM(b)$ = 30 CPM and T_w = 0.5 min.

Klein and Hunt (ref. 47) interpret the detection limit as the activity DPM (min) that must be present in the eluting peak in order to obtain a peak height count rate of twice the background count rate. Assuming a triangular peak shape, an expression for DPM (min) is derived:

$$DPM(min) = \frac{CPM(b) \times [(F_e \times S + F_s)] \times [T_w + V_d/ (F_e \times S + F_s)]}{S \times E \times V_d} \quad (8)$$

in which S is the fraction of the eluate transported through the radio-activity detector. T_d thus follows from $T_d = V_d/[F_e \times S + F_s]$.

At this point it should be noted that in practice, T_d is the only parameter which the analytical chemist can vary at will to improve precision in the measurement. Methods other than off-line counting that are intended to improve precision via T_d have been compiled in sections 3.4 and 3.5.

2.4.2 DATA ACQUISITON, PROCESSING AND PRESENTATION

In order to adequately represent the shape of an eluting peak it is normally advised to collect 20-30 data points (samples) in the peak. For peak widths of about 0.5 min, this requires sample frequencies of 1 s to be used.

However, the total number of counts accumulated per sample decreases with decreasing sample frequency, which results in noisy signals.

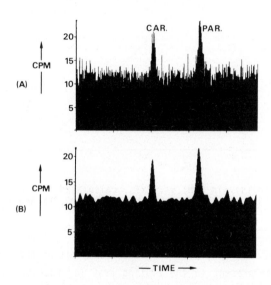

Fig. 14 Radiograms of the ^{14}C-labeled pesticides carbaryl (1.0 Bq) and parathion (0.8 Bq). Chromatographic conditions, see Fig. 3; raw (A) and filtered (by Fast Fourier Transform; B) data. (Reproduced from ref. 7; see also refs. 89-93).

Therefore, the use of data filtering techniques is of prime importance in low-level flow-through radioactivity counting in order to be able to recognize small peaks recorded at short sampling frequency. An electronic filter for radiochromatography has been described (ref. 52). With the introduction of data collection, storage and handling by personal computers, the use of filtering algorithms has eliminated much of the need for electronic filtering. Filtering can thus be performed in real time during the collection of data samples or, alternatively, after storage of the raw data. The latter is preferred, mainly because it allows more sophisticated filtering programmes such as Savinsky-Golay moving average (ref. 53) or Fast Fourier Transform (FFT) (ref. 54) to be used. Fig. 14 illustrates the effectiveness of FFT on a detection limit determination of ^{14}C-labeled pesticides with flow-through LSC. No peak broadening occurs after filtering, allowing a more simple peak detection algorithm to be used.

2.5. COMMERCIALLY AVAILABLE FLOW-THROUGH RADIOACTIVITY DETECTORS

At present, at least six detectors for flow-through ß-counting are commercially available (Table IV). Most of these systems can be purchased with an (optimal) LS-pump, eluent-splitter system, built-in micro processor, software packages for controlling data collection and data handling of digital and analogue signals, flow cells with volumes ranging from 0.05 to about 1.5 ml and different types of solid scintillators.

TABLE IV Commercially available flow-through radioactivity detectors

Type(s)	Manufacturer
MODEL 171/170	Beckman Instruments, Inc., Fullerton, CA, USA
LB 506 (A, B, D)	Berthold Lab., Wildbad, FRG
FLO-ONE (IC, CR, CT, BD)	Radiomatic Instruments and Chemical Co., Inc. Tampa, Fl, USA
ISOFLO	Nuclear Enterprizes Limited, Edinburgh, GB
BETACORD 1208	LKB WALLAC, Turku, Finland
RAMONA radioactivity detectors	Isomess Isotopenmessgeräte, Straubenhardt, FRG
ß-MAT	IN/US, Fairfield, NJ, USA

Depending on the configuration, the costs of these instruments vary between Hfl. 20,000-25,000* for basic set-ups (without LS-pump), splitter and computer system, and one analogue output from a ratemeter provided) to about Hfl. 40,000-60,000 for more complex systems. These prices are, of course, subject to variation and give rough indications only.

*(Hfl. 1,000 = about $500)

3. OPTIMIZATION PARAMETERS FOR FLOW-THROUGH ß-COUNTING
3.1 INTRODUCTION

In this section, the optimization of E, CPM(b) and T_d is considered in more detail. It should be stressed that most of the variables that can be used to improve the counting results are to some extent interrelated; this requires careful selection of the operating conditions.

In section 3.5, some alternative methods for monitoring of ß-labeled compounds in column liquid chromatography are discussed, most of which deal with the increase in sensitivity via T_d by decoupling separation from counting.

3.2 THE COUNTING EFFICIENCY E; GENERAL ASPECTS

In flow-through SC, E is determined by the product of three indepen-dent efficiencies: the efficiency with which the scintillator/eluate mixture converts the ß-energy into photons (the intrinsic efficiency), the efficiency with which these photons reach the photocathode of the PM tubes (the geometric efficiency), and the efficiency of the electronic circuit (including the PM tubes) to convert the photons into electrical pulses that pass through the coincidence circuit.

The coincidence time and the settings of the lower and upper pulse height discriminators may be chosen to improve E. However, increasing the coincidence time or decreasing the lower pulse height discriminator may adversely affect the background count rate. To characterize a given set-up, it therefore is not sufficient to quote E but rather the Figure of Merit, FM, defined as:

$$FM = E^2/CPM(b) \tag{9}$$

E can be specified under dynamic flow conditions or at zero flow. Differences in dynamic and static E values have been reported (refs. 35, 55-57). Dynamic efficiencies appear to be somewhat smaller. One explana-tion for this phenomenon is the decrease in effective flow cell volume with increasing flow rate, as observed by Van Nieuwkerk et al. (ref. 58); in other words, the actual dynamic and static efficiencies then are about equal.

For the determination of dynamic efficiencies, it is more convenient to perform plug injections (without column installed) for greater choice in radioactive standard. Possible radiochemical impurities then do not interfere with the measurement. This also facilitates the recording of gradient curves yielding E as a function of the gradient parameters which are normally needed when combining gradient elution with flow-through LSC.

It turns out that the strongest influence the user can have on E is via the intrinsic efficiency, and experimental data from flow-through LSC and HSC on this subject are given in the next two sections.

3.2.1 E in LSC

Scintillator cocktails, especially formulated for dynamic flow counting have become available only recently. Their main properties are fast sample incorporation and high sample hold capacity. With these scintillators, LS/eluate rations of about 3-5 will normally give satisfactory performance. Nitrogen purging of the scintillator may reduce oxygen quenching by about 50% (ref. 59).

Some precautions with respect to the eluent composition should be taken. Halogenated solvents such as chloroform and dichloromethane act as strong quenchers of the LS fluorescence and must be avoided. The applicability of types of solvents to LSC may be judged from table V, which gives the relative quenching properties of some aliphatic substituents.

TABLE V Relative quenching properties in LSC of aliphatic groups[1]

Diluter	Mild quencher	Strong quencher
-H	-CH=CHR	-OCO.COR
-F	-Cl	-I > Br
$(OR)_3PO$	$-NH_2$	-NHR
-CN	-OH	$-NO_2$
-COOR	-COOH	-CHO
		-SH=SR
		$-Cl_2$
		$-Cl_3$

[1]Reproduced by permission of Friedr. Vieweg & Sohn, ref. 59.

E can further be affected by gradient elution. Roberts and Fields (ref. 35) determined dynamic counting efficiencies for ^{3}H with acetonitrile-water gradients at an LS/eluent ratio of 3. Values of 0.21, 0.13 and 0.22 were found for 0.60 and 100% (v/v) acetonitrile, respectively. Webster and Whaun (ref. 57) examined the effect of methanol or salt gradients on static efficiencies for ^{14}C in RP-HPLC. No significant influence on E was found for the methanol-water gradient (0-35% methanol in 0.01 M phosphate buffer, pH 5.6, which is in accordance with the observation made by Causey et al. (ref. 60). Salt gradient elution resulted in a slight decrease in E from 91 to 77%, with a 0.01-0.7 M phosphate buffer at pH values of 3.4 and 4.3, respectively.

Finally E can be influenced by the injected sample. This was illustrated in ref. 61, where preconcentration of a 900 ml pool of hamster urine on a conical pre-column resulted in heavy quenching by endogeneous material in parts of the radiogram. These extreme conditions, though, are not likely to be met in routine work.

3.2.2 E IN HSC

The influence of scintillator particle size on E has been studied for various materials (refs. 21, 62-64). Some results are given in Table VI. Mutual comparison of the scintillators is hampered because of differences in mesh sizes (or mesh-size distributions) and packing densities of the cells.

TABLE VI Counting efficiencies of solid scintillators for ^{3}H and [14]C

Type	Particle size Mesh	μm	[3]H[1]	[14]C[1]	Ref.
Ce(Li)		90-125		0.44	62
		63-90		0.57	62
		38-63		0.62	62
		250-350		0.16(D)	67
		250-350		0.40(D)	68
	60-80		0.018-0.014	0.38(S)	64
	140-160		0.041-0.044	0.62	64
	160-180		0.069-0.062	0.71	64
	150			0.20(S)	21
EuCaF$_2$	150-250			0.57(S)	21
	45-100		0.035(S)		63
	100-150		0.086(S)		63

[1]D = dynamic, S = static

Nevertheless, it can be concluded that E increases with decreasing mesh sizes. At the same time, however, the pressure build-up, ΔP, across the cell increases. ΔP can be approximated from the well-known equation:

$$\Delta P = [\eta \times L^2]/[K_o \times T_d \times d_p^2] \text{ (bar)} \tag{10}$$

In which L is the cell length (in m), η the eluent viscosity (in Kg/(m x s)), d_p the scintillator particle size (in m) and K_o the permeability constant (with values of 0.001-0.002). Substituting typical values of L = 0.04 m, η = 0.001 kg/(m x s), K_o = 0.0015, T_d = 10 s and d_p = 25x10^{-6}m ΔP is 1.7 bar.

The lower limit of d_p is further determined by the adsorption behaviour of the scintillator material, which becomes especially evident at increasing specific surface areas. At present, counting efficiencies of over 0.80 and 0.11 have been claimed by some manufacturers for ^{14}C and ^{3}H, respectively. For ^{3}H, apart from few exceptions (refs. 39, 55, 65), such impressive figures have not been confirmed in the literature as yet.

Although with HSC it is normally assumed that E is not affected by the eluent composition, some evidence of the opposite is available. Giersch (ref. 66) determined the counting efficiencies of ^{14}C-labeled sugars and corresponding esters with a lithium glass scintillator. With an aqueous salt gradient of 5-400 mM phosphate, E was 0.22, which increased to 0.39 for acetonitrile/water (80/20, v/v, pH = 3.0). The difference was attributed to alterations in the partition coefficients of the analytes between mobile phase and scintillator surface, and relativates HSC counting efficiencies calculated from equation (1).

Finally, compounds in the eluent may absorb photons emitted from the scintillator. Colour quenching by p-nitrophenol has been observed by Mackey et al. (ref. 62). Mori reported colour quenching from reagents used in the post-column ninhydrin derivatization of amino acids (ref. 67) and the bromocresolpurple derivatization of carboxylic acids (ref. 68).

3.3   THE BACKGROUND COUNT RATE CPM(b); GENERAL ASPECTS

Ideally, a stable background countrate should be observed, which originates from the thermally generated PM dark current, cosmic rays and radioactivity present in the glass envelope of the PM tubes (^{40}K, ^{232}Th and ^{238}U) (ref. 69). In practise, a number of user-determined sources contribute to CPM(b) as well. These include the laboratory environment and interfering processes in the flow cell of the radioactivity monitor.

Examples from the first category are external radioactive sources, stray light, spikes from electric apparatus and temperature fluctuations. Proper lead shielding can give a 3-fold reduction in CPM(b) (ref. 70). Connections to and from the detector should be light-tight. For this the use of stainless-steel or blackened PTFE capillary is recommended.

Dark counts from the PM tubes can be reduced substantially by cooling, but only one example of this effect is found in the literature on radio-column liquid chromatography (ref. 63). Contributions to CPM(b) arising from cross-talk of two opposite PM tubes can be diminished by using paper masks (ref. 62).

The second category is perhaps of more importance because it refers more to the daily routine. Radioactive contamination in the flow cell is the most abundant example in this class, but chemiluminescence and phosphorescence have also been shown to add to CPM(b). Although one-photon events, these processes may pass the coincidence circuit if the pulse rate exceeds the coincidence resolving time and thus be counted (ref. 59). Differences in the pulse-height distribution of luminescence and scintillator fuorescence enable corrections to be made for pulses not originating from scintillator fluorescence (ref. 71) and some radio-activity monitors have provisions to perform such correction.

From the above reflections it can be argumented that CPM(b) should be specified under realistic conditions, i.e. between successive mea-surements with the flow cell filled with the scintillator-eluate mixture. It is good practice to run 'blank' chromatograms, prefereably at short sample frequency to observe spikes and thus be able to identify possible background sources.

The following sections summarize some observations described in the literature on background contributions arising in the flow cell.

3.3.1 CPM(b) in flow-through LSC

Chemiluminescence may occur after mixing alkaline solvents or ethers with liquid scintillators (refs. 19, 30, 36, 55, 59). The effect is especially pronounced for dioxane-based cocktails. Peroxides in ether or dioxane are believed to play a major role. It can be suppressed by cooling of the LS/eluate mixture to 10 ^{o}C before entering the flow cell (refs. 30, 36) or electronically corrected for after pulse height analy-sis. Because of the relatively short lifetimes of most chemiluminescence processes, increasing the time lapse between mixing and counting may also be used.

Radioactive contamination of the flow cell is not expected to be a ma-jor problem, yet adsorption of radiolabeled peptides (ref. 35), pesticides (ref. 72) and anions (refs. 73, 74) on PTFE-capillary flow cells has been observed.

3.3.2. CPM(b) IN FLOW-THROUGH HSC

Contamination of packed cells has been observed for a large variety of solutes, some of which are collected in Table VII. These observations are not always consistent. For instance, Schutte (ref. 77) explained the observed adsorption of ^{14}C-labeled nucleotides on calcium fluoride by the low solubility of calcium-salts of nucleotides, whereas Nakamura and Koizumi (ref. 21) found no adsorption for the same types of materials.

TABLE VII   Observations of adsorption on solid scintillator particles

Scintillator	Sample	Remedy	Ref.
Yt_2SiO_5	phospholipids	HNO_3 flush (20%)	75
CeLi	sugars	hot detergent and 0.1 M HCl flush	66
CeLi	carboxylic acids	repacking after 50 inj.	68
CeLi	phosphates, acetates	repacking	62
$EuCaF_2$	polymers	--	63
$EuCaF_2$	glucuronides	methanol, water or detergent flush	76
$EuCaF_2$	ribonucleotides	silanization	44
$EuCaF_2$	nucleotides	--	77

In general, HSC is not compatible with solutes having molecular weights larger than about 600, such as proteins and polymers. The increase in CPM(b) observed with these solutes is explained both by filtering processes and the adsorption of radioactivity to the scintil-lator due to the relatively large number of reactive sites in these molecules (ref. 78). As a general rule one should avoid injecting radio-active samples of unknown identity or samples with radiolabeled solutes with large differences in polarity. Before starting flow experiments, the adsorption behavior of the solutes should be established using test-tubes filled with the scintillator powder (ref. 39). These test tubes also facilitate developing washing procedures for the packed cell. Other pre-cautions that can be taken are presaturation of the scintillator with non-labeled analogs or silanization of the scintillator surface (ref. 44).

Some of the disadvantages in HSC may be circumvented by the principle introduced by Rucker et al. (ref. 78), who employed flow cells filled with axially aligned 0.1 mm i.d. Ce(Li) glass fibers. Computer models were developed for the prediction of the geometric counting efficiencies of ß-emitters. Except for ^{3}H, the predicted values agreed well with the experimental counting efficiencies. For ^{14}C and ^{32}P, E values of 0.55 and 0.93 were found, respectively. For ^{3}H, E was only 0.001 (predicted 0.10). The system was compared with a HSC cell packed with Yt_2SiO_5 powder (with $d_p < 25$ μm). As compared to the packed cell, substantial reductions in pressure build-up and contamination were observed using the glass fiber cell (Table VIII). No radiograms demonstrating the flow characteristics of the glass-fiber cell were given as yet.

3.4   OTHER PARAMETERS
3.4.1 SELECTING THE FLOW CELL VOLUME V_d IN LSC

Some authors simply relate the maximum permitted V_d to the minimum time ΔW, observed between two neighbouring peaks of interest in the reference detector (refs. 47, 55). V_d(max) is then calculated according to:

$$V_d(\text{max}) = \Delta W \times [F_e \times S + F_s] \text{ (ml)} \tag{11}$$

Obviously, only those peaks are detected which refer to radiolabeled products. For instance, in metabolism studies only the parent compound and its metabolites are labeled and therefore detected in the radiogram. It may then be worthwhile to optimize resolution of radioactive peaks, which allows larger cell volumes to be used. Alternatively, V_d(max) may be related to the effective base peak width T_w (ref. 30). For most practical purposes, a T_w/V_d ratio of about 10 is acceptable with respect to extra column peak broadening (ref. 79). V_d(max) then becomes:

$$V_d(\text{max}) = [0.4 \times V_o \times (1 \times k')]/[X \times N^{0.5}] \text{ (ml)} \tag{12}$$

in which X is the eluate fraction in the eluate/scintillator mixture. Substituting typical values of N = 3000, V_o = 0.5 ml and X = 0.25, maximum permitted flow cell volumes of 0.030 and 0.073 ml are calculated for k' = 1.0 and 4.0, respectively.

The criteria given above do not include losses in resolution caused by poor flow dynamics in the mixing-T, connecting capillaries or flow cell. To minimize the extra-column peak broadening, eluate segmentation techniques can be used. These are treated in section 3.5.

3.4.2 SELECTING THE FLOW CELL VOLUME V_d IN HSC

The criteria for V_d(max) given in section 3.4 may be adapted to flow-through HSC if additional band broadening from the packed cell can be neglected.

To some extent the packed cells can be treated as packed-bed reactors. For analytes that do not adsorb to the scintillator particles (no retention), the added variance in peak volume σ_w arising from axial molecular diffusion and convective mixing in a cylindrical packed cell can in that case be approximated by:

$$\sigma_w = [(h \times d_p \times T_d^2)/L]^{0.5} \text{ (ml)} \tag{13}$$

in which h is the reduced plate height ($2 < h < 6$). From this equation it follows that it is advantageous to use long, small-bore capillaries rather than short, large-bore ones in constructing the cell. Substituting typical values of $T_d = 1.0$ s, $h = 4$, $d_p = 25 \times 10^{-6}$ m and $L = 0.04$ m, the added variance is 0.05 s. A 20-fold decrease in F_e($T_d = 20$ s) gives $\sigma_w = 1.0$ s, which still seems quite acceptable for most applications.

Large differences between experimental and theoretical values for σ_w may be explained by adsorption of the analyte on the scintillator material, resulting in asymmetric tailing peaks (ref. 80).

3.4.3 REPRODUCIBILITY AND LINEARITY

Limited data are available on reproducibilities and linear dynamic ranges in flow-through ß-counting. For high levels of activity, statistical and background fluctuations can be neglected (refs. 46, 60, 81) and reproducibility is expected to be determined primarily by E. This has not been thoroughly evaluated yet. Both for E and T_d, pulse-free pumping of solvents and (with LSC) efficient mixing of eluent and scintillator are probably the most important requirements. RSD values, as determined from repetitive injections of high ^{14}C-levels, can be as low as 1-2% for the LSC mode (refs. 57, 60) and 2.2% for the HSC mode (ref. 82).

The effect of replacing packed cells on reproducibility must also be considered; repacking a single HSC cell resulted in static [14]C E of 0.716

$\pm$ 0.003 (n = 3), while packing of five HSC cells of about the same geometric dimensions resulted in static ^{14}C E values of 0.688 $\pm$ 0.024 (ref. 62).

The linear dynamic response is determined primarily by the dead time of the electronic circuit and the maximum number of counts per sample frequency which the computer can accumulate (ref. 83). A linearity of up to about 4×10^3 Bq/peak can easily be obtained (ref. 44).

TABLE VIII Adsorption (%) on granular Yt_2SiO_5 and fiber scintillators[1]

Compound	Eluting solvent	Retained % Yt_2SiO_5	Fiber
Glycine	0.5 M acetate	0.5	0.7
Glucose	50% ethanol	1.6	0.2
Cholesterol	benzene	2.1	1.4
UDP	50% ethanol	38[2]	1.0
ATP	0.01 M phosphate	55[2]	9.9
Inulin	0.01 M phospahte	1.4	0.1
Insulin	50% ethanol	83[2]	24
Cytochrome C	0.01 M phosphate	86[2]	84

[1] Reproduced by permission of Friedr. Vieweg & Sohn, ref. 78
[2] Irreversible sorption

3.5   SPECIAL METHODS

Bakay (refs. 74, 84-86) suppressed peak broadening in flow-through LSC of ^{14}C-labeled amino acids by using post-column gel segmentation. For this, a polyacrylamide gel is pumped by motor-driven high-pressure syringes to the scintillator/eluate mixture. The benefits of segmentation on the clearance of the flow cell is clearly demonstrated in Fig. 15. Schutte (ref. 77) used air segmentation, but data of its effect on suppresion of peak broadening were not given.

Snyder (ref. 87) suggested the usage of post-column segmentation and buffer storage systems for decoupling separation and counting steps. In addition, in comparing post-column reactors (open tubular, liquid- or gas-segmented flow and packed bed), Scholten et al. (ref. 88) showed that when dealing with capillaries with internal diameters of over 0.3 mm and residence time of over about 20 s, segmentation techniques should be applied in order to prevent peak broadening. Van Nieuwkerk (ref. 89) adapted the principle of solvent segmentation and extraction of aqueous column eluates for subsequent storage of the segmented stream in a capillary storage loop (Fig. 16). During the separation, the analytes are extracted from the aqueous eluate into water-immiscible LS plugs. The LS plugs act both as detection medium and segmentator. The segmented flow is transported through a flow-through ß-detector for recording the direct radiogram and subsequently stored in a capillary storage loop. After storage of the complete chromatogram the contents of the loop are re-introduced into the detector at low flow rates by turning the switching valve. These flow rates can be selected according to the counting time T_d and, therefore, the sensitivity needed, independently from the separation (reverse radiogram).

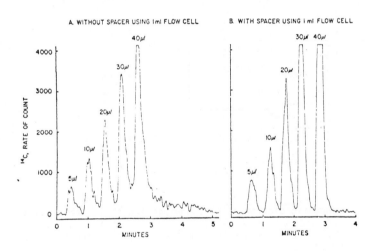

Fig. 15 Tracings of radioactivity of samples containing ^{14}C-labeled amino acids. Samples were injected at 0.5 min intervals, without (A) and with (B) gel segmentation. Total flow rate of column eluate and liquid scintillator is 2.5 ml/min to which the gel spacer is added at 0.044 ml/min (B). (Reprinted with permission from Anal. Biochem., Academic Press. ref. 84)

The principle was applied in the determination of ^{14}C-pesticides (refs. 7, 90) and ^{14}C-amino acids (ref. 91). For extractable analytes, a 0.75 mm i.d. stainless-steel capillary was used as storage loop, and it was shown that segmentation effectively suppressed peak broadening in the

172

capillaries, permitting T_d values of over 5 min to be used (see also Figs. 3 and 14).

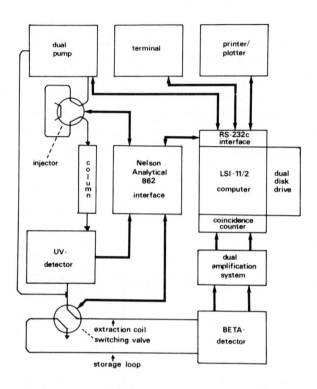

Fig. 16 Schematic diagram of HPLC equipment with flow-through ß-detector and extraction/segmentation/storage system. (Reproduced with permission from Chromatographia, Friedr. Vieweg and Sons, ref. 90.)

For non-extractable analytes, the stainless-steel capillary led to significant peak broadening, which was explained by aqueous wetting on the inner wall of the stainless-steel capillary and transport of the non-extracted analytes through the aqueous film. This effect was suppressed by using a 0.80 mm i.d. PTFE loop instead. For the non-extracted analytes, as compared to corresponding direct measurements, a significant decrease in E was observed in the reverse measurements. This could be explained by the increase in the volumes of the aqueous and scintillator segments during the transport and storage in the capillaries. As a consequence, less ß-particles reach the scintillator segments in the reverse mode. In this case, reverse counting efficiencies could be improved by homogenizing the contents of the storage loop just before reintroduction into the ß-detector while still maintaining the peak broadening suppression effect. A second scintillator pump was installed for this purpose for the addition of water-miscible LS to the contents of the storage loop (Fig. 17). For some selected amino acids the performance of the system for non-extractable analytes is illustrated in Fig. 18. Alternatively, the principle can be adapted for non-extractable analytes by using post-column ion-suppression or ion-pair extraction techniques as shown by Veltkamp et al. (refs. 92, 93), thereby avoiding the need for a second scintillator pump. This was demonstrated in the determination of the ^{14}C-labeled, amine-containing pharmaceuticals remoxipride (ref. 92) and urapidil (ref. 93). Urapidil and its main metabolites (Fig. 19) were separated on a cyano-bonded phase, using an aqueous eluent consisting of acetonitrile/water (12/88, v/v, pH = 2.2).

Under these conditions, the analytes are non-extractable because of protonation of the amine substituents, which resulted in counting efficiencies for ^{14}C of less than 0.05 for the direct radiogram. Addition of the ion-pair reagent sodium dodecylbenzenzesulphonate to the water-immiscible LS increased the counting efficiency to over 0.80 (Figs. 20, 21). The principle was applied in the determination of urapidil and its main metabolites in rat plasma.

An additional advantage of using water-immiscible LS for flow-through LSC in reversed-phase systems is that water is excluded from the scintillator segments, and that the LS/eluate ratio may be chosen such as to optimize the sensitivity $E \times T_d$. Extraction into the scintillator plugs, and thus E, improves at high ratios. At the same time, however, the total flow rate F_t increases, and T_d decreases proportionally (and vica versa). This is shown in Fig. 21 for remoxipride. Repetitive injections of the analyte were made at F_e = 1.0 ml/min while varying the scintillator flow rate F_s. For extractable compounds, the sensitivity was normally at its maximum at F_s/F_e = 0.2 (ref. 31). This compares favourably to water-miscible scintillators for which ratios of over 3 are recommended in order to obtain a homogeneous phase (refs. 30, 59).

Baba et al. (refs. 48, 94, 95) developed a flow-through synchronized accumulating radioisotope detector for GC and HPLC. It consists of five flow cells connected in series. The signals from each cell are synchronized and accumulated. As a result, a five-fold increase in sensitivity is obtained. Additional peak broadening in the fifth cell was negligible as compared to the first cell with V_d = 1.1 ml each, F_e = 1.5 ml/min and F_s = 8.3 ml/min (Fig. 22). In the HSC mode considerable peak

174

broadening took place with yttrium silicate packed cells with volumes of 0.39 ml each (ref. 48), which hampered synchronization of the signals. The high cost of the counting equipment required for data collection of five different cells will probably prevent other workers to adapt this principle.

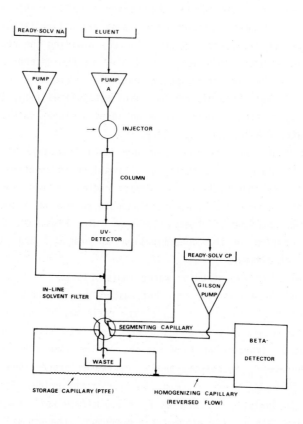

Fig. 17 Schematic diagram of HPLC equipment with flow-through ß-detector and segmentation/storage system, adapted for the determination of radiolabeled, non-extractable analytes. (Reproduced from J. Chromatogr., Elsevier, ref. 91).

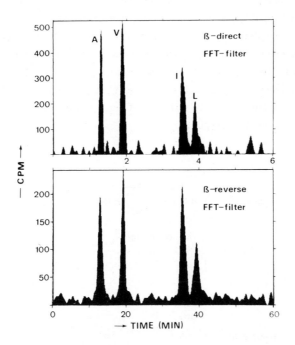

Fig. 18 Radiograms of ^{14}C-labeled amino acids. Chromatographic conditions, see Fig. 7; sample, alanine (65 Bq), valine (77 Bq),
isoleucine (82 Bq) and leucine (41 Bq); upper trace, radiogram
recorded during separation at T_d = 0.03 min; lower trace,
radiogram recorded after storage of the complete chromatogram by
reintroduction of the contents of the storage loop into the
ß-detector at lower flow rate (T_d = 0.15 min); both traces
filtered by FFT. (Reproduced from J. Chromatogr., Elsevier, ref.
91).

Karmen et al. (ref. 96) adapted an automatic micro-fraction collector
for fractionation of the column eluate on filter paper and subsequent
autoradiography. Fractions with volumes of up to 0.30 ml are collected
into wells, formed in non-wetting fluorocarbon film. After evaporation to
near-dryness, the remaining spots are transferred to a filter paper
placed over the wells by using a vacuum technique.

Depending on the volatility of the fractions, the time needed to
evaporate and quantitatively transfer the spots to the paper is about
5-10 min. The volume of the remaining spots was controlled by the addition of 0.0005% glycerol to the eluent. The performance of the system was
tested in the determination of selected ^{14}C-labeled amino acids, with
0.125 ml fractions and at F_e = 1.0 ml/min (Fig. 23). The sensitivity and
the linear dynamic range was strongly determined by the exposure time of

the filter paper to the X-ray photographic film. The widest linear range (250-5000 DPM/spot) was obtained for a 6 h exposure time. Exposure for 72 h permitted the determination of 5-80 DPM/spot. With this method, a high sample throughput may be obtained even for extended exposure times at moderate costs. The reproducibility was satisfactory. Since autoradiography is normally used for qualitative purposes only (ref. 97), care should be taken when quantitatively interpreting autoradiographs with large differences in activity/spot.

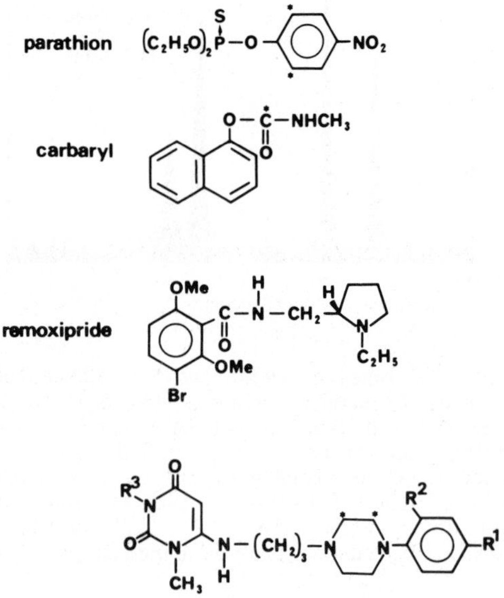

	R³	R²	R¹
urapidil	CH₃	OCH₃	H
m1	CH₃	OCH₃	OH
m2	CH₃	OH	H
m3	H	OCH₃	H

Fig. 19 Structural formulae of ^{14}C-labeled pesticides (parathion and carbaryl) and pharmaceuticals (remoxipride and urapidil and main metabolites). The positions of the labels are indicated by the asterisk.

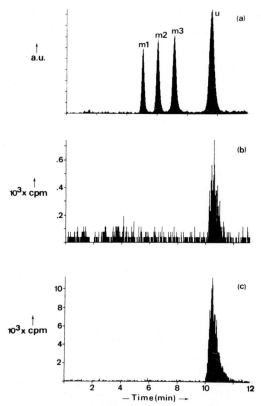

Fig. 20 Representative UV chromatogram (269 nm) of urapidil and its main metabolites (A) and radiograms of ^{14}C-urapidil (730 Bq) under non-extractive (B) and extractive (C) conditions. Column, Pierce Cyano Spheri-5 (100 x 4.6 mm); eluent, acetonitrile/water (15/85, v/v), pH = 2.2 with phosphoric acid, 1.0 ml/min; water-immiscible liquid scintillator, Ready-Solv NA, 0.2 ml/min. For Fig. 20C, 1.0 mM sodium dodecylbenzenesulphonate was added to the scintillator.

4. APPLICATIONS

4.1 PREPARATION, PURIFICATION, IDENTIFICATION

In most experiments involving the use of radioisotopes, the validity of the results is strongly dependent on the radiochemical purity of the compound used, which is defined as the total radioactivity present as the nuclide of interest in a specific chemical form. Furthermore, knowledge of the specific activity is normally also required.

Repeated analysis just prior to the use of the stored material is recommended because, apart from its expected chemical and microbiological decomposition, its radiochemical purity decreases upon storage due to radiolysis. In particular ß-labeled compounds are prone to radiolysis because the ß's emitted form high density spurs of reactive radicals in the surrounding matrix. A detailed discussion on radiolysis can be found in ref. 98.

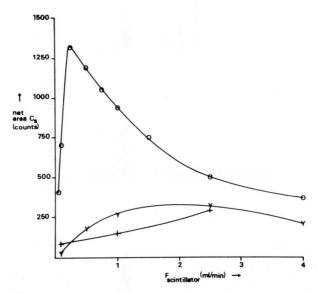

Fig. 21 ^{14}C-peak area (in counts) as function of the scintillator flow rate. Chromatographic conditions, see Fig. 20; sample, ^{14}C-remoxipride (600 Bq); liquid scintillator, water-immiscible (+), water-immiscible with 1.0 mM sodium dodecylbenzenesulphonate (O) and water miscible (Y). (Reproduced by permission from Eur. Chromatogr. News, John Wiley & Sons, Ref. 31.)

Of the various methods for purification, isolation and identification, HPLC is the most convenient because of its superior selectivity, easy operation (which is important from the standpoint of safety in handling high levels of radioactivity), and speed of analysis which even allows the purification of compounds labeled with shortlived isotopes. Examples have been collected in Table IX, some of which are described below in more detail, with emphasis on the benefits of using HPLC in the procedures.

Chasko and Thayer (ref. 99) considerably simplified isolation of cyclotron-produced ^{13}N-labeled nitrite or nitrate from water targets by using a reductor, a concentrator and two analytical columns in series, packed with copperized cadmium and Partisil-10 SAX anion exchanger, respectively (Fig. 24A, B). As compared to the alternative method of rotary evaporation, an increase in concentration of the radiolabeled material of at least 10-fold is obtained, with the additional advantages of less sample handling and higher radiochemical purities of the material. Boothe et al. (ref. 100) used RP-HPLC with mobile phases containing the ion-pairing reagent n-octylamine for separation of anions labeled with ^{11}C, ^{13}N or ^{18}F. A significant influence of carrier addition on the elution patterns was observed. For example, NCA $^{18}F^-$ only eluted form the column after the addition of carrier F^-. These observations are consistent with results by other workers (refs. 8-10).

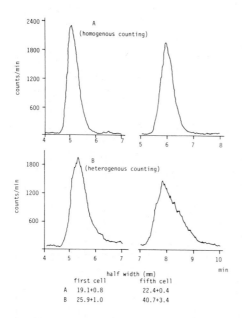

	half width (mm)	
	first cell	fifth cell
A	19.1+0.8	22.4+0.4
B	25.9+1.0	40.7+3.4

Fig. 22   Peak broadening using a synchronized accumulating radioisotope detector in liquid (upper trace) and heterogeneous (lower trace) scintillation counting. For explanation, see text. (Reprinted from J. Chromatogr., Elsevier, ref. 48).

In the determination of specific activities of radiohalogenides, Kloster and Laufer (ref. 101) used pre-column derivatization of the halides with 2-naphthol in the presence of the oxidizing agent chloramine-T and subsequent HPLC. The detection limit of the halonaphthol reaction products was 0.5 pmol (at 220 nm), which allowed the determination of specific activities of NCA radioiodine or radiobromine using only 5% of the total radioactivity.

TABLE IX   Preparation, purification and identification in radio-CLC

Sample(s)	Column and mobile phase	Detection[1]	Ref.
^{13}N-anions ($^{13}NO_3^-$, $^{13}NO_2^-$, $^{13}NH_4^+$)	Partisil-10 SAX, 30 mM phosphate buffer (pH = 3.0)	f.t.	99
radioanions (^{11}C, ^{13}N, ^{18}F, ^{82}Br, ^{131}I, ^{99m}Tc)	C_{18}, 0.01 M octylamine (aqueous; 4.5 < pH < 6.5)	f.t.NaI(Tl)	100
radio-halogenides (^{38}Cl, ^{80}Br, ^{82}Br, ^{128}I, ^{131}I)	TSK LS-222, acetone/sodiumnitrate (1.5 N) (1/1,v/v)	f.t.GM	108
^{125}I$^-$, ^{125}I-proteins	Sephadex G-25	external LSC	24
radioanions (^{13}N, ^{77}Br, $^{128}IO_3^-$, mixture)	YEW AX-1, 4 mM Na_2CO_3/$NaHCO_3$	f.t.Ge(Li)	29
^{125}I-monoiodoglucagon	C_{18}, n-propanol/phosphate buffer (10mM, pH = 2.5)	external plastic	26
^{125}I/^{14}C-peptides	C_{18}, acetonitrile/trifluoroacetic acid (0.05%),	o.l.	103
^{99m}Tc-diphosphonates	Aminex A28, 0.7 M sodium acetate	f.t.NaI(Tl)	106
^{14}C-S-adenosylmethionine	C_{18}, methanol/phosphate buffer (45/55, v/v)	f.t.HSC	109
^{14}C-chloroform, ^{14}C-dibromoethane	C_{18}, methanol/water (75/25, v/v)	f.t.LSC	110
^{14}C-proteinhydrosylate	SCX (preparative), lithium/sodium citrate, gradient f.t.HSC		111
^{14}C-butoprozine	C_8, methanol/dichloromethane/water (100/35/15, v/v/v) + 0.5% diethylamine (v/v)	o.l.	112

1 f.t. = flow-through; o.l. = off-line

181

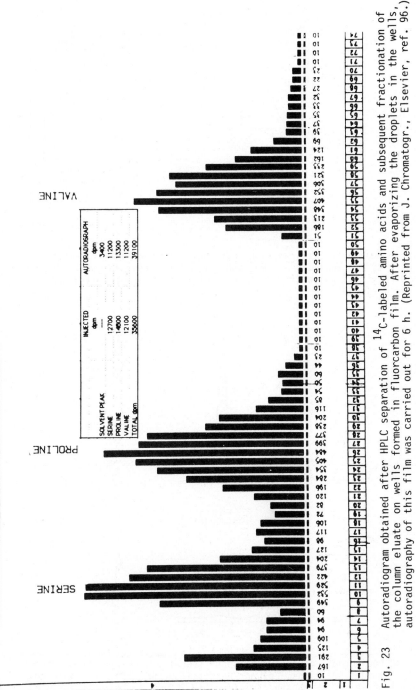

Fig. 23 Autoradiogram obtained after HPLC separation of ^{14}C-labeled amino acids and subsequent fractionation of the column eluate on wells formed in fluorcarbon film. After evaporizing the droplets in the wells, autoradiography of this film was carried out for 6 h. (Reprinted from J. Chromatogr., Elsevier, ref. 96.)

182

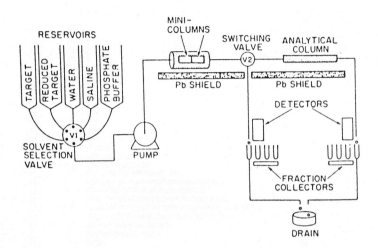

Fig. 24A Schematic diagram for concentration and purification of cyclotronproduced $^{13}NO_3^-$ and $^{13}NO_2^-$ from water targets. (Reprinted by permission from Int. J. Appl. Radiat. Isot., Pergamon Press, ref. 99).

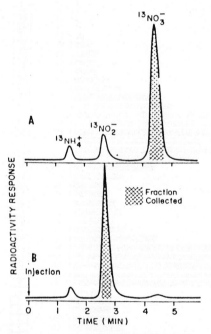

Fig. 24B Radiograms of unreduced (upper trace) and reduced (lower trace) target solutions. Column, Partisil-10 SAX (250 x 4.6 mm): mobile phase, 30 mM phosphate buffer, pH 3.0, 3.0 ml/min. (Reprinted by permission from Int. J. Appl. Radiat. Isot., Pergamon Press, ref. 99).

Radioiodine can be introduced via covalent binding to tyrosine-groups into peptides and proteins by reaction of the radiolabel with the biomolecule in the presence of chloramine-T or lactoperoxidase (ref. 102). However, if the biomolecule contains two or more tyrosine groups, heterogeneous mixtures may be formed, even if single-labeled compounds are obtained. Pingoud (ref. 26) analyzed the intramolecular distribution of ^{125}I incorporated into the 29-amino acid polypeptide hormone glucagon which contains two tyrosine residues (at positions 10 and 13). Two monoiodinated products were obtained that could be separated by RP-HPLC (Fig. 25). Receptor binding ability of these isomers to isolated intact rat hepatocytes was shown to differ by a factor of two. The intra-molecular distribution of ^{125}I was confirmed after enzymatic cleavage of the isolated products, followed by reinjection into the chromatograph. Judd and Caldwell (ref. 103) studied the level of ^{125}I-incorporation after in-gel chloramine-T iodination into fragments of the major outer membrane proteins of chlamydia trachomatis strain L2/434. After enzymatic cleavage of the proteins, RP-HPLC with radioactivity determination was used for peptide mapping (Fig. 26). Since peptides could not be detected by UV-monitoring, the L2 strain was intrinsically labeled with a mixture of ^{14}C-amino acids to ensure that each peptide fragment would be radiolabeled and could be detected. Dual isotope detection of the eluting fragments then gives the level of radioiodination. Most of the fragments contained the ^{125}I label, which indicated that amino acid groups other then tyrosine are also iodinated.

To overcome some of the drawbacks involved in radioiodination of pro-teins, such as the need to expose the protein to oxidizing agents and the in vivo instability of the radiolabeled material, the use of metallic radionuclides has been suggested (ref. 104). They can be incorporated into proteins by complexation with chelating agents covalently attached to the protein. For this purpose, Hnatowich et al. (ref. 105) used di-ethylenetriaminepentaacetic acid (DTPA) as chelating group. To determine the coupling efficiency of DTPA to IgG antibody, a radio-HPLC method was developed for the separation of the 'free' and 'protein-bound' DTPA-metal complex. ^{111}In was used as radiotracer (Fig. 27). Since DTPA forms strong complexes with a large number of metals, the principle can be extended to other metallic radionuclides. However, this necessitates the use of ultrapure reagents and deionized water in the preparation of the labeled proteins in order to avoid the introduction of non-radioactive metals.

Finally, radio-column liquid chromatography is used for the characterization of complex mixtures of radiopharmaceutical preparations.

184

The work on ^{99m}Tc-diphosphonate complexes may serve as an example because the carrier itself (^{99}Tc) is radioactive which hinders handling the amounts of material needed for alternative methods, such as X-ray crystallography. These complexes are widely used as bone-imaging agents. Their synthesis involves the reduction of ^{99m}Tc-pertechnetate electrochemically or with reductants such as SN^{2+} or BH_4^- in the presence of the diphosponate ligand. The preparations normally result in a mixture of ^{99m}Tc-complexes. Because these complexes may exhibit different in vivo distribution, the usage of a single complex is recommended in order to obtain a clear picture of the distribution. After sodiumborohydride reduction of ^{99}Tc in the presence of hydroxyethylidene diphosphonate, at least twelve complexes could be separated by anion exchange HPLC on Aminex A-29 (ref. 106), which were shown to interconvert on standing. The charges on eluting species could be established by correlation of the retention times with the activity of the counter ion used for the anion-exchange separation. In addition, relative molecular sizes were determined by size exclusion chromatography.

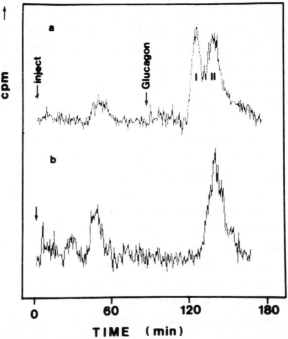

Fig. 25 Radiogram of [mono-^{125}I-Tyr^{10}]- and [mono^{125}I-Tyr^{13}]glucagon (upper trace) and commercially available [mono-^{125}I-Tyr^{10}]-glucagon (lower trace). Column, LiChrosorb C$_{18}$ (250 x 4.6 mm); eluent, n-propanol/phophate buffer (10 mM, pH 2.5) (19/81, v/v); detection was by flow-through external scintillation counting, using plastic scintillator. (Reprinted from J. Chromatogr., Elsevier, ref. 26).

When using tin as reductant, ^{113}Sn can be used to determine whether tin incoorporates into the Tc-diphosphonate complexes. The absence of tin as tracer from peaks containing ^{99}Tc (or ^{99m}Tc) then indicates that no tin is incoorporated (Fig. 28). Finally, the metal-ligand ratio can be established by using ^{32}P-labeled ligands and dual isotope determination in eluting peaks, as was illustrated in ref. 107.

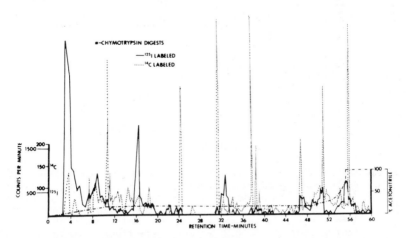

Fig. 26 HPLC elution profiles of ^{125}I- and ^{14}C-labeled peptides of C. trachomatis L2 MOMP, generated by α-chymotrypsin digestion. Column, Bondapak C_{18} (300 x 3.9 mm); gradient elution with acetonitrile/water, containing 0.05% trifluoroacetic acid (v/v); dual isotope detection was performed by off-line scintillation counting techniques on 0.2 ml fractions. (Reprinted by permission from J. Liq. Chromatogr., Marcel Dekker Journals, ref. 11103).

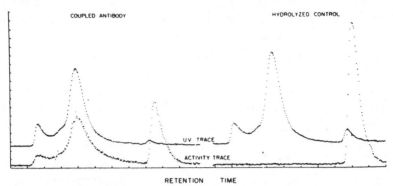

Fig. 27 HPLC-UV (280 nm) and radioactivity trace of a DTPA-coupled antibody sample (human IgG) after labeling with ^{111}In and its corresponding hydrolyzed control. Separation was performed using a protein column (I125 or I250, Waters Associates). Early eluting peaks contain protein material. (Reprinted from J. Immun. Methods, Elsevier, ref. 105)

186

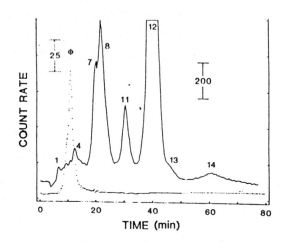

Fig. 28   Separation of tin-reduced technetium-hydroxyethylidene diphosphonate complexes, prepared at pH 10 in the presence of
^{99m}Tc and ^{113}Sn. The solid line represents the ^{99m}Tc responce,
while the dotted line indicates the ^{113}Sn responce. Column,
Aminex A28 anion exchange (150 x 4 mm); mobile phase, 0.70 M
sodium acetate at 0.2 ml/min; detection, flow-through dual-
channel detection by NaI(T1) scintillation counting. (Reprinted
by permission from Anal. Chem., American Chemical Society, ref.
106).

4.2   DISTRIBUTION AND METABOLISM OF EXOGENIC COMPOUNDS

From the numerous examples in the field of drug and pesticide analysis
some that employ flow-through radioactivity determination have been
summarized in Table X. ^{3}H-, ^{14}C- or ^{35}S-labeled parent compounds are
readily available from commercial sources, or can be supplied as custom
preparation. One example is quoted below.

In the study on the metabolism of the chemotherapeutic agent
Pt-diaminocyclohexane (Pt-DACH), the ligand was labeled to high specific
activity by reductive tritiation of 1,2-diaminocyclohexene (ref. 113).
The sensitivity of off-line LSC was better by about one order of
magnitude as compared to atomic absorption determination.

Metabolism took place by complexation of the agent with amino acids.
By using ^{14}C-or ^{35}S-labeled amino acids it was possible to determine the
stoichiometry of some major amino acid-platinum biotransformation
products (Fig. 29).

TABLE X   Radio-CLC of exogenic compounds

Sample(s)	Column and mobile phase	Detection[1]	Ref.
14C-doxylaminesuccinate	CN bonded phase, methanol/phosphate buffer	f.t.HSC	6
14C-omeprazole	C8, SI-60 or amberlite XAD	"	114
14C-felopidine	column switching techniques	"	115
3H-and 14C-labeled drugs	C18, aqueous acetonitrile mostly	f.t.HSC/LSC	116
Pt-3H-diaminocyclohexane	C18 and SCX subsequently	o.l.LSC	113
14C-MDL 257	C18, acetonitrile/acetate (0.01 N), gradient	f.t.HSC	117
14C-glyceryltrinitrate	C18, methanol/water (50/50, v/v)	f.t.LSC	94
14C-carbaryl and 14C-parathion	CN bonded phase, acetonitrile/water (70/30, v/v)	"	90
14C-deltametrin	SI-60, hexane/pentane/acetonitrile/dioxane/ 2-propanol/lumaflow	"	118
14C-ethyldipropylthiocarbamate	C18	f.t.HSC	119
14C-caffeine	CN bonded phase, acetonitrile/methanol	"	120
14C-nicotine	CN bonded phase, acetonitrile/methanol/water gradient	f.t.HSC	121
3H-benzo[a]pyrene	C8, methanol/tetrabutylammonium bromide (0,04 M), gradient	o.l.LSC	122
33S-glutathionoconjugate of 2-bromohydroquinone	C18, methanol/water/acetic acid (9/90/1, v/v/v)	"	123
13N-nitrite/-nitrate	Partisil-10 SAX, 30 mM phosphate buffer (pH: 3.1)	f.t.NaI(Tl)	124

[1] f.t. = flow-through; o.l. = off-line

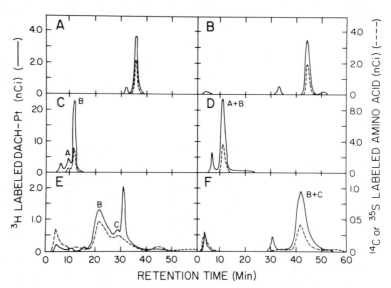

Fig. 29 Use of dual isotope detection and cation exchange HPLC in the determination of the stoichiometry of amino acid complexes of the chemotherapeutic agent Pt-1,2-diaminocyclohexane. The complexes were prepared by incubation of $[4,5-^3H](cis-1,2-$diaminocyclohexane)malonateplatinum(II) with ^{14}C-methionine (A, B), ^{14}C-serine (C, D) or ^{35}S-glutathione (E, F) at 37 °C. After purification by ion-pair reversed-phase chromatography, peak fractions were concentrated and subsequently chromatographed on Partisil-10 SCX (250 x 4.5 mm), with aqueous mobile phases containing acetonitrile (15%, v/v) and 50 mM phosphate (at pH 4 in Figs. A, C, and E and pH 2.3 in Figs. B, D and F). Radioactivity was determined in 1 ml fractions. (Reprinted by permission from Anal. Biochem., Academic Press, ref. 113).

4.3   DISTRIBUTION AND METABOLISM OF ENDOGENIC COMPOUNDS

Isotopic techniques often are the only resource by which studies on the distribution and biotransformation of endogeneous materials at physiological concentration can be undertaken. Examples in this area deal with amino acids, peptides, catecholamines, nucleotides, sugars, fatty acids, lipids, steroids and vitamins. A broad range of high-purity radiolabeled endogeneous compounds are available from various commercial sources. These also include simple precursors, such as ^{32}P-pyrophosphate or $^{14}CO_2$ which can be used for incubation studies.

Some recent examples on the use of flow-through radioactivity determination in metabolism studies of endogeneous materials have been collected in Table XI. Nieves et al. (ref. 25) studied the distribution of ^{13}N-glutamate in rats and its metabolic products in rat liver. Separation of radiolabeled products was evaluated using cation-, and anion exchange and amino columns (Fig. 30), or RP-18 columns following pre-column derivatization. For the latter O-phthaldialdehyde (OPA) derivatization was chosen because it could be carried out under the time constraints imposed by the short-lived isotope (T1/2 = 9.96 min). Problems were observed, however, in the reaction of OPA with glutamate and ammonia.

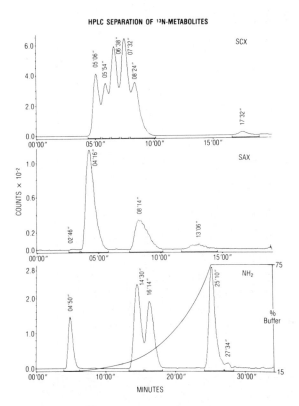

Fig. 30 Separation of ^{13}N-labeled metabolites in rat liver formed 1 min after hepatic portal vein injection of L-^{13}N-glutamate. Columns used were Partisil-10 SCX (upper), Particil-10 SAX (middle) and RSIL-NH$_2$ (lower) (250 x 4.6 mm, all). In this case, detection was by flow-through external scintillation counting using a plastic scintillator. Urea and amino acid metabolites were identified enzymatically and by coelution with unlabeled standards. (Reprinted from J. Chromatogr., Elsevier, ref. 25).

4.4 RADIOASSAYS

In radioassays, radiolabeled analytes are used in order to determine a property of an unlabeled compound, such as the concentration, localization or biological activity. Examples in this category are enzyme activity determinaitons and the ^{32}P post-labeling of DNA adducts (Table XII).

TABLE XI   Radio-CLC of endogeneous compounds

Sample(s)	Column and mobile phase	Detection[1]	Ref.
^{3}H/^{14}C-steroids, -nucleotides, -proteins	various	f.t.vs. o.l. LSC	34
^{3}H-arachidonic acid metabolites	C$_{18}$, acetonitrile/0.005% phosphoric acid (gradient)	f.t. HSC	75
^{14}C-fatty acid methyl esters	C$_{18}$, acetonitrile/water (90/10, v/v)	f.t. LSC	125
^{14}C-steroid metabolites	C$_{18}$, "	"	126
^{3}H-cholesterol/^{14}C-pregnenolone	C$_8$, acetonitrile (95/5, v/v)	"	127
^{3}H/^{14}C-purine nucleotides	Partisil-10 SAX	"	57
^{3}H/^{14}C-purine nucleotides and bases	C$_{18}$	"	57
^{3}H-enkephalin metabolites	C$_{18}$, methanol/phosphate buffer (gradient)	"	40
^{14}C-amino acids	Durrum-Dc 4A resin, Lithium/citrate buffer	f.t.HSC	67
^{14}C-amino acids	Aminex A9	"	42
^{14}C-carboxylic acids	Diaion-CK08S, 3mM HClO$_4$ (aqueous)	"	68
^{32}P-pyro (PP$_i$)- and orthosphophate (P$_i$)	NH$_2$bonded phase, citric acid/Mg^{2+} (gradient)	f.t.cérenkov	45
^{14}C-sugar degradation	ion-exchange, water	f.t. HSC	128

[1] f.t. = flow-through; o.l. = off-line

TABLE XII   Radioassay techniques and enzyme activity determinations

Radioassay	Detection[1]	Ref.
Characterization of deoxyribonucleoside phosphates in DNA adducts by ^{32}P-postlabeling	f.t.LSC	131
Determination of: 5'-methyldeoxycytidine in DNA by ^{32}P-postlabeling	o.l.LSC	134
estradiol-17-fatty acid esters by acylation with ^{3}H-acetic anhydride		135
ADP-ribose by ^{3}H-borohydride reduction of corresponding ketones		132
in vitro and in vivo receptorbinding with ^{3}H-steroids		136
anions in column eluates by indirect detection with ^{35}SO$_4$$^{2-}$ mobile phases		133

Enzyme	Substrate(s)		Ref.
Purine (in intact and lysed cells)	various	f.t.HSC	137
α-Amylase	^{14}C-starch	f.t.LSC	138
Aromatic-L-aminodecarboxylase	^{14}C-L-3, 4-dihydroxyphenylalanine		139
Dog-liver N-methyltransferase	^{14}C-amines		140
Brain adenylate cyclase	^{3}H-adenosinetriphosphate (ATP)	o.l.LSC	141
ATP-sulfurylase	^{35}SO$_4$$^{2-}$		142
UDP-glucuronosyltransferase	UDP-[U-^{14}C]glucuronic acid and ^{3}H-substrates		130

[1]f.t. = flow-through; o.l = off-line

TABLE XIII   Tracer applications in radio-CLC

Sample(s)	Application[1]	Ref.
[3]H-steroids	i.s. for determination of: selected steroids in human urine	143
[3]H-dopamine O-sulphate (DAS)	DAS in human plasma	144
[3]H-vitamin D	vitamin D in various milk powders	155
[14]C-benzo[a]pyrene (BaP)	BaP in refined coal- and petroleum fuels	156
[14]C-carboxylic acids	conversion into 4'-bromophenacyclester derivatives	157
[3]H/^{14}C-phospholipids	determination of retention behaviour on: normal-phase packing	33
[14]C-methylated aprotin	C_{18}	145
[125]I-albumin	C_8	146
[125]I-cytochromec and -ovalbumin	reversed-phase packings	147
alkali-and transition metal ions	ion-exchanger	151
[153]Gd(DTPA)	C_{18}	148
[140]La	study of memory effects from components of the LC set-up	152
[59]Fe-ferritin	determination of intracellular, ferritin associated iron	153
[14]C-hypoxanthine	study of hypoxantine transport in human erythrocytes	154

[1] i.s. = internal standard

Abeijon et al. (ref. 129) measured the enzymatic activity of cytidine monophosphate-N-acetylneuraminic acid (CMP-NeuAc) synthetase. In the experiment, this enzyme catalyzes the reaction

$$CTP + [^3H]NeuAc <---> CMP-[^3H]NeuAc + PP_i$$

(CTP = cytidine triphosphate). The labeled products were separated by anion-exchange HPLC with off-line LSC. The specific activity of $[^3H]NeuAc$ (0.8 TBq/mmol) enabled the detection of about 160 fmol CMP-NeuAc, which is approximately three orders of magnitude more sensitive than alternative methods such as colorimetry or fluorimetry.

The use of flow-through SC in radioenzyme assay is rather limited. Coughtrie et al. (ref. 130) describe a method for the assay of UDP-glucuronosyltransferase activity towards various xenobiotic and endogeneous substrates. Metabolites could be identified by double labeling analysis, using ^3H-labeled substrates with $UDP[^{14}C]$-glucuronic acid (GA) as co-substrate:

$$^3H-substrate + UDP[^{14}C]GA --> [^3H-^{14}C]glucuronide-metabolite$$

HPLC on a polar amino-cyano bonded phase column was used in separating the radiolabeled glucuronide conjugates from $UDP[^{14}C]GA$, with flow-through LSC. An example is given in Fig. 31, where ^3H-androsterone was used as substrate. For most of the substrates tested, the corresponding glucuronide conjugates eluted between 8-10 min after injection, and could be detected at the 100-200 pmol level. Flow-through radioactivity determination considerably simplified the procedure, and the authors therefore suggest that the assay can be extended in the investigation of many other compounds.

Relatively few papers are concerned with the use of radioisotopes in pre- or post-column derivatization reactions for HPLC, probably because of the hazards involved in handling high levels of radioactivity and the small masses of radiochemicals available, which make it difficult to obtain satisfactory reaction yields at at high specific activities. Dietrich et al. (ref. 131) adapted the method of Randerath for the characterization of DNA adducts by ^{32}P-postlabeling (Scheme 2). It is based on the reaction of DNA with the chemical (i.e. carcinogenic agent) under study followed by enzymatic hydrolysis of the resulting modified DNA to form deoxyribonucleoside-3'-phosphates, which are subsequently radiolabeled by reaction with ^{32}P-ATP in the presence of T4 poly-nucleotide kinase. The resulting [3',5'-^{32}P]biphosphates are charac-terized by HPLC either directly or after enzymatic hydrolysis with Nuclease 1 to [5'-^{32}P]monophosphates. Figs. 32A-C give chromatograms of products obtained after reaction of 3'-dGMP with the ethylating agent diethylsulfate (DES) followed by ^{32}P-postlabeling with UV and radio-activity detection by flow-through LSC.

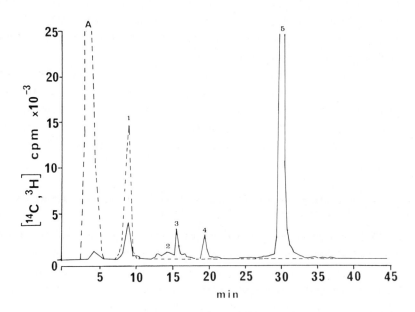

Fig. 31   HPLC assay of UDP-glucuronosyltransferase activity towards androsterone, using UDP-glucuronic acid (UDP-GA) as co-substrate. Column, Partisil-5 PAC (Polar Amino-Cyano) (260 x 4.5 mm); gradient elution from 100% acetonitrile to 100% 0.01 M TBAHS; detection by flow-through liquid scintillation counting; incubations, performed separately, contained either unlabeled androsterone and UDP-^{14}C-GA or ^{3}H-androsterone and unlabeled UDP-GA (---); peak identification, (A) = androsterone; 1 = androsterone glucuronide; 2 = glucuronic acid; 3 and 4, unidentified; 5 = UDP-GA (UDP = uridine diphospho). Reprinted by permission from Anal. Biochem., Academic Press, ref. 130).

Hakam et al. (ref. 132) used ^{3}H-labeling in the simultaneous determi-nation of mono- and poly-ADP-ribose. The cis-diols of these molecules are selectively oxidized by periodate treatment followed by reduction with sodium-^{3}H-borohydride. The tritiated products were separated by RP-HPLC with off-line LSC (Fig. 33), and could be detected at the subpicomol level. In principle, this result can be improved by using borohydride with higher specific activity.

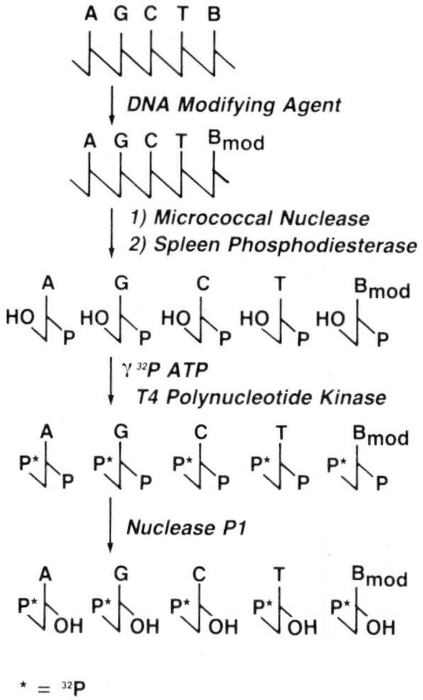

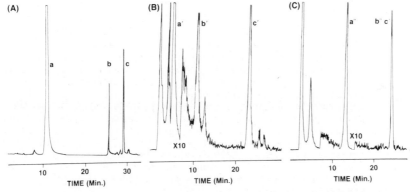

Scheme 2   Outline of the DNA postlabeling method. (Reprinted with permission from Chromatographia, Friedr. Vieweg and Sons, ref. 131).

Fig. 32.   HPLC characterization of products obtained from the reaction of dGMP with diethyl sulphate and ^{32}P postlabeling (Scheme 2) as deoxyribonucleoside-5'-phosphates (A), as deoxyribonucleoside-3,5'-phosphates (B) and as deoxyribonucleoside-5'-phosphates (C). Column, Spherisorb ODS-2 (250 x 4.6 mm); gradient elution from 0-40% (v/v) methanol in 0.075 M potassium phosphate (pH 3.0); detection by UV (at 260 nm, A) and flow-through liquid scintillation counting (B, C); peaks: a = 3'-dGMP, a' = 3',5'-dGbisP, a'' = 5'-dGMP; b = N-7 ethyl-3'-dGMP, b'= N-7 ethyl-3',5'-dGbisP, b'' = N-7 ethyl-5'-dGMP; c = ethyl ester of 3'-dGMP, c' = ethyl ester of 3',5'-dbisP.(Reprinted by permission from Chromatographia, Friedr. Vieweg & Sohn, ref. 131).

196

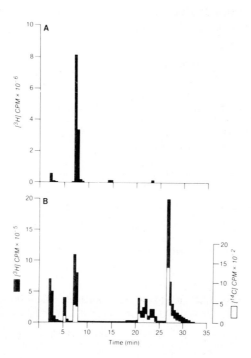

Fig. 33   RP-HPLC analysis of products obtained after ^{3}H-borohydride reduction of poly(ADP-ribose) (B). Before reduction, the cis-diols in the ribose are selectively oxidised by using periodate. In (B), in vitro-generated ^{14}C-labeled poly(ADP-ribose) was used to determine whether all ribose compounds are tritiated. In (A), the products obtained after ^{3}H-borohydride reduction of AMP is given, which is used as primary external standard. Column, Ultrasphere ODS (250 x 4.6 mm); gradient elution from 100% 0.10 M potassium phosphate (pH 4.25) to 100 % methanol/0.10 M potassium phosphate (20/80, v/v) to 100% acetonitrile/1.0 M urea (50/50, v/v); dual isotope detection by off-line liquid scintillation counting of 0.75 ml fractions. (Reprinted from J. Chromatogr., Elsevier, ref. 132).

Banerjee and Steimers (ref. 133) developed a method for indirect detection of anions in ion chromatography, based on the use of ^{35}S-sulfate in the eluent. When an anion elutes, a simultaneous decrease in sulfate concentration occurs; hence, the decrease in radioactivity signal in collected fractions can be used to quantitate eluting anions. Chloride was used as test analyte with plug injections. With a mobile phase containing 3.4 nM sulfate (13.4GBq/mmol), the detection limit of

Cl⁻ was about 0.1 μg. In principle, the detection limits can be brought down to the low nanogram region when using higher specific activities. For cation analysis, the use of ^{45}Ca is suggested. It is very unlikely, however, that the principle will be adapted by other workers because alternatives for ion determination which are not based on radioactivity detection are readily available.

It is worth mentioning here that the selectivity of many radioassay techniques can be significantly improved by using column liquid chromatography in sample clean up prior to the actual assay of the pooled fractions, for example in activation analysis and HPLC-RIA (refs. 159-163).

4.5 MISCELLANEOUS

In this section applications on radio-column liquid chromatography have been collected in which radiolabeled compounds are used as tracers for which no change in the chemical from of the compound is to be expected (Table XIII).

Radiotracers are excellent internal standards that are identical to the compound to be determined except for the radiolabel. They are most frequently applied in optimizing sample preparation steps, based on one radiolabeled tracer only. The advantages are best exploited, however, by using mixtures of radiolabeled internal standards for radio-CLC. Depending on the specific activities, the mass of the standard can normally be neglected as compared to the mass to be determined. The added signal in the reference detector may then be neglected. The amounts of activity may be chosen as high as to allow satisfactory quantitation of the radioactivity by flow-through radioactivity counting. This was illustrated by Frey and Frey (ref. 143) in the simultaneous determination of prednisone, prednisolone and 6-hydroxyprednisolone in urine samples using RP-HPLC, with ^{3}H-labeled prednisolone and 6-hydroxycortisol as internal standards. The radioactivity in the HPLC effluent was monitored by flow-through HSC (Fig. 34). Benedict (ref. 144) developed a clean-up procedure for the determination of catecholamines in urine and plasma samples. ^{14}C-labeled catecholamines were used to correct for losses during the sample treatment step and to identify eluting peaks after HPLC.

198

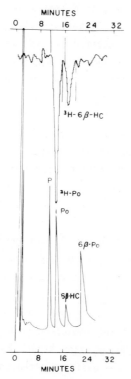

Fig. 34 HPLC chromatogram of prednisone (P), prednisolone (Po), 6ß-
hydroxycortisol (6ß-HC) and 6ß-hydroxyprednisolone (6ß-Po), with
³H-Po and ³H-6ß-HC) as internal standards. Column LiChrosorb
SI-60 (250 x 3.2 mm); mobile phase, hexane/diethyl ether/tetra-
hydrofuran/ethanol/glacial acid (55.9/31/6.5/2.3/0.3); detection
by UV (254 nm; lower trace) and flow trough heterogeneous scintillation counting (upper trace). (Reprinted from J. Chromatogr.,
Elsevier, ref. 143).

A second type of radiotracer work for CLC is the study on the chromatographic behaviour of analytes, i.e. the study of retention characteristics of proteins (refs. 145-147) and Gd-DTPA (ref. 148) on reversed-phase
packings, phospholipids on normal-phase packings (ref. 33), pro-
gesteronesuccinyltyrosine methyl esters on Sephadex LH-20 (refs. 149,
150) or alkali and transition metal ions on a silica-supported zirconium
phosphate ion exchanger (ref. 151). In all examples, radionuclides were
used to facilitate detection. ^{140}La was used by Cassidy et al. (ref. 152)
to identify the sources of memory effects in the HPLC of lanthanides on
dynamically generated ion exchangers. Apart from the normal peak
broadening, major sources for tailing found in eluting peaks were
sorption on metal surfaces and retention within sample loops (Fig. 35).

By using PTFE loops and tubing, most of the tailing could be eliminated, which resulted in cross-contamination of collected fractions of less than 0.006%.

Finally, examples of radio-column liquid chromatography can be found in biochemical transport studies. Josic et al. (ref. 153) determined the kinetics of intracellular uptake of iron by ferritin, for which ^{59}Fe was used as tracer. In this case, ferritin-bound and free radioactivity were separated by ion-exchange HPLC (Fig. 36). Boulieu et al. (ref. 154) studied the transport of the purine base hypoxanthine in human erythrocytes using ^{14}C-hypoxanthine. This enabled the determination of the analyte at physiological concentrations in erythrocytes and incubation medium. Radio-HPLC with flow-through LSC confirmed that no metabolism of the analyte took place during the experiments.

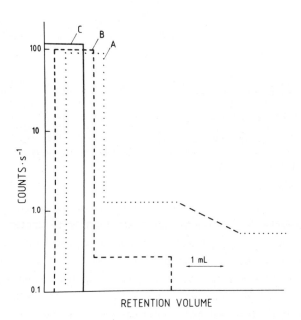

Fig. 35 Removal rate of [140]La from sample loops:(A) metal loop and 0.01N HNO$_3$ eluent, (B) metal loop and 0.01 N α-hydroxyisobutyric acid at pH 4.6, (C) teflon loop and 0.01 N HNO$_3$ eluent; flow rate, 2 ml/min. Curves (B) and (C) are shifted from origin for visibility (Reprinted by permission from Anal. Chem., American Chemical Society, ref. 152).

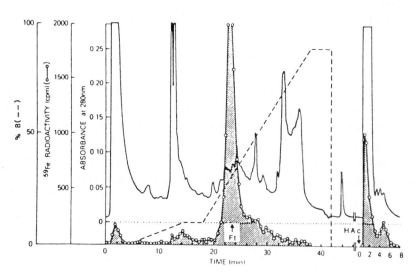

Fig. 36 Ion-exchange HPLC of K562 cell lysate after incubation with
^{59}Fe-transferrin. Column, Mono Q (50 x 5 mm); salt gradient
elution from 5 mM Hepes (pH 6.0) (buffer A) with 1 M NaCl
(buffer B); after the salt gradient, 0.5 ml of 60% acetic acid
was injected (arrow HAc); Ft = ferritin-bound radioactivity;
detection by UV (280 nm) and off-line NaI(Tl) scintillation
counting of 0.5 ml fractions. (Reprinted by permission from
Anal. Biochem., Academic Press, ref. 153)

5.   CONCLUDING REMARKS

Compared to the flow-through mode, off-line counting in radio-column
liquid chromatography is inherently more sensitive, with better control
over the precision in counting results and the background count rate.

The sensitivity of the off-line procedure can be obtained by using
different methods which are also based on the decoupling of the separation and counting step, for example by the deposition of the column eluate
on fluorocarbon films followed by autoradiography (ref. 96), or by using
flow-programming as in refs. 89-93. The principle of the latter is based
on solvent segmentation of the column eluate and temporary storage of the
segmented stream in a capillary storage loop. After reintroduction of the
contents of the loop and flow-programming, counting times of over 5 min
can be used in selected regions of the radiogram, which is comparable
with counting times commonly used in off-line determinations. The
chromatographic integrity, however, is somewhat better preserved in
flow-through counting.

When comparing flow-through LSC and HSC, in the authors'view, the liquid mode is the more versatile because of lower risks of (radioactive) contamination of the flow cell and the absence of restrictions on the type of samples. In addtion, for ^{3}H the counting efficiency normally is better by a factor of 3-5 for the liquid mode. The radiolabeled solutes can sometimes be recovered from the liquid scintillator. For example, Devine (ref. 158) used Sep-pak cartridges for the recovery of ^{14}C-androstenedione and carotine: Quantitative recovery could be obtained provided the polarity of the scintillator compounds and the solutes of interest were different and no detergents had been used in the scintillator.

The main drawback of flow-through LSC originates from the high scintillator/eluate ratios of at least three normally required with corresponding high flow rates (short counting time) in the radioactivity detector; this results in high volumes of radioactive waste. For extractable analytes, however, it has been shown that ratios of about 0.2 can be applied when using post-column extraction techniques with a water-immiscible liquid scintillator (ref. 31).

The applications presented in section 4 show that column liquid chromatography is of great value in the preparation and characterization of radiolabeled solutes. In addition, this chromatographic technique becomes increasingly important as sample pretreatment step in order to further improve selectivity of well-established radio-analytical methods (refs. 159-163). Examples are found in activation analysis (Fig. 1) and HPLC-RIA. Continuous attention is however, also given to the development of non-radio HPLC procedures as an alternative to radioanalyses (refs. 166-171). For example, Stevens et al. (ref. 170) studied the sorption and desorption of the preservative chlorhexidine on soft contact lenses by radioassay or, alternatively, by HPLC-UV. The former procedure involves the catalytic oxidation of the lens matrix which contains ^{14}C-chlorhexidine to ^{14}C-carbon dioxide and water which are trapped in a cocktail and analyzed by liquid scintillation counting. Although the sensitivity, accuracy and precision of both methods were satisfactory, the HPLC procedure has the advantages of being more selective, easily automatable and allowing monitoring of analyte degradation.

The combination of column liquid chromatography and radioactivity detection has proven to be an indispensable tool in many areas, in particular in the elucidation of complex biomedical problems, such as the distribution and metabolism of endogenous compounds in biological samples (ref. 173). Future work should be focussed on the combination of on-line preconcentration of radiolabeled solutes for radio-column liquid chromatography, thereby rendering flow-through quantitation feasible. There can be no doubt that the use of flow-through radioactivity detectors - which are now commercially available from several manufacturers - will stimulate the utilization of radioisotopes in the various areas of research mentioned above.

6. ACKNOWLEDGEMENTS

The author would like to thank Prof. Dr. U.A.Th. Brinkman, Prof. Dr. Ir.H.A. Das and Prof. Dr. R.W. Frei for stimulating discussions and their many helpful suggestions in the preparation of this chapter.

REFERENCES

1   H.J.M. Bowen, Chemical Applications of Radioisotopes, Muthuen & Co. Ltd, London (1969).
2   W. Geary, Radiochemical Methods, published on behalf of ACOL, London, by John Wiley & Sons, Chichester (1986).
3   D.D. Breimer, Rational Selection of Methods for Therapeutic Drug Monitoring. In: Therapeutic Relevance of Drug Analysis, F.A. DeWolff, H. Mattie and D.D. Breimer (Eds.), Martinus Nijhoff, The Hague (1979) pp. 9-21.
4   H.C. Dorn, Anal. Chem., 56 (1984) 747A-758A.
5   H.F. Haas and V. Krivan, Fresenius Z. Anal. Chem., 324 (1986) 13-18.
6   C.L. Holder, H.C. Thompson, Jr, A.B. Gosnell, P.H. Siitonen, W.A. Korfmacher, C.E. Cerniglia, D.W. Miller, D.A. Casciano and W. Slikker, Jr., J. Anal. Toxicol., 11 (1987) 113-121.
7   A.C. Veltkamp, H.A. Das, U.A. Th. Brinkman and R.W. Frei, poster presented at the 9th International Symposium on Column Liquid Chromatography, Edinburgh (1985).
8   G. Kloster and P. Laufer, Int. J. Appl. Radiat. Isot., 35 (1984) 545.
9   D.S. Wilbur, Development of No-Carrier-Added Radiopharmaceuticals with the Aid of Radio-HPLC. In: Analytical and Chromatographic Techniques in Radiopharmaceutical Chemistry, D.M. Wieland, M.C. Tobes and T.J. Mangner (Eds.), Springer-Verlag, New York, 1986, Ch. 11.
10 T.J. Manger, Potential Artifacts in the Chromatography of Radiopharmaceuticals. In: Analytical and Chromatographic Techniques in Radiopharmaceutical Chemistry, D.M. Wieland, M.C. Tobes and T.J. Manger (Eds.), Springer-Verlag, New-York, (1986), Ch. 13.
11 T.J. Beugelsdijk and D.W. Knobeloch, J. Liq. Chromatogr., 9 (1986) 3093-3131.
12 T.R. Roberts, Radiochromatography, the Chromatography and Electrophoresis of Radiolabelled Compounds, Elsevier, Amsterdam, (1978), Ch. 6.
13 P.C. White, Analyst, 109 (1984) 976-979.
14 R.P.W. Scott, Liquid Chromatography Detectors, Elsevier, Amsterdam (1986).
15 H.G. Barth, W.E. Barber, C.H. Lochmüller, R.E. Majors and F.E. Regnier, Anal. Chem., 58 (1986) 211R-250R.
16 H.G. Barth, W.E. Barber, C.H. Lochmüller, R.E. Majors and F.E. Regnier, Anal. Chem., 60 (1988) 387R-435R.
17 D.M. Wieland, M.C. Tobes and T.J. Mangner (Eds.), Analytical and Chromatographic Techniques in Radiopharmaceutical Chemistry, Springer-Verlag, New York, 1986.

18   G. Simonnet and M. Oia, Les Mesures de Radioactivité a l'Aide des Compteurs à Scintillateur Liquide, Editions Eyrolles, Paris (1977) p. 19.
19   C.T. Peng, Sample Preparation in Liquid Scintillation Counting, Review 17, The Radiochemical Centre, Amersham, Bucks., England (1977).
20   M.J. Kessler, J. Liq. Chromatogr., 5 (1982) 313-325.
21   Y. Nakamura and Y. Koizumi, J. Chromatogr., 333 (1985) 83-92.
22   C.A. Mathis, R.M. Jones and J.H. Chasko, Overall Radio-HPLC Design. In: Analytical and Chromatographic Techniques in Radiopharmaceutical Chemistry, D.M. Wieland, M.C. Tobes and T.J. Mangner (Eds.), Springer-Verlag, New-York (1986) Ch. 6.
23   J.D. Baker, R.J. Gehrke, R.C. Greenwood and D.M. Meikrantz, J. Radioanal. Chem., 74 (1982) 117-124.
24   F. Simonnet, J. Combe and G. Simonnet, Int. J. Appl. Radiat. Isotop., 38 (1987) 571-572.
25   E. Nieves, K.C. Rosenspire, S. Filc-DeRicco and A.S. Gelbard, J. Chromatogr., 383 (1986) 325-337.
26   V. Pingoud, J. Chromatogr., 331 (1985) 125-132.
27   B. Langstrom and H. Lundqvist, Radiochem. Radioanal. Letters, 41 (1979) 375-381.
28   R.E. Needham and M.F. Delaney, Anal. Chem., 55 (1983) 148-150.
29   K. Susuki, Int. J. Appl. Radiat. Isot., 35 (1984) 801-804.
30   D.R. Reeve and A. Crozier, J. Chromatogr., 137 (1977) 271-282.
31   A.C. Veltkamp, H.A. Das, R.W. Frei and U.A.Th. Brinkman, Eur. J. Chromatogr. News, 1(2) (1987) 16-21.
32   B.F.H. Drenth, T. Jagersma, F. Overzet, R.T. Ghijsen and R.A. de Zeeuw, Meth. Surv. Biochem. Anal., 12 (1983) 75-80.
33   H. Alam, J.B. Smith, M. J. Silver and D. Ahern, J. Chromatogr., 285 (1982) 218-222.
34   M.J. Kessler, J. Chromatogr. Sci., 20 (1982) 523-527.
35   R.F. Roberts and M.J. Fields, J. Chromatogr., Biomed. Appl., 342 (1985) 25-33.
36   J.A. Hunt, Anal. Biochem., 23 (1968) 289-300.
37   D. Westerlund, J. Carlqvist, M. Ovesson and B. G. Pring, Acta Pharm. Suec., 21 (1984) 93-102.
38   H.A. Dugger and B.A. Orwig, Drug. Metab. Rev., 10(2) (1979) 247-269.
39   S.W. Wunderley, poster presented at the 11th International Symposium on Column Liquid Chromatography, Amsterdam (1987).
40   H. Lentzen and R. Simon, J. Chromatogr., 389 (1987) 444-449.
41   Brochure 126L, Nuclear Enterprises, Edinburgh, Great Britain, (1980).
42   S.J. Potashner, N. Lake and J.D. Knowles, Anal. Biochem., 112 (1981) 82-89.
43   G.B. Sieswerda and H.L. Polak, J. Radioanal. Chem., 11 (1972) 49-58.
44   D. de Korte, Y.M.T. Marijnen, W.A. Haverkort, A.H. van Gennip and D. Roos, J. Chromatogr., 415 (1987) 383-387.
45   A.P.A. Prins, E. Kiljan, R.J. van de Stadt and J.K. van de Korst, Anal. Biochem., 152 (1986) 370-375.
46   G.B. Sieswerda, H. Poppe and J.F.K. Huber, Anal. Chim. Acta, 78 (1975) 343-358.
47   F.K. Klein and C.A. Hunt, Int. J. Appl. Radiat. Isot., 32 (1981) 669-671.
48   S. Baba, Y. Suzuki, Y. Sasaki and M. Horie, J. Chromatogr., 392 (1987) 157-164.
49   H.J. van Nieuwkerk, Thesis: On-line Radiometry in High-Performance Liquid Chromatography using a Storage Loop, Petten, The Netherlands, 1987, pp. 68-71.
50   L.A. Currie, Anal. Chem., 40 (1968) 586-593.
51   P.R. Bevington, Data Reduction and Error Analysis for the Physical Sciences, McGraw-Hill Book Co., New York, (1969).

52 D.J. Malcolme-lawes, S. Massey and P. Warwick, J. Radioanal. Chem., 57 (1980) 335-361.
53 A. Savitsky and M.J.E. Golay, Anal. Chem., 36 (1964) 1627-1639.
54 G.J. de Groot, Trends in Anal. Chem., 4 (1985) 134-137.
55 B.M. Frey and F.J. Frey, Clin. Chem., 28 (1982) 689-692.
56 M.J. Kessler, J. Chromatogr., 255 (1983) 209-217.
57 H.K. Webster and J.M. Whaum, J. Chromatogr., 209 (1981) 283-292.
58 H.J. van Nieuwkerk, A.C. Veltkamp, H.A. Das, U.A.Th. Brinkman and R.W. Frei, J. Radioanal. Nucl. Chem., 100 (1986) 165-176.
59 N.G.L. Harding, Y. Farid, M.J. Stewart, J. Shepherd and D. Nicoll, Chromatographia 15 (1982) 468-474.
60 A.G. Causey, B. Middleton and K. Bartlett, Biochem. J., 235 (1986) 343-350.
61 H.M. Ruijten, P.H. van Amsterdam and H. de Bree, J. Chromatogr., 314 (1984) 183-191.
62 L.N. Mackey, P.A. Rodriguez and F.B. Schroeder, J. Chromatogr., 208 (1981) 1-8.
63 K. Shirahashi, G. Izawa, Y. Murano, Y. Muramastu and K. Yoshihara, J. Radioanal. Nucl. Chem., Letters, 86 (1984) 1-10.
64 L.J. Everett, Chromatographia, 15 (1982) 445-448.
65 M.H. Simonian and M.W. Capp, Chromatogram 8 (1987) 5-6 (HPLC newsletter published by Beckman Inc., Fullerton, CA, USA).
66 C. Giersch, J. Chromatogr., 172 (1979) 153-161.
67 S. Mori, Plant & Cell Physiol., 23 (1982), 703-708.
68 S. Mori, Agric. Biol. Chem., 45 (1981) 1881-1884.
69 B.H. Candy, Rev. Sci. Instrum., 56 (1985) 183-193.
70 K. Strandgarden and P.O. Gunnarson, poster presented at the 11th International Symposium on Column Liquid Chromatography, Amsterdam (1987).
71 G. Dietzel, GIT Fachz. Lab., 28 (1984) 909-913.
72 B. Detjens, K. Figge and H. Martinen, GIT Fachz. Lab., 22 (1978) 578-585.
73 A.D. Nunn and A.R. Fritzberg, in: Analytical and Chromatographic Techniques in Radiopharmaceutical Chemistry. D.M. Wieland, M.C. Tobes and T.J. Mangner (Eds.), Springer-Verlag, New York, 1986, p. 116.
74 B. Bakay, E.Nissinen and L. Sweetman, Anal. Biochem., 86 (1978) (65-77).
75 M.A. Clark, T.M. Conway and S.T. Crooke, J. Liq. Chromatogr., 10 (1987) 2707-2719.
76 P.W. Albro, J. Lorenzo and J. Schroeder, LC. Liq. Chromatogr. HPLC Mag., 2 (1984) 310-312.
77 L. Schutte, J. Chromatogr., 72 (1972) 303-309.
78 T.L. Rucker, H.H. Ross and G.K. Schweitzer, Chromatographia, 25 (1988) 31-36.
79 J.C. Sternberg, in J.C. Giddings and R.A. Keller (Eds.), Advances in Chromatography, Vol. 2, Dekker, New York, USA, 1966, p. 205.
80 L. Nondek, U.A.Th. Brinkman and R.W. Frei, Anal. Chem., 54 (1983) 1466-70.
81 P.R. Bewington, Data Reduction and Error Analysis for the Physical Sciences, McGraw-Hill Book Co., New York, 1969, p. 78.
82 D.W. Roberts, Meth. Surv. Biochem. Anal., 12 (1983) 81-88.
83 C.A. Mathis, R.M. Jones and J.H. Chasko, Overall Radio HPLC Design. In: Analytical and Chromatographic Techniques in Radiopharmaceutical Chemistry, D.M. Wieland, M.C. Tobes and T.J. Manger (Eds.), Springer-Verlag, New York, (1986) pp. 133-137.
84 B. Bakay, Anal. Biochem., 63 (1975) 87-98.
85 B. Bakay, Clin. Chem., 21 (1975) 1212-1216.
86 B. Bakay, in: Liquid Scintillation Counting. H.A. Crook and P. Johnson, Vol. 4, p. 133, Heyden Press, London, 1977.
87 L.R. Snyder, J. Chromatogr., 149 (1978) 653-668.

88   A.H.M.T. Scholten, U.A. Th. Brinkman and R.W. Frei, Anal. Chem., 54
      (1982) 1932-38.
89   H.J. van Nieuwkerk, Thesis: On-line Radiometry in High-Performance
      Liquid Chromatography using a Storage Loop, Petten, The Netherlands
      (1987).
90   H.J. van Nieuwkerk, H.A. Das, U.A. Th. Brinkman and R.W. Frei,
      Chromatographia, 19 (1984) 137-144.
91   H.J. van Nieuwkerk, H.A. Das, U.A.Th., Brinkman and R.W. Frei,
      J. Chromatogr., 360 (1986) 105-117.
92   A.C. Veltkamp, H.A. Das, R.W. Frei and U.A.Th. Brinkman,
      J. Chromatogr., 384 (1987) 357-369.
93   A.C. Veltkamp, H.A. Das, R.W. Frei and U.A.Th. Brinkman, J. Pharm.
      Biomed. Anal. 6 (1988) 609-622.
94   S. Baba, Y. Shinohara, H. Sano, T. Inoue, S. Masuda and M. Kurono,
      J. Chromatogr., 305 (1984) 119-126.
95   S. Baba, M. Horie and K. Watanabe, J. Chromatogr., 244 (1982) 57-64.
96   A. Karmen, G. Malikin, L. Freundlich and S. Lam, J. Chromatogr., 349
      (1985) 267-274.
97   H.C. Treutler and K. Freyer, ZfI-Mitt., 103 (1985) 18-32.
98   E.A. Evans, Self-decomposition of Radiochemicals: Principles,
      Control, Observations and Effects, Review 16, Amersham International
      Ltd., Amersham, Bucks., England, 1976.
99   J.H. Chasko and J.R. Thayer, Int. J. Appl. Radiat. Isot., 32 (1981)
      645-649.
100   T.E. Boothe, A.M. Emran, R.D. Finn, P.J. Kothari and M.M. Vora,
      J. Chromatogr., 333 (1985) 269-275.
101   G. Kloster and P. Laufer, J. Lab. Comp. Radioph., 20 (1983)
      1305-1315.
102   A.E. Bolton, Radioiodination Techniques, Review 18, Amersham.
      Bucks., England (1985).
103   R.C. Judd and H.D. Caldwell, J. Liq. Chromatogr., 8 (1985) 1109-1120.
104   S.J. Wagner and M.J. Welch, J. Nucl. Med., 20 (1979) 428.
105   D.J. Hnatowich, R.L. Childs, D. Lanteigne and A. Najafi J. Immun.
      Meth., 65 (1983) 147-157.
106   M.V. Mikelsons and T.C. Pinkerton, Anal. Chem., 58, (1986) 1007-1013.
107   J.A.G.M. van den Brand, H.A. Das, B. G. Dekker and C.L. de Ligny,
      Int. J. Appl. Radiat. Isot., 33 (1982) 917.
108   D. Ishii, A. Hirose and I. Horiuchi, J. Radioanal. Chem., 45 (1978)
      7-14.
109   J. Vockova and V. Svoboda, J. Chromatogr., 410 (1987) 500-503.
110   D.A. Wells, G.A. Garbolas and G.A. Digenis, J. Chromatogr., 356
      (1986) 367-371.
111   I.Kleinmann, V. Svoboda and J. Vockova, J. Chromatogr., 411 (1987)
      335-344.
112   B.F.H. Drenth and R.A. de Zeeuw, Int. J. Appl. Radiat. Isot., 33
      (1982) 681-683.
113   S.K. Mauldin, F.A. Richard, M.Plescia, S.D. Wyrick, A. Sancar and
      S.G. Chaney, Anal. Biochem., 157 (1986) 129-143.
114   K.J. Hoffman, Drug Metab. Disp., 14 (1985) 341-348.
115   L. Weidolf, J. Chromatogr., 343 (1985) 85-97.
116   K.O. Vollmer, W. Klemisch and A. von Hodenberg, Z. Naturforsch.,
      C: Biosci., 41 (1986) 115-125.
117   J.E. Coutant, R.J. Barbuch, D.K. Satonin and R.J. Cregge, Biomed.
      Environ. Mass Spectrom., 14 (1987) 325-330.
118   C. Meinard, P. Bruneau and M. Roche, J. Chromatogr., 349 (1985)
      105-108.
119   G.L. Lamoureux and D.G. Rusness, J. Agr. Food. Chem., 35 (1987) 1-7.
120   M.J. Arnaud, Drug Metab. Disp., 13 (1985) 471-478.
121   G.A. Kyerematen, L.H. Taylor, J.D. deBethizy and E.S. Vesell,
      J. Chromatogr., 419 (1987) 191-203.

206

122   I. Plakunov, T.A. Smolarek, D.L. Fischer, J.C. Wiley Jr. and W.M. Baird, Carcinogenesis, 8 (1987) 59-66.

123   T.J. Monks, S.S. Lau, R.J. Highet and J.R. Gillette, Drug. Metab. Disp., 13 (1985) 553-559.

124   N.J. Parks, K.A. Krohn, C.A. Mathis, J.H. Chasko, K.R. Geiger, M.E. Gregor and N.F. Peek, Science, 212 (1981) 58-60.

125   R.S. Chapkin, V.A. Ziboh, C.L. Marcelo and J.J. Voorhees, J. Lipid Res., 27 (1986) 945-954.

126   P.I. Lundmo and A. Sunde, J. Chromatogr., 308 (1984) 289-294.

127   D.L. Tribble, M.R. Glover and J.D. Lambeth, J. Chromatogr., 414 (1987) 411-416.

128   G.Bonn, J. Chromatogr., 387 (1987) 393-398.

129   C. Abeijon, J.M. Capasso and C.B. Hirschberg, J. Chromatogr., 360 (1986) 293-297.

130   M.W.H. Coughtrie, B. Burchell and J.R. Bend, Anal. Biochem., 159 (1986) 198-205.

131   M.W. Dietrich, W.E. Hopkins, K.J. Asbury and W.P. Ridley, Chromatographia, 24 (1987) 545-551.

132   A. Hakam, J. McLick and E. Kun, J. Chromatogr., 359 (1986) 275-284.

133   S. Banerjee and J.R. Steimers, Anal. Chem., 57 (1985) 1476-1477.

134   V.L. Wilson, R.A. Smith, H. Autrup, H. Krokan, D.E. Musci, Ngoc-Nga-Thi Le, J. Longeria, D. Ziska and C.C. Harris, Anal. Biochem., 152 (1982) 275-284.

135   L. Janocko and R.B. Hochberg, J. Steroid Biochem., 24 (1986) 1049-1052.

136   E.W. Bergink, P.S.L. Janssen, E.W. Turpijn and J. van der Vies, J. Steroid Biochem., 22 (1985) 831-836.

137   L.D. Fairbanks, A. Goday, G.S. Morris, M.F.J. Brolsma, H.A. Simmonds and T. Gibson, J. Chromatogr., 276 (1983) 427-432.

138   R.T. Marsili and H. Ostapenko, J. Agr. Food Chem., 35 (1987) 304-308.

139   E. Nissinen, J. Chromatogr., 342 (1985) 175-178.

140   K.L.L. Fong and B.Y.H. Hwang, Biochem. Pharmac., 32 (1983) 2781-2786.

141   S.R. Childers, Neurochem. Res., 11 (1986) 161-171.

142   F.A. Hommes and L. Moss, Anal. Biochem., 154 (1986) 100-103.

143   B.M. Frey and F.J. Frey, J. Chromatogr., 229 (1982) 283-292.

144   C.R. Benedict, J. Chromatogr., 385 (1987) 369-375.

145   G. Raspi, A. LO Moro and M. Spinetti, Ann. Chim., 77 (1987) 525-532.

146   F.L. de Vos, D.M. Robertson and M.T.W. Hearn, J. Chromatogr., 392 (1987) 17-32.

147   P.C. Sadek, P.W. Carr., L.D. Bowers and L.C. Haddad, Anal. Biochem., 153 (1986) 359-371.

148   M.M. Vora, S. Wukovnig, R.D. Finn, A.M. Emran, T.E. Boothe and P.J. Kothari, J. Chromatogr., 369 (1986) 187-192.

149   G. Toth and J. Zsadanyi, J. Chromatogr., 329 (1985) 264.

150   G. Toth, J. Chromatogr., 404 (1987) 258-260.

151   L. van So and L. Szirtes, J. Radioanal. Nucl. Chem., 99 (1986) 55-60.

152   R.M. Cassidy, F.C. Miller, C.H. Knight, J.C. Roddick and R.W. Sullivan, Anal. Chem., 58 (1986) 1389-1394.

153   D. Josic, E. Mattia, G. Ashwell and J. van Renswoude, Anal. Biochem., 152 (1986) 42-47.

154   R. Boulieu, C. Bory, P. Baltassat and C. Gonnet, J. Liq. Chromatogr., 7 (1984) 1013-1021.

155   H. van den Berg, P.G. Boshuis and W.H.P. Schreurs, J. Agr. Food Chem., 34 (1986) 264-268.

156   B.A. Tomkins and W.H. Griest, J. Chromatogr., 386 (1987) 103-110.

157   S.T. Ingalls, P.E. Minkler, C.L. Hoppel and J.E. Nordlander, J. Chromatogr., 299 (1984) 365-376.

158   P.C. Devine and B.V. Milborrow, J. Chromatogr., 325 (1985) 323-326.

159   J.J. Fardy, G. D. McOrist and T.M. Florence, Anal. Chim. Acta, 159 (1984) 199-209.

160   G.V. Iyengar, J. Radioanal. Nucl. Chem., 110 (1987) 503-517.

161   E. Gelpi, Trends in Anal. Chem., 4 (1985) XIII-XIV.
162   O.B. Holland, M. Risk, H. Brown, K. Komes, P. Dube and C. Swann,
      J. Chromatogr., 385 (1987) 393-396.
163   G.H. Fridland and D.M. Desiderio, Life Sci., 41 (1987) 809-812.
164   J.J. Pratt, Ann. Clin. Biochem., 23 (1986) 251-276.
165   P.K.F. Yeung, J.W. Hubbard, B.W. Baker, M.R. Looker and K.K. Midha,
      J. Chromatogr., 303 (1984) 412-416.
166   J. Krska, G.M. Addison and S.D. Soni, Ann. Clin. Biochem., 23 (1986)
      340-345.
167   O. Lantto, Clin. Chem., 28 (1982) 1129-1132.
168   G.J. Peters, E. Laurensse, A. Leyva and H.M. Pinedo, Anal. Biochem.,
      161 (1987) 32-38.
169   P. Hjemdahl, M. Daleskog and T. Kahan, Life Sci., 25 (1979) 131-138.
170   L.E. Stevens, J.R. Durrwachter and D.O. Helton, J. Pharm. Sci., 75
      (1986) 83-86.
171   F. Overzet and R.A. de Zeeuw, J. Pharma. Biomed. Anal., 2 (1984)
      3-17.
172   J.A. van der Krogt, C.F.M. van Valkenburg and R.D.M. Belfroid,
      J. Chromatogr. Biomed. Appl., 427 (1988) 9-17.
173   E. Buncel and J.R. Jones (Eds.), Isotopes in the Physical and
      Biomedical Sciences, Volume 1 Labelled Compounds (Part A), Elsevier,
      Amsterdam, 1987.

CHAPTER V

MODERN POST-COLUMN REACTION DETECTION IN HIGH-PERFORMANCE
LIQUID CHROMATOGRAPHY

H. JANSEN and R. W. FREI

1.   GENERAL INTRODUCTION

In many analytical laboratories all over the world, high performance
liquid chromatography (HPLC) is used for the analysis of a wide variety
of samples. Applications include, e.g., environmental analysis, clinical
analysis, food analysis, pharmaceutical analysis and quality control. In
many cases, trace level concentrations of the analytes of interest are
encountered. Because of the often inadequate sensitivity and/or
selectivity in the detection process in HPLC needed for trace analysis in

complex matrices (biological fluids, heavily contaminated environmental samples) methods have been developed to overcome this problem. Two powerful approaches can be distinguished:

- The use of small precolumns for selective on-line sample handling. With the proper selection of sorbents, clean-up and trace enrichment can be achieved. An excellent review showing the possibilities of this method was published recently (ref. 1).
- The use of chemical derivatization techniques with the goal of improving the detectability of the compounds of interest.

Currently, the best and most reliable detectors for HPLC that are widely used are UV-VIS, fluorescence and electrochemical detectors. It is desirable to use these detectors for trace analysis which is possible when suitable chemical derivatization techniques are incorporated to convert the analytes of interest with their originally poor detection properties into compounds that can be detected with high sensitivity with these detectors. Apart from an increase in detectability, the derivatization step can also improve the selectivity of the total analytical method. The derivatization can be carried out in the precolumn mode, i.e., before the separation takes place and usually off-line, although recent advances in automation of pre-column labelling has caused a revival of pre-chromatographic techniques. However, the majority of on-line derivatization procedures are carried out in the post-column mode. Both approaches have their inherent advantages and disadvantages which have been discussed previously (ref. 2) and the choice for one of the two modes will generally depend a lot on the analytical problem that is to be solved. This paper will deal with post-column derivatization.

A schematic diagram of an HPLC-system with an on-line reaction detector can be found in Fig. 1. Pump 1 is used for the mobile phase supply. After separation of the compounds in the sample on the analytical column, the reagent is added with pump 2 via a suitable mixing device. The combined streams are passed through the post-column reactor, providing the desired c.q. required holdup time for the reaction to occur, and finally through the detector. Several advantages of post-column reaction can be given:

- As opposed to derivatization prior to the separation, the reaction does not have to yield a single detectable species, since the reaction products are not separated anymore.
- The reaction does not have to go to completion and the reaction products do not have to be stable. As long as the reaction is reproducible, it can, in principle, be used.

- The technique can be used with different detectors in series, i.e., an UV-detector directly at the column outlet followed by a fluorescence detector after a post-column reactor.
- Because the compounds are separated in their original form it is possible to use separation procedures from the literature.

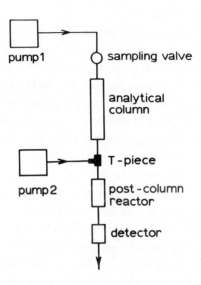

Fig. 1   General schematic of an HPLC system with reaction detector.

There are also distinct disadvantages of the technique:
- Additional pumps are needed for reagent supply. A stable, non-pulsating flow is required.
- Mixing problems can be encountered, e.g., when the mobile phase and the reagent have very different viscosity.
- The addition of reagent gives a dilution of the mobile phase.
- Frequently, the mobile phase needed for the separation is not the best medium for the derivatization reaction. Therefore, compromises must be made for those cases.
- The post-column reactor gives a contribution to total band broadening with a resultant loss in chromatographic resolution.
- In some cases, the excess of reagent interferes in the detection process.

When using post-column reaction detection, it is necessary to have a reactor suitable in terms of chemical resistance, holdup time, band broadening etc. for the reaction involved. In principle, three basically different reactor types can be distinguished, viz. straight, coiled or knitted open tubular reactors (OTRs), packed bed reactors (PBRs) and segmented stream tubular reactors (SSRs). Band broadening can, as far as OTRs and PBRs are concerned, be predicted by the use of simple equations as will be outlined below. Huber et al. have published a good comparison between OTRs and PBRs (ref. 3). Special attention was given to the loss of resolution caused by the reactor and to the mixing process.

2. TYPES OF POST-COLUMN REACTORS

2.1 OPEN TUBULAR REACTORS

The simplest post-column reactor is a piece of stainless steel or PTFE tubing. Coiling or knitting is attractive since it effects an increase in radial mass transfer resulting in a decrease in band broadening (refs. 4, 5). A good discussion of different types of these deformed OTRs can be found in ref. 6. Band broadening in time units in open tubes is described by

$$\sigma_t = (x \; \frac{t_r \cdot d_t^2}{96 \cdot D_m})^{0.5} \tag{1}$$

In this equation, x $(0 < x < 1)$ accounts for the reduced band broadening for coiled or knitted as compared to straight ($x = 1$) tubes, t_r ist the mean residence time in the reactor, d_t is the tube inner diameter and D_m is the molecular diffusion coefficient in the carrier (mobile phase + reagent) stream. A thorough investigation of the influence of coiling has been given (refs. 5, 6). To demonstrate the influence of coiling of open tubular reactors, x is calculated to be 0.14 by using the equations given (ref. 5) for the following set of parameters, which are typical for a post-column reactor coupled to a 4.6 mm I.D. chromatographic column: p (solvent density) = $10^3 kg/m^3$, η (viscosity) = 10^{-3} kg/m.s, λ (coil to tube diameter ratio) = 0.01, D_m = 10^{-9} m^2/s, F (total volumetric flow rate) = 30 . 10^{-9} m^3/s, and d_t = 500 . 10^{-6} m. For a 30 s residence time, σ_t is calculated to be 3.3 s which is lower by 62 % when compared with the value for a straight tube. It is possible to get less band broadening by using smaller inner diameter tubing but a larger pressure drop is the price that must be paid. Pressure drop (Δp) in the tubular reactor is described by the poiseuille equation

$$\Delta p = \frac{512 \cdot \eta \quad F^2 \cdot t_r}{\pi^2 \cdot d_t^6} \tag{2}$$

For the set of parameters given above, p is calculated to be 0.9 bar, which is fully acceptable but a reduction in tube diameter to $200 \cdot 10^{-6}$ m would lead to a pressure drop of over 200 bar for the same residence time.

2.2 PACKED BED REACTORS

Packed bed reactors are glass or stainless-steel columns that are packed with small, inert nonporous glass beads. They can be considered as chromatographic columns operated under non-retention conditions. Therefore, band broadening in these reactors essentially follows the theory for HPLC columns, which yields:

$$\sigma_t = \left(\frac{t_r^2 \cdot h \cdot d_p}{L} \right)^{0.5} \tag{3}$$

h stands for the reduced plate height with values of between 2 and 6 roughly dependent on the packing quality, d_p is the particle diameter and L is the reactor length. From Eq. 3 it can be concluded that reducing the particle size and increasing the reactor length while keeping the residence time unchanged will result in less band broadening. It is important not to have excessive pressure drop over the reactor. The pressure drop is given by

$$\Delta p = \frac{\cdot L^2}{K_o \cdot t_r \cdot d_p^2} \tag{4}$$

where K_o is the permeability constant which has a value of 0.001 to 0.002.

2.3 SEGMENTED STREAM TUBULAR REACTORS

Using this type of post-column reactor, the mixed mobile phase + reagent stream is segmented either with gas (air) bubbles or with non-miscible solvent plugs. If there is no carry-over from one segment of another, which can be achieved by the proper choice of the reactor material (ref. 7), the axial dispersion is determined by the size of the segment. With sufficiently small segments, band broadening can be kept very low. Band broadening using this type of reaction detection system is usually determined mainly by the phase separator or debubbler, needed to create a homogeneous flow through the detector, rather than the reactor itself. The theoretical treatment of band broadening in segmented stream reactors is rather complex. Pressure drop over a gas segmented stream reactor should be low due to compressibility of the gas bubble hence, relatively large inner diameter tubing (0.8 to 1 mm I.D.) is commonly used in combination with 4.6 mm I.D. analytical columns. For solvent segmented systems higher back pressures can be tolerated hence, a higher flexibility exists for the choice of coil geometries and some degree of miniaturization is possible (ref. 9).

3.   CHOICE OF REACTION DETECTOR

For the effective use of post-column reaction detection systems, it is of outmost importance to have a reactor attuned to the particular reaction conditions. The choice of one of the reactor types given above is mainly influenced by the reaction time needed. One can start calculating band broadening and pressure drop for open tubular and packed bed reactors using Equations 1 through 4. For open tubular reactors, band broadening will be rather high for longer reaction times (above ca. 1 min.) when pressure drop should not be too high. Therefore, this reactor type is commonly recommended for shorter reaction times. i.e., below about 30 s. With a knitted reactor, somewhat longer reaction times, up to several minutes according to Engelhardt (refs. 4, 6), are feasible. For reactions with intermediate kinetics, packed bed reactors are to be prefered. They are mainly used for reaction times between about 0.5 and 5 minutes. In this regions, they offer an attractive compromise between band broadening and pressure drop. For reaction times longer than 5 minutes also as an alternative to packed bed reactors segmented stream reactors are used. Reaction times up to 20 minutes have been reported. This reaction time can be considered as an upper practical limit in post-column reaction detector systems. Apart from band broadening and pressure drop considerations, other aspects might have their influence on the reactor choice. For instance, an open tubular reactor can be prefered for its simple construction, for the fact that clogging problems are almost never observed or because an alkaline reaction medium would cause the glass particles in a packed bed reactor to dissolve. On the other hand, segmented stream reactors must be used when reactions on liquid-liquid interfaces are used for the derivatization. As an example, ion-pair formation can be carried out. With this technique an ion-pair is extracted to one phase while the excess of reagent, i.e., the counter ion, remains in the other phase. This post-column ion-pairing is part of the so-called post-column extraction detection techniques. Another example of post-column extraction is extracting the analyte to the segmenting solvent thereby separating it from interferring compounds. After phase separation, the segmentation solvent phase is guided through the detector. Post-column extraction and ion-pairing techniques are in their appearance closely related to post-column reaction techniques and have been reviewed in the recent past (refs. 8, 9).

Several papers dealing with the optimal choice of reaction detectors and giving a more detailed theoretical treatment of the various reactor types have appeared during the last years (e.g. refs. 3, 5, 6, 8, 10) and are suggested for additional reading.

4. APPLICATIONS OF POST-COLUMN REACTION DETECTION

In Table I, a list is given of some of the possibilities of post-column reaction detection. The table is intended to give an idea of the broad applicability in various analytical fields and problem areas rather than to give a complete review of all the applications published during the past few years. Examples of the use of the different reactor types discussed above and coupled to different detection modes are included in the table. Though not always the best choice in terms of band broadening, it appears from the literature that the open tubular reactor is most frequently used for post-column derivatization. This is most probably a result of the simplicity of construction and of the reliability during operation.

TABLE I Some applications of post-column reaction detectors in HPLC

Compounds	Reagent(s)	Reactor	Detection*	Reference
Aflatoxins	I_2	OTR	F	11
Aldehydes and ketones	semicarbazide hydrochloride	PBR	E	12
Amino acids	o-phthalaldehyde	OTR	F	13
		PBR	F	5
Anthraquinones	sodium dithionite/ sodium hydroxide	OTR	VIS	14
Barbiturates	borate buffer (pH 10)	no reactor	UV	15
Carbamates	NaOH, o-phthalaldehyde	OTR	F	16
Catecholamines	Ethylenediamine	SSR	F	17
Chloroanilines	Fluram^R	OTR, PBR and SSR	F	18
Phenols	potassium ferricyanide 4-aminoantipyrine	no reactor	VIS	19
Formaldehyde	acetylacetone/ ammonia	knitted OTR	VIS and F	20
Guanidines	NaOH, phenanthrenequinone	OTR	F	21
Hydroperoxides	NaI	PBR	UV	22
ß-Lactam antibiotics	o-phthalaldehyde	OTR	F	23
Organosulphur compounds	Pd(II)-calcein	SSR	F	24
Penicillins	sodium hydroxide/ mercury(II) chloride/ethylene-diaminetetraacetic acid	OTR	UV	25
Prostaglandins	Br_2	OTR	E	26
Reducing carbohydrates	2-cyanoacetamide	OTR	E	27
		OTR	F	28
	various reagents	OTR	UV and VIS	29
Thiamine and thiamine phosphate esters	potassium ferricyanide/ sodium hydroxide	OTR	F	30
Vitamin K_3	sodium borohydride	SSR	F	31

* F: Fluorescense   E: Electrochemical   VIS: Visible light absorption
  UV: Ultraviolet absorption

5.   NEW APPROACHES TO POST-COLUMN REACTION DETECTION

5.1   INTRODUCTION

Post-column reaction detection in its straight forward form as discussed above has developed into a powerful method of detection for various analytes and is used in many analytical laboratories. Since the additional hardware needed is quite simple, problems in operating an HPLC/reaction detection system are seldom encountered. However, there are situations in which conventional reaction detection is not successful or where other approaches yield better results, lower operation cost, simpler systems or other advantages. These facts have stimulated researchers to develop new techniques in post-column reaction detection for liquid chromatography. There are a number of distinct trends in research being done at present times. These trends, demonstrated with several applications, will be discussed below and include a) the use of immobilized enzymes in post-column reactors; b) the use of solid-phase chemistries (other than reactions on immobilized enzymes), i.e., catalysis and the use of solid supported reagents; c) the use of electro-chemical reagent production; d) the use of photochemical and thermo initiated reactions; e) miniaturization, as a logical necessity when a microbore column is preferred or when expensive and/or highly toxic chemicals and solvents are used, and f) the use of hollow fibers as post-column reactors.

5.2   THE USE OF IMMOBILIZED ENZYMES IN POST-COLUMN REACTORS

5.2.1 INTRODUCTION

Enzymes are attractive for use in reaction detection systems because of their inherent selectivity. When the product formed in the enzymatic reaction is well detectable, an interesting highly specific analytical method can be obtained. Much research is being done on the use of im-mobilized enzymes since, as cited below, there are some distinct advantages as compared to enzymes in solution:

- the elimination of enzyme solution pumps and mixing units and, hence, the reduction of cost and the absence of mixing and dilution problems
- the possibility of working with enzymes that would otherwise interfere in the detection process
- immobilized enzymes can be re-used.
- immobilization often improves the storage properties and the pH stability of the enzymes.

There are many methods for immobilizing enzymes on a wide variety of supports (refs. 32-34). Supports that are widely used include agarose, cellulose, polyacrylamide, nylon, glass, silica and ion exchangers. Binding can take place either by physical adsorption or by covalent attachment. Physical adsorption has the disadvantage that loss of enzyme due to desorption can take place. For this reason, covalent attachment is generally preferred.The binding capacity of a support is an important parameter to consider. High surface area is advantageous. On the other hand, this should not be the result of very small pore diameters. Relatively large pore diameters (> ca. 40 nm) are needed for enzymes to enter into the pores.

5.2.2 ENZYME IMMOBILIZATION

Two types of enzyme reactors can be found in the literature, viz. a wall-coated open tubular reactor and a reactor packed with enzyme immobilized on beads. A major disadvantage of the first reactor type is its unfavourable ratio of active surface to mobile phase volume, leading to low activity per unit volume. For this reason, the latter reactor type is primarily used in combination with HPLC. The first type is sometimes used in flow injection analysis. Good mechanically stable packed-bed reactors can be obtained when the enzyme is bound on glass, silica or alumina particles. Other materials are quite soft and can easily be deformed in flow systems, leading to high pressure drop and excessive peak broadening due to voids. Small particles should be used if possible for reasons of band broadening as discussed before. Besides the band broadening that occurs in any packed bed reactor, the reaction itself on the enzymatic layer may give rise to band broadening. This phenomenon has been discussed in the literature (refs. 35-37). No additional band broadening due to the reaction is observed either when reactant and product move at the same velocity through the reactor or when the reaction occurs instantaneously. When using enzymes in a post-column reactor, it is necessary to be careful with the mobile phase composition. Since enzymes are usually active in only a relatively small pH range and usually can not withstand high organic modifier concentration, development of a separation with a mobile phase directly compatible with the enzyme reactor is mandatory. If this is not possible, an adaptation of the mobile phase by adding a make-up flow, e.g., a buffer, prior to entering the reactor is possible. However, this has the disadvantage that an additional pump is needed and that dilution occurs. Therefore, direct coupling is prefered if possible.

5.2.3 APPLICATIONS OF IMMOBILIZED ENZYMES
5.2.3.1 EARLY WORK

Preliminary work in the use of immobilized enzymes in post-column reaction detection systems was done by Schlabach and Regnier (ref. 38). A column with glucose-6-phosphate dehydrogenase immobilized on glass beads was used to react glucose-6-phosphate with the formation of NADH which is detected fluorimetrically.

5.2.3.2 UREASE

A schematic diagram of an HPLC-system equipped with an enzymatic reactor is given in Fig. 2. The system is used for the determination of urea and ammonia, which is based on ion-pair HPLC with on-line post-column derivatization on immobilized urease (ref. 39). In the urease solid-phase reactor (SPR), urea is quantitatively converted into ammonia, which reacts with o-phthalaldehyde and is detected by fluorescence monitoring. The method is very specific due to the inherent selectivity of the enzyme combined with the selectivity of the OPA reaction. This is shown in Table II which gives the response for compounds related to urea that are present in the samples from an urea plant for which the method was developed. Later, the same method was adapted to the determination of urea and ammonia in urine and in serum Fig. 3. An off-line sample pre-treatment procedure with an anion exchanger was needed to eliminate interfering amino acids (ref. 40).

TABLE II Response of by-products relative to response of urea
(Reprinted from ref. 39)

Compound	Structural formula	Relative response
Urea	$H_2N - \underset{\underset{O}{\|\|}}{C} - NH_2$	100
Biuret	$H_2N - \underset{\underset{O}{\|\|}}{C} - \underset{\underset{H}{\|}}{N} - \underset{\underset{O}{\|\|}}{C} - NH_2$	0.29
Guanidine	$H_2N - \underset{\underset{NH}{\|}}{C} - NH_2$	0
Cyanamide	$H_2N - C{\equiv}N$	1.5
Dicyandiamide	$H_2N - \underset{\underset{NH}{\|}}{C} - \underset{\underset{H}{\|}}{N} - C{=}N$	0.64

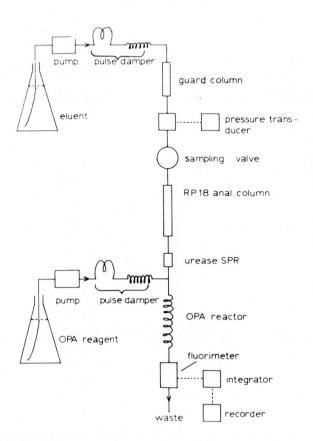

Fig. 2   Schematic diagram of an HPLC/urease reactor/OPA reactor/
fluorescence detector system. Reprinted from ref. 39.

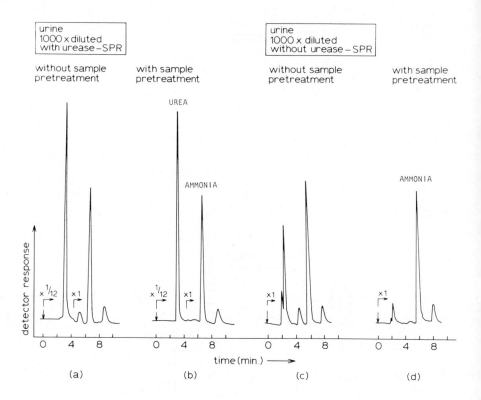

Fig. 3   (a) Chromatogram of a 1000-fold diluted urine sample without pretreatment procedure. Column, 5-μm Spherisorb ODS-2; eluent, 0.05
M potassium phosphate buffer (pH 6.9) with 0.005 M sodium octyl-
sulphonate. Urease-SPR, stainless-steel column packed with
immobilized urease. Fluorescence detection. Note: X 1/12 means
that the actual peak height is twelve times the peak height as
indicated in the figure. (b) As Fig. 3a , but with a urine
sample cleaned by pretreatment procedure. (c) As Fig. 3a, but
without urease-SPR. (d) As Fig. 3b, but without urease-SPR.
Reprinted from ref. 40.

5.2.3.3 HYDROXYSTEROID DEHYDROGENASES

Several papers discuss the use of 3α-hydroxysteroid dehydrogenase in a
post-column reactor for the determination of bile acids. After the early
work by Okuyama et al (ref. 41) methods for the routine determination of
bile acids in serum were developed. Bile acids react with NAD in the
enzyme column with the formation of NADH; NADH is reacted with phenazine
methosulphate solution which is followed by electrochemical detection at
+ 0.1 V (ref. 42). Direct oxidation of NADH was not successful due to the
higher potential needed (+ 0.33 V) which considerably decreased the

selectivity of the method. Another approach is the fluorescence detection of NADH (ref. 43). The authors succeeded in the use of the same alkaline pH (9.7) for both separation and enzyme reaction. Besides the eluent pump the system needs a second pump for NAD addition, but compared to electro-chemical detection (ref. 42) one pump less is required. Spherical cellulose beads were chosen as support because of their chemical stability at alkaline pH. Later, the system was equipped with an on-line sample pretreatment system to eliminate tedious manual precessing steps and make automatic controll possible (ref. 44). Now, NAD was added to the mobile phase so that a single pump system was obtained. Hayashi et al. (ref. 45) succeeded in the separation of the bile acids in rat bile. Rats possess several peculiar bile acids which complicate the separation. A C_8 column was used for separation and a second pump was used for addition of NAD and for pH adaptation after the separation. A chromatogram of a rat bile extract is given in Fig. 4. Another hydroxysteroid dehydrogenase was used in a similar analytical system (ref. 46). A column packed with 3ß, 17ß-hydroxysteroid dehydrogenase immobilized on glass beads was used to react Δ^5-3ß-hydroxysteroid sulphates with NAD with the formation of NADH which was detected by fluorescence monitoring. The method was successful for serum analysis and good correlation between the HPLC method and radio immuno assay was found.

5.2.3.4 ACETYLCHOLINE ESTERASE AND CHOLINE OXIDASE

A method for determination of choline and acetylcholine in neuronal tissue was introduced by Damsma et al. (ref. 47). Choline and acetyl-choline are separated by means of cation exchange HPLC. Without adaptation, the mobile phase enters the reactor packed with acetyl-cholinesterase and choline oxidase co-immobilized on sepharose beads. The latter enzyme releases hydrogen peroxide from choline, the former converts acetylcholine to choline. The hydrogen peroxide formed is detected electrochemically. A very similar method was described by Asano et al. (ref. 48), the difference being the use of glass beads as the support for immobilizing the enzyme. Instead of detecting the evolved hydrogen peroxide electrochemically, there is also the possibility to use peroxyoxalate chemiluminescence (ref. 49). Three pumps were needed to provide for the optimal media for separation, enzymatic reaction and chemiluminescence detection. Only standard mixtures were analyzed. Later, a more elegant system with chemiluminescence detection was published (ref. 50). The system has an enzyme reactor coupled directly to the analytical column and uses solid phase chemistries in the chemi-

222

luminescence detection step. Determination of choline and acetylcholine in urine and in serum (Fig. 5) was reported. Due to the selectivity of the chemiluminescence detection system, the sample pretreatment was simplified. An alternative immobilization method is the use of adsorption on an anion exchange cartridge (refs. 51, 52). This is a very simple way of preparing a reactor; just by injection of an enzyme solution on the cartridge at the proper pH. If low ionic strength mobile phases are used the reactor can be used for ca. 3 weeks before it needs to be reloaded. Problems arise when the reaction product that is to be detected is strongly retained on the ion exchanger. Strong retention of hydrogen peroxide formed in a reactor containing glucose oxidase immobilized on an ion exchanger was reported by Van Zoonen et al. (ref. 53).

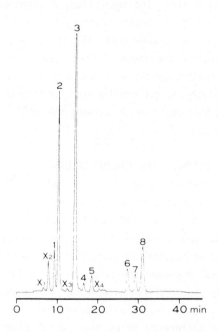

Fig. 4   Typical chromatogram of bile acids extracted from normal rat bile. Peaks: 1 = α-muricholic acid; 2 = ß-muricholic acid; 3 = cholic acid; 4 = ursodeoxycholic acid; 5 = hyodeoxycholic acid; 6 = chenodeoxycholic acid; 7 = deoxycholic acid; 8 = 5ß-andro-stane-3α, 11α, 17ß-triol (internal standard); X_{1-4} = unknown. Fluorescence detection. Reprinted from ref. 45.

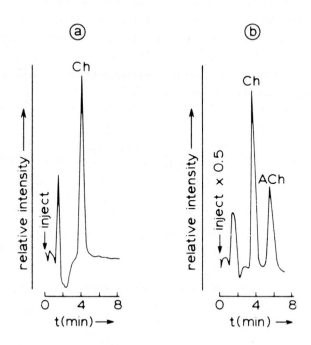

Fig. 5   (a) Chromatogram of a deproteinated pooled serum sample, (b)
serum sample spiked with 200 pmole of Choline and Acetylcholine.
Chemiluminescence detection. Reprinted with permission from ref.
50.

5.2.3.5 GLUCURONIDASE AND GLYCOSIDASE

End products from metabolic pathways in plants and animals are
detectable by means of immobilized enzyme post-column reaction detection
systems. ß-glucuronidase from bovine liver was immobilized on agarose and
on glass beads and was used for post-column cleavage of phenolic
glycosides with electrochemical detection of the phenolic compounds
formed (ref. 54). After enzymatic cleavage, lower detection potentials can
be used. Though the system can work with a single pump, the sensitivity
can be improved by changing the pH prior to the enzymatic reaction.
Almost the same system was used for the determination of glucuronide
conjugates for fenoldopam, an antihypertensive agent in human plasma and
urine (ref. 55). The diastereomeric fenoldopam glucuronides are cleaved
by the enzyme and the resultant catechol moiety ca be detected electrochemically. Chromatograms obtained with the system are given in Fig. 6.

The reactor containing the enzyme immobilized on glass beads is coupled directly between the analytical column and the detector. Dalgaard and Brimer (refs. 56,57) developed a method of analysis for xyanogenic glycosides. Glycosidase from Helix pomatia immobilized on glass beads was used in the post-column reactor. After the enzymatic reaction, base is added and the evolved cyanide is detected amperometrically at a silver electrode. The reactions involved are

Enzymatic reaction:

$$R_1 - \underset{\underset{O-Gly}{|}}{\overset{\overset{R_2}{|}}{C}} - CN + H_2O \xrightarrow{\text{Glycosidase}} R_1 - \underset{\underset{OH}{|}}{\overset{\overset{R_2}{|}}{C}} - CN + GlyOH$$

Cleavage by base:

$$R_1 - \underset{\underset{OH}{|}}{\overset{\overset{R_2}{|}}{C}} - CN + OH^- \longrightarrow R_1R_2CO + CN^- + H_2O$$

Anodic reaction:

$$Ag + 2CN^- \longrightarrow Ag(CN)_2^- + e^-$$

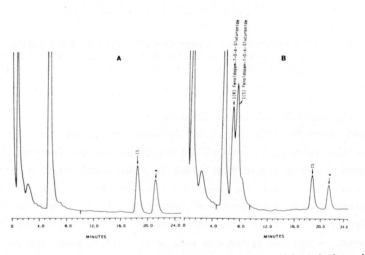

Fig. 6   Chromatograms of plasma extracts from a human subject before (A) and 0.5 h after (B) oral administration of fenoldopam. The concentrations of 1(R)- and 1(S)-fenoldopam-7-O-ß-glucuronide are 156 and 160 ng/nl, resprectively. Electrochemical detection. Reprinted from ref. 55.

Introduction of a split between the analytical column and the reactor was used to create a longer residence time in the reactor, useful for slower reacting substrates. The method was used for the determination of cyanogenic glycosides in human urine, serum and in crude plant tissue extracts (see Fig. 7).

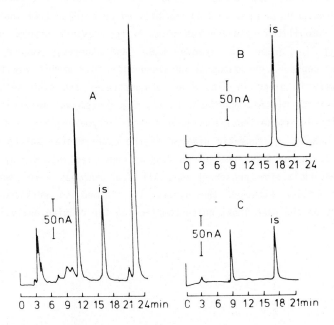

Fig. 7   Chromatograms of cyanogenic glycosides in plant extracts after suitable dilution. (A) S. nigra L (dilution factor 50); (B) P. Lauracerasus L.; (C) T. baccata L. (dilution factor 50). Amygdalin added as internal standard (is). Electrochemical detection. Reprinted from ref. 57.

5.2.3.6 XANTHINE OXIDASE

The simultaneous determination of hypoxanthine (Hyp) and Xanthine (Xan) in biological fluids is important for various pharmacological and physiological reasons. An HPLC method with on-line post-column enzymatic derivatization was developed (refs. 58, 59). Hyp and Xan were oxidized by the enzyme:

$$Hyp + H_2O + O_2 \xrightarrow{\text{xanthine oxidase}} Xan + H_2O_2$$

$$Xan + H_2O + O_2 \xrightarrow{\text{xanthine oxidase}} \text{uric acid} + H_2O_2$$

226

The resulting uric acid was detected by UV-absorption at 290 nm. However this detection mode did not offer the selectivity and sensitivity needed for serum analysis. An alternative was found by using fluorescence monitoring of the product formed in the reaction between H_2O_2 and p-hydroxyphenylacetic acid on immobilized peroxidase (ref. 60). A reactor with immobilized catalase was used in the reagent stream prior to the mixing in order to remove H_2O_2 and thereby reduce background fluorescence. A chromatogram obtained with this method applied to urine analysis is given in Fig. 8 and demonstrates the high selectivity and sensitivity of this approach. Later, the system was modified to be able to also determine inosine (ref. 61). Four enzyme reactors and three pumps were needed. A schematic diagram of the experimental set-up is given in Fig. 9. In the first reactor inosine is converted into Hyp by immobilized purine nucleoside phosphorylase. All four enzymes were immobilized on glass beads. Although the system is claimed to work reliably, the detection approach looks quite complicated for routine analysis.

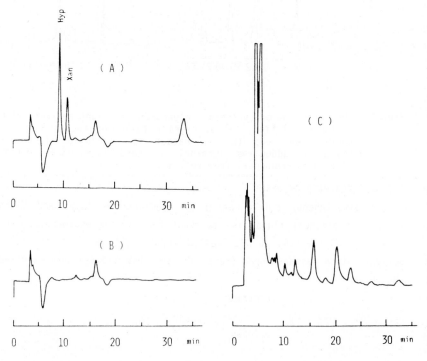

Fig. 8   Chromatograms of a urine extract from a normal subject obtained by the present method (see text) (A), when the immobilized xanthine oxidase reactor is removed from the system (B), and with UV absorbance at 254 nm when all reactors are removed (C). Reprinted from ref. 60.

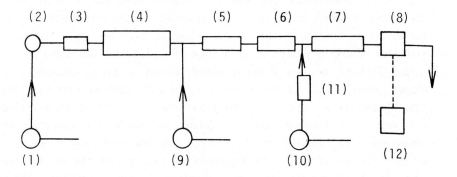

Fig. 9   Flow diagram of an HPLC system for Hyp, Xan and inosine. (1)
Eluent pump, (2) injector, (3) guard column, (4) analytical
column, (5) immobilized purine nucleoside phosphorylase reactor,
(6) immobilized xanthine oxidase reactor, (7) immobilized peroxidase reactor, (8) fluorescence detector, (9) pump for pH
adaptation, (10) reagent pump, (11) immobilized catalase
reactor, (12) integrator. Reprinted with permission from ref. 61.

5.2.3.7 L-AMINO ACID OXIDASE

Another system in which H_2O_2 is detected via homovanillic acid
reaction on coimmobilized peroxidase was introduced for the selective
detection of L-amino acids (ref. 62). The immobilized L-amino acid
oxidase catalyses the stereo-selective deamination of L-amino acids:

$$L\text{-amino acid} + O_2 + H_2O \longrightarrow 2\text{-keto acid} + NH_3 + H_2O_2$$

D-amino acids do not react and will not be detected. Flavine adenine
dinucleotide was added in the immobilization process in order to obtain a
more stable and highly active reactor.

The use of stereo-selective enzymes in the detection process offers an
attractive alternative to chiral separations. Unfortunately, in this
system the fact that the mobile phase should be compatible with the
enzyme reactor resulted in insufficient resolution of some of the amino
acids. An alternative in the detection of amino acids after the enzymatic
reaction is to monitor the conductivity of the reactor effluent (ref.
63). The System works with a single pump but, compared with fluorescence
detection (ref. 62), limits of detection are inferior.

5.2.3.8. CHOLESTEROL OXIDASE

Cholesterol oxidase immobilized on glass beads was used in the detection of cholesterol and some auto-oxidation products (ref. 64). Cholesterol has only moderate UV absorption at a short wavelength whereas the oxidation product formed in the enzymatic reaction can be detected with high absorptivity at 241 nm. A high flow of buffer was added to the column effluent to reduce the ethanol content prior to entering the enzyme reactor. The optimum dilution ratio was found by balancing the loss of sensitivity due to the dilution versus the decreased conversion efficiency with increased ethanol content in the reactor. In order to reduce the flow rate through the reactor for longer reaction time, part of the flow was split off. UV-monitoring at 241 nm with the reactor was found to yield ca. 4 times higher sensitivity than at 211 nm without reactor.

5.2.3.9 ALKALINE PHOSPHATASE

Inorganic phosphate resulting from inositol bis- and triphosphates and other organic phosphates in a reactor packed with immobilized alkaline phosphatase was detected by reacting with a molybdate solution as described by Meek and Nicoletti (ref. 65). The enzyme was immobilized by simple adsorption on a hydrophobic support. No mobile phase adaptation between the anion exchange analytical column and the enzyme reactor was needed. The method was demonstrated with test mixtures but the authors claim sufficient sensitivity for tissue analysis.

5.2.4   CONCLUDING REMARKS

Up to now, immobilized enzymes in post-column reactors are almost uniquely used for the derivatization of mostly polar analytes that can be separated on the analytical column with a highly aqueous mobile phase. In this way, the enzymatic reaction can take place under conditions that are similar to those in nature. Although immobilized enzymes are reported to be more stable than enzymes in solution they can usually not withstand high modifier concentration. On the other hand, there are examples of active enzymes in solutions with a relatively high content of organic modifier. In the system described by Okuyama for the determination of bile acids (ref. 41) a 22% acetonitrile solution flows through the reactor and a 80% acetonitrile solution was used in a flow injection analysis system with immobilized glucose oxidase (ref. 53). Ethanol at high concentration (17.3%) was pumped through the reactor with immobilized cholesterol oxidase (ref. 64). However, most analytical systems with an immobilized enzyme reactor contain only a few per cent of organic modifier and preferably alcohols rather than acetonitrile since otherwise the enzymes lose activity in either a reversible or in an irreversible way. The influence of various organic solvents in different concentrations on reaction kinetics was investigated by Bowers and Johnson for immobilized ß-glucuronidase (ref. 66). When p-nitrophenylglucuronide is used as the substrate, increasing the methanol content results first in an increase in activity up to ca. 10% methanol followed by a steep decrease whereas with estriol-3-glucuronide as the substrate there is only a decrease in reaction rate upon increasing the methanol content. The reaction rate on the enzyme was also investigated with ethanol, acetonitrile and ethylene glycol as the organic cosolvent (see Fig. 10). The above results reveal that an enzyme can react in different ways when it is brought in contact with non-natural compounds. Also the substrate to be converted is an important parameter. Other enzymes being investigated in a similar way can show very different behaviour. This makes it impossible to predict whether a certain mobile phase needed for a particular separation will be compatible with the immobilized enzyme reaction. It is very questionable whether such a prediction will be possible in the future. For the time being, in each particular analytical system involving an immobilized enzyme some trial and error work will be needed. However, the compiled applications above (see Table III) clearly reveal that there are distinct possibilities for successfully coupling HPLC with immobilized enzyme reactors and it is likely that more applications will be published in the coming years.

TABLE III   The use of immobilized enzymes in post-column reactors

Analyte(s)	Enzyme(s)	Immobilized on	Product detected	Detection*	Ref.
Glucose-6-phosphate	gucose-6-phosphate dehydrogenase	glass	NADH	F	38
Urea	urease	silica	NH_3, after derivatization with OPA reagent	F	39, 40
Bile acids	3 α-hydroxysteroid dehydrogenase	glass	NADH	F	41
Bile acids	3 α-hydroxysteroid dehydrogenase	glass	NADH, after derivatization with phenazine methosulphate	E	42
Bile acids	3 α-hydroxysteroid dehydrogenase	cellulose	NADH	F	43, 44
Bile acids	3 α-hydroxysteroid dehydrogenase	glass	NADH	F	45
Δ^5-3ß-hydroxysteroid sulphates	3ß,17ß-hydroxy-steroid dehydrogenase	glass	NADH	F	46
Choline and acetylcholine	acetylcholinesterase and choline oxidase	sepharose	H_2O_2	E	47
Choline and acetylcholine	acetylcholinesterase and choline oxidase	glass	H_2O_2	E	48
Choline and acetylcholine	acetylcholinesterase and choline oxidase	silica	H_2O_2	CL	49
Choline and acetylcholine	acetylcholinesterase and choline oxidase	sepharose	H_2O_2	CL	50
Choline and acetylcholine	choline oxidase and cholinesterase	ion exchange resin	H_2O_2	E	51
Acetylcholine	choline oxidase and cholinesterase	ion exchange resin	H_2O_2	E	52
Phenolic glycosides	ß-glucuronidase	agarose, glass	phenolic compounds	E	54
Glucuronide conjugates of fenoldopam	ß-glucuronidase	glass	fenoldopam	E	55

TABLE III   (Continued)

Analyte(s)	Enzyme(s)	Immobilized on	Product detected	Detection*	Ref.
Cyanogenic glycosides	glycosidase	glass, silica	cyanide, formed after basic cleavage of product from enzyme reaction	E	56, 57
Hypoxanthine and xanthine	xanthine oxidase	glass	uric acid	UV	58, 59
Hypoxanthine and xanthine	xanthine oxidase	glass	H_2O_2, after reaction with p-hydroxyphenyl-acetic acid on immobilized peroxidase	F	60
Hypoxanthine, xanthine and inosine	xanthine oxidase and purine nucleoside phosphorylase	glass	H_2O_2, after reaction with p-hydroxyphenyl-acetic acid on immobilized peroxidase	F	61
L-amino acids	L-amino acid oxidase	glass	H_2O_2, after reaction with homovanillic acid on co-immobilized peroxidase	F	62
L-amino acids	L-amino acid	glass	change in the ionic strenght created by the reaction	C	63
Cholesterol and autooxidation products	cholesterol oxidase	glass	oxidized cholesterol	UV	64
Inositol bis- and triphosphates and other organic phosphates	alkaline phosphatase	phenoxyacetyl-cellulose	Inorganic phosphate after reaction with a molybdate solution	UV	65

*F: Fluorescence
E: Electrochemical
CL: Chemiluminescence
UV: Ultraviolet absorption
C: Conductivity

232

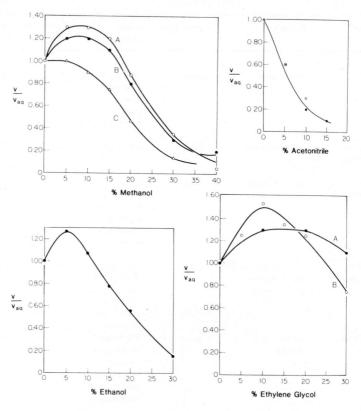

Fig. 10 Normalized reaction rate as a function of vol. percent organic cosolvent. In the methanol panel, curves A and B represent the behaviour observed for immobilized and soluble enzyme, respectively, with p-nitrophenyl glucuronide substrate, while curve C represents the reaction of estriol-3-glucuronide with immobilized enzyme. The acetonitrile and ethanol panels are the results of p-nitrophenyl glucuronide hydrolysis by immobilized ß-glucuronidase. The ethylene glycol panel illustrates the difference observed between the immobilized (A) and soluble (B) enzyme with p-nitrophenyl glucuronide substrate. In all cases the buffer was 0.1 mol/l phosphate, pH. 6.7. Reprinted from ref. 66.

5.3   OTHER SOLID-PHASE CHEMISTRIES

5.3.1 INTRODUCTION

In common reaction detection, reagent solution is added to the column effluent by means of a pump as outlined before. An alternative is to introduce the chemicals needed for the derivatization reaction in a heterogeneous way. The reactor, usually a short column packed with active material, can act as a catalyst or provide reagents for the reaction. In the latter case the reactants can come directly in contact with the solid reagent or reagent is dissolved gradually before the reaction takes place. The reactor has to be recharged periodically. The former approach is interesting since the reactor keeps its activity without the need for reloading, provided no catalyst poisoning takes place. The use of immobilized enzyme in a post-column reactor as outlined in the previous section is an example of this approach. This section will discuss the other uses of solid-phase chemistries in the field of post-column reaction detection.

5.3.2 APPLICATIONS OF SOLID-PHASE CHEMISTRIES
5.3.2.1 EARLY WORK

The first applications of solid-phase chemistry in a post-column reactor were published by Studebaker et al. (refs. 67, 68). The method is designed for the detection of thiols, disulfides and proteolytic enzymes. In each case, the compound of interest releases a detectable species from the packing material in a column downstream from the analytical column. The detection system for disulfides is outlined in Fig. 11. The disulfides release thiols in the upper reactor and the thiols release the detectable species which is initially bound to the polymer in the lower reactor. Later, the system was modified by Millot et al. (ref. 69), the main difference being the use of silica instead of a polymer as the support in order to minimize band broadening.

5.3.2.2 CATALYTIC SOLID-PHASE CHEMISTRIES

The concept of solid-phase catalysis in post-column derivatization was first introduced for the determination of non-reducing carbohydrates (refs. 29, 79, 71). Several strongly acidic cation exchangers were used as the catalyst for conversion of the non-reducing carbohydrates into reducing carbohydrates which are detectable by several means. Best results were obtained using 4% cross-linked polystyrene resins. The reactor (6 cm in length) was operated at 85 $^\circ$C to ascertain 100% conversion. The separations were carried out at a sulphonic acid type (Ca^{2+}) cation exchanger with water as the mobile phase. A thorough investigation of band broadening in this reactor type was published by Nondek et al. (ref. 35). Apart from the band broadening that occurs in any packed bed, additional band broadening is observed when reactant and product have different retention in the reactor column. A mathematical model was proposed the validity of which was tested with the decomposition of diacetone alcohol on alumina and the catalytic hydrolysis of 1-naphthyl-N-methylcarbamate/Carbaryl) on a strongly basic anion ex-

234

changer. Later, the reactivity of other N-methylcarbamates on the ion exchange resin and the applicability to residue analysis were investigated (refs. 36, 72). Upon decomposition of the N-methylcarbamates, methylamine is splitt off which, after labelling with o-phthalaldehyde (OPA), is detected by fluorescence monitoring. The reactor is operated at high temperature (100 $^\circ$C or slightly above) in order to keep reaction band broadening low and to obtain high conversion. The method was applied to the analysis of river water samples (ref. 36) and, after having been combined with on-line preconcentration and clean-up, to the analysis of heavily polluted water samples with detection limits well below 1 ppb (ref. 72).

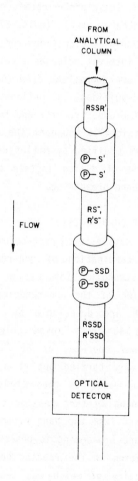

Fig. 11 Diagram of the solid-phase apparatus for detection of disulfides. RSSR' represents a disulfide in the eluate. P-S is a polymer with a bound thiol, P-SSD is a polymer with a bound detectable species. Reprinted from ref. 68.

5.3.2.3 NON-CATALYTIC SOLID-PHASE CHEMISTRIES

The use of non-catalytic solid-phase chemistries has resulted in different approaches in the field of post-column reaction detection. The earliest publication dealt with the use of a column packed with zinc powder (ref. 73). By pumping an acidic mobile phase through this column, hydrogen in statu nascendi is produced that splits off iodine from iodinated thyronines. The iodide ion was detected by means of a catalytic principle based on the iodide-catalyzed reaction of chloramine-T and N, N'-tetramethyldiaminodiphenylmethane in an air segmented tubular reactor. Chromatograms of plasma extracts are given in Fig. 12. The zinc column should be repacked daily. Another interesting use of a reactor packed with zinc was introduced by Sigwardson and Birks (ref. 74). Nitropoly-aromatic hydrocarbons (nitro-PAHs) were on-line reduced to the corresponding amino-PAHs that could be detected with high sensitivity using peroxyoxalate chemiluminescence detection. The reactor can be placed either before or after the analytical column so that the analytes elute either as the amino-PAHs or as the nitro-PAHs. This was found to be useful for identification. The method was applied to the analysis of carbon black extracts.

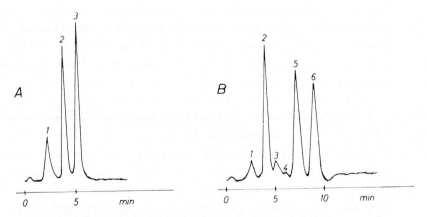

Fig. 12 Chromatographic determination of iodinated thyronines with catalytic detection. Column, C-18; mobile phase, methanol-water (67:33) plus 0.05% of methanesulphonic acid; flow-rate, 0.5 ml/min; detection wavelength, 600 nm. (A) Determination of total T_4 (thyroxine). Injection volume, 50 ml of ethanolic serum extract. 1, Free halide ions; 2, 10 ng of T_3 (tri-iodothyronine, internal standard); 3,8 ng of T_4. (B) Stereo-specific determination L-T_4 in serum after derivatization. Injection volume, 30 ml. 1, Free halide ions; 2, 10 ng of T_3 (internal standard); 3 and 4, not identified; 5, L-Leu-L-T_4 corresponding to 7 ng of L-T_4); 6, L-Leu-D-T_4 (corresponding to 7 ng of D-T_4). Reprinted from ref. 73.

Krull et al. (refs. 75-78) described various reactors containing solid-supported reagents for on-line reduction and oxidation. Most of the applications, e.g., the on-line reduction of aldehydes to the corresponding alcohols using solid supported borohydride (ref. 75) and the oxidation of primary and secondary alcohols, aldehydes and ketones using an anion exchanger in the permanganate form (ref. 78), deal with pre-column derivatization using difference chromatography to study the extent of reaction. This is mainly done for identification purpose since no significant changes in detector response are reported which limits, though possible and described (e.g. for the post-column reduction of a variety of aldehydes, ketones and acid chlorides on solid-supported borohydride in normal-phase HPLC) (ref. 76), the applicability in post-column derivatization. All reactors described by Krull et al. can be used for hundreds of analyses before they loose activity.

Solid-phase reactors containing either lead dioxide or manganese dioxide precipitated on silica were compared for use in the post-column oxidation of catecholamines to the respective adrenochromes (ref. 79). This was followed by homogeneous reduction to the fluorescent trihydroxyindoles. The two reactors yield similar results in all respects and allow for a simpler, more economic and more reliable post-column reaction system as compared to an all homogeneous approach. The number of analyses possible before depletion of the reagent occurs is more than 300 for both reactors. The same lead dioxide reactor was used to oxidize lower oxidation state chromium ions to chromate which is detected by complexation with 1,5-diphenylcarbazide (ref. 80). As applications, wastewaters and samples from a steel company were analyzed.

A solid-phase reactor used in a sulphate selective post-column derivatization system was developed for the analysis of wastewaters of the potato starch industry (ref. 81). The solid-phase reactor is packed with a mixture of silica and barium chloranilate. The solubility of barium sulphate is less than that of barium chloranilate. The sulphate eluting from the ion exchange analytical column will precipitate as barium sulphate and an equivalent amount of the highly coloured acid chloranilate ion is released:

$$SO_4^{2-} + BaC_6Cl_2O_4 + H^+ \longrightarrow BySO_4 \downarrow + HC_6Cl_2O_4^-$$

Irth et al. (ref. 82) developed a post-column solid-phase derivatization for the selective detection of thiram and disulfiram. These thiuram disulfides undergo complexation in a very short (4 mm in length) cartridge-type reactor packed with finely divided metallic copper to form a coloured copper complex with an absorption maximum at 435 nm.

The post-column complexation enhances the selectivity of the analytical method with almost the same sensitivity as UV detection at 254 nm. The post-column reactor was found to cause only little additional band broadening since no retention takes place and the reactor is short. The method was demonstrated with the determination of thiram in surface water and with the determination of disulfiram in urine (Fig. 13). A short precolumn was used for on-line trace enrichment. When nanogram amounts of analyte are injected in the analytical system, the copper reactor can be used for more than 200 analyses.

Another use of solid-supported reagent was introduced by Jansen et al. (ref. 83). The analytical system contains an anion exchange column in the hydroxy form inserted parallel to the injector and analytical column (Fig. 14). One part of the acetate-containing mobile phase flows through the injection valve and analytical column to achieve the separation, the other part flows through the anion-exchange column where the acetate ion causes the release of ion exchanger-bound hydroxide ions. Finally, the alkaline stream from the anion-exchange column is recombined with the analytical column effluent. The resultant alkaline detection medium is favourable for the UV detection of barbiturates at 254 nm. Only one pump is needed for the separation and the post-column pH modification. A low-cost large particle ion exchanger was used since the anion exchange column does not contribute to band broadening. Both the determination of barbiturates in urine and in plasma were shown. The anion-exchange column should be regenerated after approximately 17 hours use. Basically the same analytical system was adapted for aflatoxin determination (ref. 84). Now, the parallel column was packed with solid iodine and a knitted open tubular reactor was mounted between the mixing T-piece and the fluorescence detector. Iodine is only partly soluble in the aqueous mobile phase; therefore, the parallel column can be used in the analytical system for the delivery of a saturated iodine solution over long periods of time before refilling is necessary. For optimum response, a splitting ratio of ca. 30 to 1 was used. Iodine attachment to the double bond of aflatoxin B_1 and G_1 makes them about as fluorescent as aflatoxin B_2 and G_2. As in the system with the parallel ion exchanger, only one high-quality pump is needed making low-cost post-column derivatization possible without the need for a daily preparation of the iodine solution as was necessary before (ref. 11). The method was successfully applied to

238

the analysis of peanut butter extracts as shown in Fig. 15. The parallel column approach was also found to be applicable in chemiluminescence detection (ref. 85). Bis-2,4,6-trichlorophenyloxalate (TCPO) is added from a solid reagent bed and the fluorophore is immobilized on glass beads packed in a flow cell. Hydrogen peroxide generated photochemically by quinone analytes is measured.

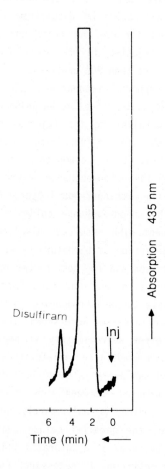

Fig. 13   Determination of disulfiram in urine. HPLC conditions:
analytical column packed with 5μm Hypersil ODS; pre-column packed with 5-μm LiChrosorb RP-18; eluent, acetonitrile-aqueous acetate buffer (10 mM, pH 5.0) (65:35); detection wavelength, 435 nm (0.02 a.u.f.s.). Pre-concentration of 1.0 ml of urine spiked with 87 ppb of disulfiram (sample stabilized with 10 mM EDTA-citrate). Reprinted from ref 82.

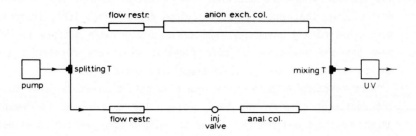

Fig. 14 Experimental set-up of a parallel column derivatization system. Reprinted from ref. 83.

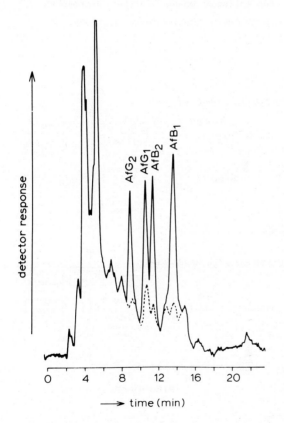

Fig. 15 ———— Chromatogram of an extract of peanut butter spiked with AfB$_1$ (8 ppb), AfG$_1$ (8ppb), AfB$_2$ (4 ppb) and AfG$_2$94 ppb); -------- Chromatogram of an unspiked peanut butter extract. Conditions: analytical column packed with 5-μm LiChrosorb RP-18, parallel column packed with solid iodine, knitted open tubular reactor, fluorescence detection. Reprinted with permission from ref. 84.

240

Two systems with TCPO addition from a solid reagent bed, a dual pump design (Fig. 16a) and a parallel column design (Fig. 16b), respectively, were compared and advantages and disadvantages were discussed. The dual pump has the advantage of more flexible flow rate regulation but the split-flow has the advantage of simplicity and economics.

Much research in post-column reaction detection has as goal to render the equipment simpler and less costly. The development of immobilized enzyme reactors and reactors based on solid-phase chemistries as outlined in the previous and in this section are clear examples of this trend since all these systems contain at least one pump less than their classical homogeneous reaction analogues or permit reaction types which are impossible by conventional means. Further approaches for pumpless reaction units are discussed in the following sections.

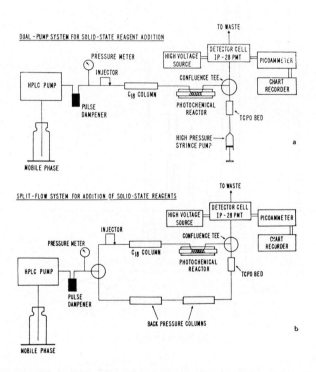

Fig. 16   Schematic diagram of the dual-pump system (a) and the split-flow system (b) for TCPO addition. Reprinted with permission from ref. 85.

5.4   THE USE OF ELECTROCHEMICAL REAGENT PRODUCTION

Electrochemical techniques can be used for on-line production of the reagent. One example is the use of copper electrodes or other suitable metal electrodes at which metal ions are generated. Complexing analytes (amino acids, dicarboxylic acids) can then be detected by amperometric techniques (refs. 86, 87). Another example of on-line electrochemical reagent production is the use of a microcoulometric production cell downstream from the analytical column in which the production of bromine or iodine from KBr or KI dissolved in the mobile phase is effected. This technique, introduced by King and Kissinger (ref. 88), involves reaction of bromine or iodine with suitable groups of compounds such as un-saturated organics, phenolics, methoxysubstituted aromatics and or-ganosulphur compounds and the decrease in reagent concentration is detected amperometrically (refs. 89, 90). A schematic diagram of the post-column system is shown in Fig. 17 and an application to the determination of ampicillin in plasma is given in Fig. 18. The bromine or iodine produced in the microcoulometric cell could also be used for the direct oxidation of the analyte to yield a fluorescent derivative. This principle was demonstrated with the determination of thioridazine in plasma after oxidation with bromine (ref. 91) and with the determination of aflatoxins in cattle feed following iodine addition (ref. 92). A more detailed discussion on electrochemical reaction techniques for detection in HPLC can be found in ref. 93.

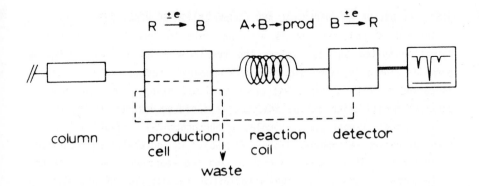

Fig. 17   Scheme of the on-line reagent production system: (A) analyte, (B) reagent (bromine), (R) precursor of the reagent (bromide). Reprinted from ref. 89.

242

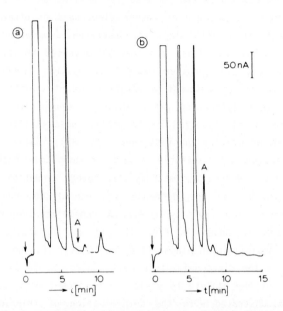

Fig. 18 Chromatograms obtained with deproteinized plasma; (a) blank
plasma; (b) plasma spiked with ampicillin (8 μg/ml). Column:
LiChrosorb RP-18; reagent production system as in Fig. 17.
Reprinted with permission from ref. 90.

The extension of this principle to other reagents and particularly
reactions involving noxious or unstable reagents offers interesting
possibilities for the future.

5.5 THE USE OF PHOTOCHEMICAL AND THERMO INITIATED REACTIONS

Instead of reagents, it is also possible to "pump" photons into a
detection system and to explore the resulting reactions for detection
purpose. Generally, some piece of UV-transparent tubing, usually teflon,
is coiled or knitted (ref. 94) around a lamp. Also in this approach, a
chemical modification of the analytes is achieved without having to use
an additional reagent pump. Examples of photochemical derivatization have
been discussed elsewhere (refs. 8, 95). In general, photochemical
derivatization is not easy to control and the reaction products are not
fully known. However, if the irradiation results in an increase in
sensitivity and/or selectivity, this is not a drawback for use in
a post-column system. In the more recent literature examples of better
controlled photochemical derivatizations can be found. One way of
improving the situation is to add small amounts of reagent to the mobile

phase to be able to conduct the reaction in the desired direction. An example is the photochemical reduction of Vitamin K_1 (ref. 96). To ensure the conversion into the desired hydroquinone, a small amount of ascorbic acid is added to the mobile phase and oxygen is carefully removed. The use of sensitizers is another means to expand the usefulness of photochemical reactions. A quinone can be used as sensitizer for the photochemical detection of reducing species (sugars, alcohols, aldehydes, etc.) (ref. 97). The resulting hydroquinone is detected fluorimetrically. Verbeke and Vanhee (ref. 98) used photochemical derivatization for the selective determination of diethylstilbestrol (DES). The photochemical reaction was followed by on-line oxidation to highly fluorescent products. The experimental set-up used is shown in Fig. 19. The method was successfully applied to the selective quantitation of DES residues at the 1 ppb level in extracts of urine and animal tissues (Fig. 20).

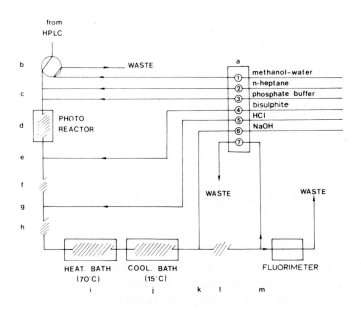

Fig. 19   Post-column reactor arrangement for the detection of DES: (1) methanol-water (65:35) (0.70 ml/min); (2) n-heptane (0.4 ml/min): (3) phosphate buffer, 0.02 M (0.1 ml/min); (4) hydrogen sulphite solution, 0.05 M (0.1 ml/min); (5) 3.7 HCl (0.2 ml/min); (6) 4.8 M NaOH (0.2 ml/min: (7) organic solvent from phase separator (0.9 ml/min). Reprinted from ref. 98.

244

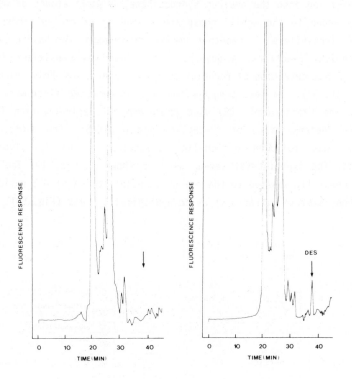

Fig. 20 (a) Chromatogram of a DES-negative urine extract. (b) Chromatogram of an urine extract containing 1 ppb of DES. For details see Fig. 19. Reprinted from ref. 98.

An interesting extension of the use of photochemical techniques in HPLC was developed by De Ruiter et al. (ref. 99). Dansylated chlorophenols were, after separation on a RP-column, irradiated to knock off the dansyl group. This resulted in enhanced fluorescence since now, the fluorescence is not reduced anymore by inductive and intramolecular heavy atom effects originating from the chlorophenols. After solid-phase extraction and precolumn dansylation, chlorophenols can be determined at the ca. 100 ppt level in river water samples.

Post-column thermolysis followed by photolysis and fluorescence detection was introduced for the trace-level determination of ciprofloxacin and its metabolites in urine, serum/plasma, bile, faeces and tissue (ref. 100). Both the thermolysis step and the photolysis were optimized yielding fluorescence gain factors up to 130 as compared to direct fluorescence detection, i.e., without the thermolysis and photolysis step. Reaction times were very short, 2 s for thermolysis and 0.6 s for photolysis. Using this detection mode, only minimal sample preparation, extraction and/or dilution, is required.

Here, it is interesting to mention that another example of the use of a temperature jump after the separation was reported by LePage and Rocha (ref. 101). They used the ninhydrin reaction for amino acids and other amines which is kinetically slow at ambient temperature. The reagent was added to the mobile phase with no reaction taking place during the separation. By applying a temperature jump after the column reaction will proceed at a reasonable rate without necessitating post-column addition of reagents.

5.6   MINIATURIZATION

In the past decade, an increasing interest has developed in the use of narrow-bore separation columns (< 2 mm I.D.) in HPLC as evidenced by several review publications and books in this field (refs. 102 - 107). These columns require a much lower volumetric flow rate than conventional HPLC columns and savings in solvents, reagents and packing material are obvious. In order to keep the chromatographic integrity of the system, the sample volume must also be reduced in comparison to conventional scale HPLC, which can be an advantage when the sample volume is limited. Also the volumes of the peaks eluting from a narrow-bore column are smaller than in the case of conventional scale HPLC. This results in the necessity to use low volume detectors. When a post-column reaction detection system is used in conjuction with narrow-bore HPLC, a careful design is needed in order to avoid excessive band broadening and loss in resolution. Special attention should be given to the mixing units used for the reagent addition and to connections. These difficulties resulted in only a few applications of post-column reaction detection in miniaturized HPLC described in the literature up to now.

Hirose et al. (ref. 108) were the first to describe a tubular post-column reactor coupled to a 0.5 mm I.D. analytical column for the determination of rare earth metals by a colour reaction with xylenol orange. Today this post-column system can be considered as an obsolete device since a several-meter-long tubular reactor of relatively large inner diameter and a wide-bore mixing unit were used. Unfortunately, no data on band broadening and detection limits were given.

Kucera and Umagat (ref. 109) described tubular post-column reactors

for the fast derivatization of primary amines with o-phthalaldehyde
(OPA). A 30 cm x 1 mm I.D. analytical column was used. The influences of
the mixing unit design and the pumps on mixing noise were carefully
examined. The contribution to total band broadening orginating from the
tubular reactor was investigated for different reactor lengths and inner
diameters. The authors' recommendation was to zigzag the reactor tubing
in order to enhance radial mixing due to the secondary flow phenomenon
resulting in lower band broadening. Knitting the reactor as proposed by
Engelhardt and Neue (ref. 4) will also be very useful for this purpose.
The analysis of a mixture of primary amino acids and a mixture of
homologous n-alkylamines were presented as applications.

Apfel et al. (ref. 110) used the same OPA reaction in the evaluation
of tubular and packed bed reactors for narrow-bore HPLC from both a
theoretical and an experimental point of view. In the model system,
catecholamines were separated on a 1 mm I.D. analytical column followed
by OPA derivatization and fluorescence detection. In comparison with the
measurement of the natural fluorescence, the detection limits were
lowered by more than an order of magnitude when using the post-column
derivatization. The band broadening caused by the post-column system was
acceptable but could have been improved with a better mixing unit.
Because of the relatively high band spreading in the mixing unit, a Valco
T-piece, no difference in performance between a system with a packed bed
reactor (particle size 5 μm) and one with a 100 μm I.D. tubular reactor
could be found.

Takeuchi et al (ref. 111) described a solid-phase reactor containing
3α-hydroxysteroid dehydrogenase immobilized on glass beads for the
detection of bile acids after separation on a 0.26 mm I.D. analytical
column. The 3α-hydroxy group in each bile acid was oxidized in the
enzymatic reaction, while NAD was reduced to NADH which was subjected to
fluorescence monitoring. Two experimental set-ups were compared (Fig.
21). Addition of NAD to the mobile phase prior to the column avoided the
need of post-column reagent addition and this was found to improve
baseline stability. A chromatogram obtained with the latter system is
shown in Fig. 22. Band broadening in the immobilized enzyme reactor was
relatively large because sufficiently small glass beads were not
available. The same equipment was used in combination with an off-line
trace enrichment procedure for the determination of bile acids in serum
(ref. 112).

(A)

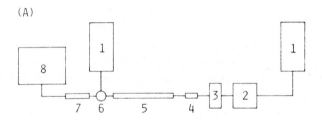

(B)

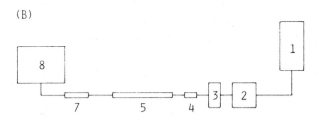

Fig. 21 Schematic diagrams of the systems (see text): (A) post-column
         mixing system; (B) pre-mixing system. 1 = Pump (Micro Feeder); 2
         = gradient equipment; 3 = micro valve injector; 4 = guard
         column; 5 = separation column; 6 = T-piece; 7 = immobilized
         enzyme column; 8 = spectrophotofluorimeter. Reprinted from ref.
         111.

Additional examples of the use of a solid-phase reactor in post-column
systems for narrow-bore HPLC can be found in ref. 113. A system was
described for the determination of N-methylcarbamate pesticides via
separation on a 180 mm x 1.0 mm I.D. reversed-phase column, hydrolysis on
an anion exchange resin in a solid-phase reactor kept at 90 °C, and
subsequent derivatization of the liberated methylamine with IPA in a very
short open tubular reactor (Fig. 23). Similar equipment has been used for
the rapid determination of urea and ammonia via their separation on a 40
mm x 1.0 mm I.D. column, hydrolysis of the urea on immobilized urease,
and followed by derivatization with OPA in a coiled open tubular reactor.
In this system, the analytical column and the reactor were coupled
without connecting capillary (Fig. 24). This elimination of a possible
source of band broadening is recommended if column and reactor are
operated at the same temperature. Band broadening in the mixing unit
could be kept sufficienty low by a special design (Fig. 25). This mixing
unit was constructed from two blocks; the upper one (B) contained the
entrance and exit capillaries (E) that ended in a small groove (D)
machined in a PTFE plate (C), supported by the lower part (A). The same
mixing unit was used in a miniaturized version of the split-flow

analytical system for barbiturates (ref. 83). A critical comparison between the conventional scale (3 mm I.D. column) and the narrow bore (1 mm I.D. column) system is made in ref. 83. It was found that the concentration sensitivity in narrow-bore system was lower than in the conventional scale system. Although the peak broadening in volume units is less in the narrow-bore system, resulting in more concentrated peaks in comparison to peaks eluting from the conventional scale column if the same mass is injected on the column, this does not yield more detector signal since the peak is now detected in a cell with a lower optical path length. Therefore, equal masses put on the two column types will yield roughly equal detector signals. Since the mass applied to the narrow-bore column is injected in a low volume, the concentration sensitivity is less.

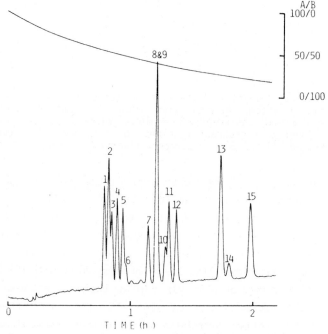

Fig. 22   Separation of bile acids (each around 5 ng) by the premixing system Column; Bilepak, 250 x 0.26 mm I.D. Mobile phase: gradient profile as indicated of acetonitrile-30 mM (A) or 10 mM (B) potassium dihydrogen orthophosphate (pH 7.8) - 10 mM potassium dihydrogen orthophosphate (pH 7.0) containing 6 mM NAD, 0.05% 2-mercaptoethanol and 1 mM EDTA, in the proportions (A) 18:52:30 and (B) 35:35:30, each containing 0.1% ammonium carbonate. Samples: 1 = ursodeoxycholic acid, 2 = cholic acid, 3 = glycoursodeoxycholic acid, 4 = glycocholic acid, 5 = tauroursodeoxycholic acid, 6 = taurocholic acid, 7 = chenodeoxycholic acid, 8 = deoxycholic acid, 9 = glycochenodeoxycholic acid, 10 = glycodeoxycholic acid, 11 = taurochenodeoxycholic acid, 12 = taurodeoxycholic acid, 13 = lithocholic acid, 14 = glycolithocholic acid, 15 = taurolithocholic acid. Wavelength of detection: excitation 365 nm, emission 470 nm. Reprinted from ref. 111.

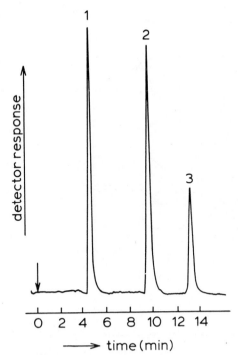

Fig. 23 Chromatogram of N-methylcarbamate pesticides. 1 = Methomyl, 2 = aldicarb, 3 = propoxur, ca. 25 ng each; 180 mm x 1.0 mm I.D. column packed with 5 μm Spherisorb ODS-2; 40 mm x 1.0 mm I.D. solid-phase reactor packed with a strong anion exchange resin; 170 mm x 0.12 mm I.D. straight open-tubular reactor for OPA-reaction. Flow rate of eluent: 35.7 μl/min; flow rate of OPA-reagent: 7.5 μl/min; fluorescence detection. Reprinted with permission from ref. 113.

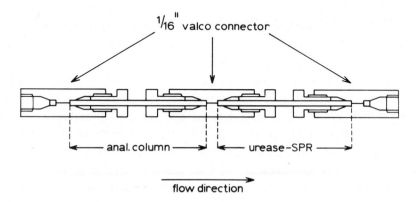

Fig. 24 Construction of the combination of short analytical column and urease-SPR (both 40 mm x 1.0 mm I.D., 1/16" O.D.) Reprinted with permission from ref. 113.

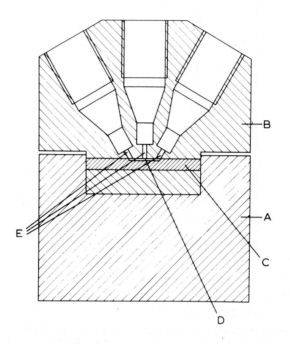

Fig. 25   Construction of the mixing unit. For explanation, see text.
Reprinted with permission from ref. 113.

A miniaturized HPLC system with a tubular reactor was used with chemiluminescence detection for dansylated amino acids (ref. 114). The
analytes are separated on a 250 mm x 1.0 mm I.D. column. The reagent, 1
mM bis(2,4,6-trichlorophenyl) oxalate in ethyl acetate-0.1 M H_2O_2 in
acetone (1:3 v/v), was added at a relatively high flow-rate, thereby also
working as a make-up flow. For this reason a cyclon-type mixing unit for
conventional scale HPLC was used after a minor modification; a 0.1 mm
I.D. stainless steel tube was inserted in the eluent inlet ((Fig. 26).
The authors claim detection limits of 0.2 fmol. A chromatogram of four
dansylated amino acids is given in Fig. 27. In a more recent publication
(ref. 115), the authors replaced the 1 mm I.D. column by a 2.1 mm I.D.
column. This was done because two pump gradient elution on the 1 mm I.D.
column was not successful due to pump irregularities, resulting in
increased baseline noise. Van Vliet et al. (ref. 116) investigated the
possibility of coupling a tubular post-column reactor to a 25 μm I.D.
open tubular separation column. Using these extremely small bore columns,
extra column volume is only tolerable on the nanoliter scale. Two

specially designed mixing units (Fig. 28) were proposed and evaluated. 25 μm I.D. fused silica tubing was used as the reactor. To make the detection in a very low volume, laser-induced fluorescence was used. Even at this extremely low volume scale, band broadening in the mixing unit and reactor could be kept at a level compatible with the column used. The OPA-derivatization of alanine was given as an example.

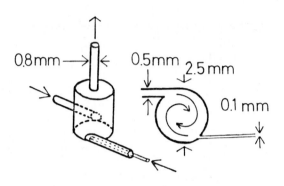

Fig. 26   Cyclon-type mixing unit for miniaturized HPLC. Reprinted from ref. 114.

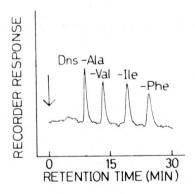

Fig. 27   Chromatogram of Dns-amino acids, obtained with the microbore HPLC-chemiluminescence detection system. The amount of each amino acid injected was 3 fmol. Eluent flow-rate 0.03 ml/min. Reagent flow-rate 0.6 ml/min. Reprinted from ref. 114.

252

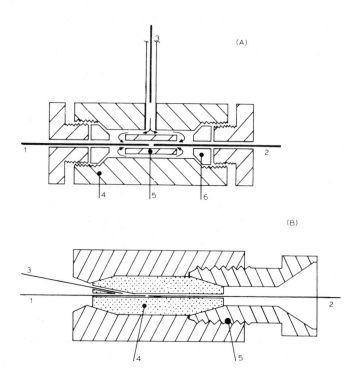

Fig. 28 Design of the mixing devices. 1 = Open-tubular column; 2 = reaction capillary; 3 = reagent delivery capillary; (A) based on a
Valco 1/16-in zero dead volume connector, 4 = connector body, 5
= stainless-steel liner, 6 = PTFE ferrules, arrows indicate
reagent flow; (B) based on Supelco butt-connector, 4 = vespel
ferrule, 5 = connector body. Reprinted from ref. 116.

5.7   HOLLOW FIBERS AS POST-COLUMN REACTORS

Hollow fiber membranes have been discussed for use as suppressors in
ion chromatography with conductometric detection and are commercialized
by the Dionex Company (refs. 117-121). They can also be used as reactor
and pumpless reagent addition device in post-column systems with
fluorescence detection (ref. 122). The reactor was made of several
parallel 325 μm I.D. sulphonated polyethylene hollow fibers, 8 inch in
length each. This reactor was suspended in a container of the appropriate
reagent. Permeation through the membrane should be small to avoid loss of
analyte. The reagent concentration in the eluent can be made sufficiently

strong by using highly concentrated reagent solution in the container. Attention should be given to the choice of solvent for the reagent as this may have a beneficial or a detrimental influence on the reagent flux through the membrane. Several model systems were discussed: the enhanced UV detection of nitrophenols and the enhanced fluorescence detection of phenols upon pH-adaptation, the fluorescence detection of amines after derivatization with fluorescamine and the ninhydrin colour formation with amino acids.

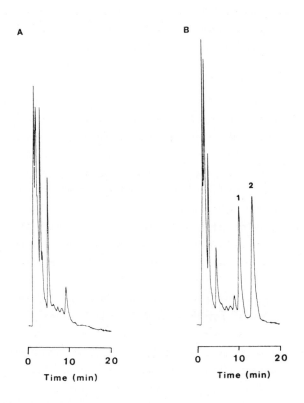

Fig. 29 Chromatograms of control saliva (A) and control saliva spiked with amobarbital (B). 1 = hexobarbital (0.5 μg/ml, internal standard); 2 = amobarbital (0.5 μg/ml). Injection volume: 50 ml. UV-detection at 240 nm, 0.008 AUFS. Reprinted from ref. 124.

Haginaka et al. (ref. 123) were the first to describe the use of a hollow fiber post-column reactor for the analysis of real samples. The determination of ß-lactamase inhibitors (clavulanic acid and sulbactam) in serum and urine based on hollow fiber pH adaptation and UV detection at 270 - 280 nm was discussed. A 1.2 m long hollow fiber (0.3 mm I.D.) suspended in a 1.0 M sodium hydroxide solution was used. The response of the analytes was optimized by changing reactor length and reagent concentration. Essentially the same analytical system was used by these authors for the determination of some selected barbiturates whereby the determination of amobarbital in saliva was given as the application (ref. 124). A 0.05 M ammonium hydroxide solution was taken as the reagent and the hollow fiber reactor was only 15 cm in length. Typical chromatograms obtained using this method are given in Fig. 29.

Since the use of hollow fiber reactors seems to give a wide range of possibilities for simple reagent introduction, it is likely that more applications will be published in the nearby future.

6. CONCLUSIONS

HPLC is a technique that is widely used in many analytical laboratories all over the world. When the HPLC separation is combined with on-line post-column reaction detection highly sensitive and selective analytical systems can be obtained, as reflected by a still growing number of publications dealing with interesting applications. The success of post-column reaction detectors depends primarily on the proper choice of the reactor itself, and where applicable, on the mixing unit used for reagent addition. If normal precautions are taken, the insertion of a reaction detector into a HPLC system does not cause a serious loss in chromatographic resolution (less than two-fold increase in peak width in most cases). Often, the loss in resolution will be compensated by an increase in selectivity, hence one should not only concentrate on keeping band broadening as low as possible since frequently chemical selectivity can be substituted for chromatographic selectivity. Sensitivity and selectivity are often improved to such an extent that sample pretreatment can be considerably simplified and automated or even omitted.

Reaction detection as such is already a relatively old technique that now, due to its mature status, is used for many analytical problems. Much interest exists for the newer developments in the field of reaction detection as discussed in the present paper. The use of immobilized enzymes and, to be more general, solid-phase chemistries has resulted in simpler and more economical analytical systems and often gave the possibility to circumvent problems on compatibility of the reagent with the detection mode chosen. With electrochemical reagent production and photochemical reaction detection post-column chemistries are possible without the need for an additional reagent pump which, again, improves the economics of the system.

Miniaturization of reaction detector has made them compatible for use with narrow-bore (1 mm I.D. or lower) analytical columns. The major advantage of miniaturized reactors being low reagent consumption hence permitting to work with expensive materials. Toxicity of samples and reagents is also better controllable. For such systems, the development of solid-phase reactor columns and other pumpless reaction units seems promising because no mixing of column effluent and reagent stream is needed. Although possible, such mixing is somewhat difficult to control in miniaturized HPLC. A very gentle approach towards the mixing problem is the use of hollow fiber reactors as introduced only a few years ago. With the ever increasing developments in membrane technologies, rapid development in this field seems to be very likely.

In conclusion, it can be stated that post-column reaction detection has now been developed to such a degree that it can be used for many different types of analyses, both in research and routine laboratories. On the other hand there are still many different aspects for researchers to work on the coming decade which will no doubt result in even more interesting applications.

REFERENCES
1   M.W.F. Nielen, R.W. Frei and U.A. Th. Brinkman, Selective Sample Handling and Detection (Editors: R.W. Frei and K. Zech), Vol. 1, Elsevier, Amsterdam, 1988.
2   R. W. Frei, J. Chromatogr., 165, 75 (1979).
3   J.F.K. Huber, K.M. Jonker and H. Poppe, Anal. Chem. 52, 2 (1980).
4   H. Engelhardt and U.D. Neue, Chromatographia, 15 403 (1982).
5   R.S. Deelder, A.T.J.M. Kuijpers and J.H.M van den Berg, J. Chromatogr., 255, 545 (1983).
6   B. Lillig and H. Engelhardt, Reaction Detection in Liquid Chromatography, I.S. Krull ed., Marcel Dekker, New York, 1987, Ch. 1.
7   L. Nord and B. Karlberg, Anal. Chim. Acta, 164 233 (1984).
8   R.W. Frei, Chemical Derivatization in Analytical Chemistry, Vol. 1 Chromatography', R.W. Frei and J.F. Lawrence eds., Plenum Press, New York, (1981), Ch. 4
9   J.F. Lawrence, U.A. Th. Brinkman and R. W. Frei, Reaction Detection in Liquid Chromatography, I.S. Krull ed., Marcel Dekker, New York, (1987), Ch. 6.
10 J.H.M. van den Berg, R.S. Deelder and H.G.M. Egberink, Anal. Chim. Acta, 114, 91 (1980).
11 L.G.M.Th. Tuinstra and W. Haasnoot, J. Chromatogr., 282, 457 (1983).
12 C.J. Little, J.A. Whatley and A.D. Dale, J. Chromatogr., 171, 63 (1979).
13 P. Böhlen and R. Schroeder, Anal. Biochem., 126, 144 (1982).
14 N. Kiba, M. Takamatsu and M. Furusawa, J. Chromatogr., 328, 309 (1985).
15 C.R. Clark and J. Chan, Anal. Chem., 50, 635 (1978).
16 H.A. Moye, S.J. Scherer and P.A. St. John, Anal. Lett., 10 1049 (1977).
17 G. Schwedt, Chromatographia, 10, 92 (1977).

256

18   A.H.M.T. Scholten, U.A.Th. Brinkman and R.W. Frei, J. Chromatogr., 218, 3 (1981).
19   F.P. Bigley and R.L. Grob, J. Chromatogr., 350, 407 (1985).
20   H. Engelhardt and R. Klinkner, Chromatographia, 20, 559 (1985).
21   M.D. Baker, H.Y. Mohammed and H. Veening, Anal. Chem., 53 1658 (1981).
22   R.S. Deelder, M.G.F. Kroll and J.H.M. van den Berg, J. Chromatogr., 125, 307 (1976).
23   M.E. Rogers, M.W. Adlard, G. Saunders and G. Holt, J. Chromatogr., 257, 91 (1983).
24   C.E. Werkhoven-Goewie, W.M.A. Niessen, U.A.Th. Brinkman and R.W. Frei, J. Chromatogr., 203, 165 (1981).
25   J.Haginaka and J. Wakai, Anal. Chem., 57, 1568 (1985).
26   W.P. King and P.T. Kissinger, Clin. Chem., 26 1484 (1980).
27   S. Honda, T. Konishi and S. Suzuki, J. Chromatogr., 299 245 (1984).
28   J.H.M. van den Berg, H.W.M. Horsels and R.S. Deelder, J. Liq. Chromatogr., 7, 2351 (1984).
29   P. Vrátny, U.A.Th. Brinkman and R.W. Frei, Anal. Chem., 57, 224 (1985).
30   M. Kimura and Y. Itokawa, J. Chromatogr., 332, 181 (1985).
31   A.J. Speek, J. Schrijver and W.H.P. Schreurs, J. Chromatogr., 301, 441 (1984)
32   L.D. Bowers and W.D. Bostick, Chemical Derivatization in Analytical Chemistry, Vol. II, Separations and Continuous Flow Techniques, R.W. Frei and J.F. Lawrence eds., Plenum, New York, 1982, Ch. 3.
33   P.W. Carr and L.D. Bowers, Immobilized Enzymes in Analytical and Clinical Chemistry, Wiley, New York (1980), Ch. 4
34   G.G. Guilbault, Analytical Uses of Immobilized Enzymes, Marcel Dekker, New York (1984).
35   L. Nondek, U.A.Th. Brinkman and R.W. Frei, Anal. Chem., 54, 1466 (1983).
36   L.Nondek, R.W. Frei and U.A.Th. Brinkman, J. Chromatogr., 282 (1983).
37   L.Nondek, Anal. Chem., 56 1192 (1984).
38   T.D. Schlabach and F.E. Regnier, J. Chromatogr., 158, 349 (1978).
39   H. Jansen, R.W. Frei, U.A.Th. Brinkman, R.S Deelder and R.P.J. Snellings, J. Chromatogr., 325, 255 (1985).
40   H. Jansen, E.G. van der Velde, U.A.Th. Brinkman and R.W. Frei, J. Chromatogr., 378, 215 (1986).
41   S. Okuyama, N. Kokubun, S. Higashidate, D. Uemura and Y. Hirata, Chem. Lett., 1443 (1979).
42   S. Kamada, M. Maeda, A. Tsumji, Y. Umezawa and T. Kurahashi, J. Chromatogr., 239, 773 (1982).
43   S. Hasegawa, R. Uenoyama, F. Takeda, J. Chuma and S. Baba, J. Chromatogr., 278, 25 (1983).
44   S. Hasegawa, R. Uenoyama, F. Takeda, J. Chuma, K. Suzuki, F. Kamiyama, K. Yamazaki and S. Baba, J. Liq. Chromatogr., 7, 2267 (1984).
45   M. Hayashi, Y. Imai, Y. Minami, S. Kawata, Y. Matsuzawa and S. Tarui, J. Chromatogr., 338, 195 (1985).
46   M.-C. Wu, K. Takagi, S. Okuyama, M. Ohsawa, T. Masahashi, O. Narita and Y. Tomeda, J. Chromatogr., 377, 121 (1986).
47   G. Damsa, B.H.C. Westerink and A.S. Horn, J. Neurochem., 45, 1649 (1985).
48   M. Asano, T. Miyanchi, T. Kato, K. Fujimori and K. Yamamoto, J. Liq. Chromatogr., 9, 199 (1986).
49   K. Honda, K. Miyaguchi, H. Nishino, H. Tanaka, T. Yao and K. Imai, Anal. Biochem., 153, 50 (1986).
50   P. van Zoonen, C. Gooijer, N.H. Velthorst, R.W. Frei, J.H. Wolf, J. Gerrits and F. Flentge, J. Pharm. Biomed. Anal., 5, 485 (1987).

51   C. Eva, M. Hadjuconstantiou, N.H. Neff and J.L. Meek, Anal.
      Biochem., 143, 320 (1984).
52   J.L. Meek and C. Eva, J. Chromatogr., 317, 343 (1984).
53   P. van Zoonen, I de Herder, C. Goojjer, N.H. Velthorst, R.W. Frei,
      E. Küntzberg and G. Gübitz, Anal. Lett., 19, 1949 (1986).
54   L. Dalgaard, L. Nordholm and L. Brimer, J. Chromatogr., 265, 183
      (1983).
55   V.K. Boppana, K.-L.L. Fong, J.A. Ziemniak and R.K. Lynn, J.
      Chromatogr., 353, 231 (1986).
56   L. Dalgaard and L. Brimer, J. Chromatogr., 303, 67 (1984).
57   L. Brimer and L. Dalgaard, J. Chromatogr., 303, 77 (1984).
58   R. Tawa, M. Kito and S. Hirose, Chem. Lett., 745 (1981).
59   M. Kito, R. Tawa, S. Takeshima and S. Hirose, J. Chromatogr., 231,
      183 (1982).
60   M. Kito, R. Tawa, S. Takeshima and S. Hirose, J. Chromatogr., 278,
      35 (1983).
61   M. Kito, R. Tawa, S. Takeshima and S. Hirose, Anal. Lett., 18, 323
      (1985).
62   N. Kiba and M. Kaneko, J. Chromatogr., 303, 396 (1984).
63   D.W. Taylor and T.A. Nieman, J. Chromatogr., 368, 95 (1986).
64   L. Ögren, I. Csiky, L. Risinger, L.G. Nilsson and G. Johansson,
      Anal. Chim. Acta, 117, 71 (1980).
65   J.L. Meek and F. Nicoletti, J. Chromatogr., 351, 303 (1986).
66   L.D. Bowers and P.R. Johnson, Biochim. Biophys. Acta, 661, 100
      (1981).
67   J.F. Studebaker, S.A. Slocum and E.L. Lewis, Anal. Chem., 50 1500
      (1978).
68   J.F. Studebaker, J. Chromatogr., 185, 497 (1979).
69   M.-C. Millot, B. Sebille and J.-P. Mahieu, J. Chromatogr., 354, 155
      (1986).
70   P. Vrátny, J. Ouhrabková and J. Capiková, J. Chromatogr., 191, 313
      (1980).
71   P. Vrátny, R.W. Frei, U.A.Th. Brinkman and M.W.F. Nielen, J.
      Chromatogr., 295, 355 (1984).
72   K.-S. Low, U.A. Th. Brinkman and R.W. Frei, Anal. Lett., 17, 915
      (1984).
73   E.P. Lankmayr, B. Maichin, G. Knapp and F. Nachtmann, J.
      Chromatogr., 224,239 (1981).
74   K.W. Sigvardson and J.W. Birks, J. Chromatogr., 316, 507 (1984).
75   I.S. Krull, K.-H. Xie, S. Colgan, U. Neue, T. Izod, R. King and B.
      Didlingmeyer, J. Liq. Chromatogr., 6, 605 (1983).
76   I.S. Krull, S. Colgan, K.-H. Xie, U. Neue, R. King and B.
      Bidlingmeyer, J. Liq. Chromatogr., 6, 1015 (1983).
77   K.-H. Xie, S. Colgan and I.X. Krull, J. Liq. Chromatogr., 6, 125
      (1983).
78   K.-H. Xie, C.T. Santasania, I.S. Krull, U. Neue, B. Bidlingmeyer and
      A. Newhart, L. Liq. Chromatogr., 6, 2109 (1983).
79   J. Rüter, U.P. Kurz and B. Neidhart, J. Liq. Chromatogr., 8, 2475
      (1985).
80   J. Rüter, U.P. Fislage and B. Neidhart, Chromatographia, 19, 62
      (1984).
81   K. Brunt, Anal. Chem., 57, 1338 (1985).
82   H. Irth, G.J. de Jong, U.A.Th. Brinkman and R.W. Frei, J.
      Chromatogr., 370, 439 (1986).
83   H. Jansen, C.J.M Vermunt, U.A.Th. Brinkman and R.W. Frei, J.
      Chromatogr., 366, 135 (1986).
84   H. Jansen, R. Jansen, U.A.Th. Brinkman and R.W. Frei,
      Chromatographia, in press.

258

85   J.R. Poulsen, J.W. Birks, P. van Zoonen, C. Gooijer, N.H. Velthorst and R.W. Frei, Chromatographia, 21, 587 (1986).
86   W.Th. Kok, U.A. Th. Brinkman and R.W. Frei, J. Chromatogr., 256, 17 (1983).
87   W.Th. Kok, G. Groenendijk, U.A. Th. Brinkman and R.W. Frei, J. Chromatogr., 315, 271 (1984).
88   W.P. King and P.T. Kissinger, Clin. Chem., 26, 1484 (1980).
89   W.Th. Kok, U.A. Th. Brinkman and R.W. Frei, Anal. Chim. Acta, 162, 19 (1984).
90   W.Th. Kok, J.J. Halvax, W.H. Voogt, U.A. Th. Brinkman and R.W. Frei, Anal. Chem., 57, 2580 (1985).
91   W.Th. Kok, W.H. Voogt, U.A.Th. Brinkman and R.W. Frei, J. Chromatogr., 354, 249 (1986).
92   W.Th. Kok, Th.C.H. van Neer, W.A. Traag and L.G.M.Th. Tuinstra, J. Chromatogr., 367, 231 (1986).
93   W.Th. Kok, R.W. Frei and U.A. Th.Brinkman, Selective Sample Handling and Detection (Editors: R. W. Frei and K. Zech), Vol. 1, Elsevier, Amsterdam (1988).
94   J.R. Poulsen, J.W. Birks, G. Gübitz, P. van Zonnen, C. Gooijer, N.H. Velthorst and R.W. Frei, J. Chromatogr., 360, 371 (1986).
95   I.S. Krull and W.R. LaCourse in Reaction Detection, Liquid Chromatography, I.S. Krull ed., Marcel Dekker, New York, 1987, Ch. 7.
96   M.F. Lefevere, R.W. Frei, A.H.M.T. Scholten and U.A.Th. Brinkman, Chromatographia, 15, 459 (1982).
97   M.S. Gandelman, J.W. Birks, U.A. Th. Brinkman and R.W. Frei, J. Chromatogr., 282, 193 (1983).
98   R. Verbeke and P. Vanhee, J. Chromatogr., 265, 239 (1983).
99   C. de Ruiter, J.F. Bohle, G.J. de Jong, U.A.Th. Brinkman and R.W. Frei, Anal. Chem., in press.
100 H. Scholl, K. Schmidt and B. Weber, J. Chromatogr., 416, 321 (1987).
101 J.N. LePage and E.M. Rocha, Anal. Chem., 55, 1360 (1983).
102 R.P.W. Scott, J. Chromatogr., Sci., 18, 49 (1980).
103 M. Novotny, Anal. Chem., 53, 1294A (1981).
104 F.J. Yang, HRC&CC, 6, 348 (1983).
105 P. Kucera, ed., Microcolumn High-Performance Liquid Chromatography, Elsevier Scientific Publishing Company, Amsterdam, 1984.
106 R.P.W. Scott, ed., Small Bore Liquid Chromatography Columns: Their Properties and Uses, John Wiley & Sons, New York, 1984.
107 H. Jansen, U.A. Th. Brinkman and R.W. Frei, J. Chromatogr. Sci., 23 279 (1985).
108 A. Hirose, Y. Iwasaki, I.Iwata, K. Ueda and D. Ishii, HRC&CC, 4, 530 (1981).
109 P. Kucera and H. Umagat, J. Chromatogr., 255, 563 (1983).
110 J.A. Apffel, U.A. Th. Brinkman and R.W. Frei, Chromatographia, 17, 125 (1983).
111 T. Takeuchi, S. Saito, D. Ishii, J. Chromatogr., 258, 125 (1983).
112 T. Takeuchi and D. Ishii, HRC&CC, 6, 571 (1983).
113 H. Jansen, U.A.Th. Brinkman, R.W. Frei, Chromatographia, 20, 453 (1985).
114 K. Miyaguchi, K. Honda and K. Imai, J. Chromatogr., 316, 501 (1984).
115 K. Miyaguchi, K. Honda, T. Toyo'oka and K. Imai, J. Chromatogr., 352, 255 (1986).
116 H.P.M. van Vliet, G.J.M. Bruin, J.C. Kraak and H. Poppe, J. Chromatogr., 363, 187 (1986).
117 T.S. Stevens, J.C. Davis and H. Small, Anal. Chem., 53, 1488 (1981).
118 T.S. Stevens, G.L. Jewett, R.A. Bredeweg, Anal. Chem., 54, 1206 (1982).
119 J. Riviello and C.A. Pohl, Dionex Technical Note 1983 and 1984, Pittsburgh Conference.

120 R.M. Cassidy and B.D. Karcher, Reaction Detection in Liquid Chromatography, I.S. Krull ed., Marcel Dekker, New York, 1987, Ch. 3.
121 D.T. Gjerde and J.S. Fritz, Ion Chromatography, 2nd Edition, Dr. Alfred Hüthig Verlag, Heidelberg, 1987
122 J.C. Davis, D.P. Peterson, Anal. Chem., 57, 771 (1985).
123 J.Haginaka, J. Wakai and H. Yasuda, Anal. Chem., 59, 324 (1987).
124 J. Haginaka and J. Wakai, J. Chromatogr., 390, 421 (1987)

CHAPTER VI

NEW LUMINESCENCE DETECTION TECHNIQUES

C. GOOIJER, N.H. VELTHORST and R.W. FREI

1.   INTRODUCTION

The use of luminescence detection techniques in HPLC has become increasingly popular over the years. Among these methods, the most frequently applied method is fluorescence detection due to its sensitivity (refs. 1-3). Furthermore it can be nicely adapted to the demands of miniaturized HPLC especially with the help of lasers as excitation sources (laser induced fluorescence, LIF). Unfortunately, the number of compounds exhibiting intense native fluorescence is limited. That is the reason why chemical reactions play an important role in HPLC fluorescence detection: non- or weakly fluorescent compounds are converted in a pre- or post-column mode into highly fluorescent products. This can be done by chemical derivatization, ion-pair formation or the use of a photochemical reactor (ref. 4).

A well known example of chemical derivatization is the reaction of o-phthalaldehyde (OPA) with primary amino functions in alkaline media in the presence of the strong reducing agent 2-mercaptoethanol. This reaction has been applied to the detection of amino acids and primary amines in biological fluids (refs. 5-8). During the past decade a large number of reagents has been introduced for the reaction detection of a wide variety of compounds. Books and reviews on this subject are numerous (refs. 9-17). With the ion-pair formation method a non-fluorescing analyte forms an ion-pair with a highly fluorescent counter-ion. Subsequently, the apolar ion-pair is extracted from the aqueous mobile phase to an apolar phase via solvent segmentation. The method can be used for basic and acidic analytes under controlled pH conditions. Several tertiary amine and quaternary ammonium type drugs (refs. 18-20), detergents (ref. 21) and alkyl-sulphates and sulphonates (ref. 22) have been determined. With a photochemical reactor compounds are irradiated post-column in a quartz or teflon coil by UV light and converted to fluorescent reaction products (refs. 23, 24). Most applications have been in the pharmaceutical area.

In principle fluorescent analytes can also be detected via chemiluminescence (CL) or bioluminescence (BL), provided that there are chemical reactions which efficiently bring them in the electronically excited state. A lot of research has been done in this field because in chemiluminescence the light source can be eliminated from the detection system (refs. 3, 25, 26). In principle, this would be very nice since the ultimate detectability in fluorescence detection is limited by background light and its noise coming from the excitation source (ref. 27). For a number of compounds impressive results have been reported. However, it is obvious that not in all cases the CL detection limits are significantly more favourable than those obtained via fluorescence detection. In CL detection generally background due to competitive reactions or impurities for instance in the eluent determine the detection limits that can be reached (ref. 28).

In the present paper the main attention is to the detection of non-fluorescent compounds via luminescence techniques other than fluorescence, i.e., phosphorescence and chemiluminescence (Fig. 1). Regarding chemiluminescence detection in liquid chromatography, especially the so-called peroxyoxalate reaction has got a lot of attention in the literature: since the first paper of Kobayashi and Imai in 1980 (ref. 29) more than 100 papers appeared on this subject. A major constraint in applying CL in routine analysis is the relative complexity of the equipment despite of the fact that the detector itself is quite simple. In most cases more than two reagents have to be added to the analyte before CL can be observed, so that problems due to mixing and to limited solubility and stability of reagents may arise. In this chapter the solid-state reactor approach to avoid these drawbacks is discussed. Use will be made of solid oxalate (refs. 30, 31), immobilized fluorophore (ref. 32) and immobilized enzymes (ref. 33, 34). Illustrative for the eventual applicability of the method is the detection of choline and acetylcholine in complex biological matrices (as undiluted urine and the deproteinated serum) without sample pretreatment (ref. 34).

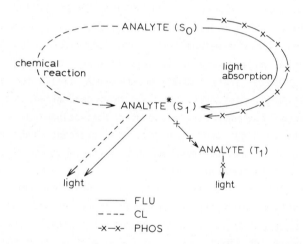

Fig. 1   Simplified diagram showing the Fluorescence, Chemiluminescence and Phosphorescence transitions. S_0 and S_1 are the lowest electronic singlet states and T_1 is the lowest electronic triplet state of the analyte.

Unlike fluorescence and chemiluminescence, phosphorescence in the liquid state is a rare phenomenon. At a first sight, this fact severely limits the potential applicability of phosphorescence detection in HPLC. Nevertheless, interesting progress has been advanced following two approaches (ref. 35). The first makes use of organized media, such as micellar solutions to extend the number of compounds that do emit phosphorescence. Such media are interesting since they become increasingly important in chemical separation problems. The second approach which will be discussed extensively in this chapter is focussed on phosphorescence of normal liquids which can be utilized as an indirect detection method (refs. 2,36,37). Two modes can be distinguished, i.e., the sensitized (refs. 38-41) and the quenched mode (refs. 42-50). In most applications developed thus far, the phosphorophore biacetyl (2,3-butanedione) is added to the eluent which needs to be deoxygenated. Compounds able to quench the biacetyl phosphorescence cause a decrease of the monitored signal. The amount of quenching is dependent on the analyte, which implies that the method has an inherent selectivity in ion chromatography, since the UV absorption properties of the analytes do not play any role in the quenching process. Furthermore, unlike indirect UV absorption- and fluorescence detection in ion chromatography, quenched phosphorescence is not based on the displacement of eluent ions by analyte ions. Instead it is a dynamic process: the efficiency of the phosphorescence is reduced by the analyte. This difference has interesting consequences.

The sensitized phosphorescence technique can be used for analytes that do absorb UV radiation but do not fluoresce because they decay to the first excited triplet state. In presence of biacetyl triplet-triplet energy tranfer from the analyte to biacetyl takes place after which the phophorescence of the latter is monitored. The sensitized phosphorescence method should be considered complementary to UV and fluorescence detection.

In this chapter the applicability of both sensitized and quenched phosphorescence will be shown, in the organic as well as the inorganic field. An interesting aspect is the relative ease compared to fluorescence to utilize time resolution to improve the ratio of the phosphopescence signal and the background radiation. This is not too complicated since the liftimes of the phosphorescence signal in the systems under consideration are in the 0.1-1 msec range. Furthermore alternative phosphorophores to overcome some of the disadvantages of biacetyl will be treated. The main disadvantages are that biacetyl is part of the eluent and, more important, that the eluent should be deoxygenated. In one alternative method use is made of an immobilized phosphorophore packed in the detector cell (ref. 49). In another approach, rare earth ions as Eu^{3+} and Tb^{3+} are utilized as luminophores (ref. 50). They emit long-living luminescence in liquid solutions without the need of deoxygenation. Although in a strict sense luminescence of rare earths is no phosphorescence, it is appropriate to discuss this technique here because there are various experimental similarities.

2. CHEMILUMINESCENCE DETECTION WITH SOLID STATE REACTORS

2.1 DETECTION BASED ON CL AND BL

Various chemiluminescence (CL) and bioluminescence (BL) reactions have been and are applied in analytical chemistry (ref. 25). Of course they commonly require more or less defined reaction conditions which generally are not equivalent with the optimal chromatographic circumstances in LC. Hence for the development of CL (and BL) as detection technique in HPLC, compatability is a crucial point.

Within this context three CL reactions are the most important: the (iso)luminol reaction, the lucigenine reaction and the peroxyoxalate reaction; these will be discussed more extensively. In addition, incidentally use has been made of the firefly BL reaction, i.e., for the isoenzymes of creatine kinase (ref. 51) and of the luciferase bacterial BL reaction, i.e., for bile acids producing NADH from NAD, which can be measured via bacterial CL (ref. 25). Furthermore, an areosol spray detector for LC based on ozon and singlet oxygen induced CL has been reported (refs. 52, 53), a cell for electrogenerated CL (ref. 54) and a thermal energy analyzer CL detector (refs. 55, 56).

To discuss the potential of the (iso)luminol, the lucigenine and the peroxyoxalate reaction for detection in LC it should be realized that the first two reaction types differ essentially from the third one. Whereas for luminol and lucigenine an energy-rich intermediate is formed in the early stage of the reaction which itself emits light, for peroxyoxalate the formation of such an intermediate is followed by energy transfer to a substance that eventually emits light. This implies that the peroxyoxalate in principle can be invoked to detect compounds with native fluorescence or carrying fluorescent labels. Detection of fluorescent compounds via the (iso) luminol or the lucigenine reaction is not appropriate.

The (iso)luminol reaction is schematically depicted as follows:

luminol : $R_1 = NH_2$ $R_2 = H$

isoluminol : $R_1 = H$ $R_2 = NH_2$

+ light (425 nm)

The reaction requires strong alkaline conditions; the (iso)luminol reacts with an oxidant while metal ions catalyze the reaction. This implies that in principle it can be used to detect oxidants (possibly generated in a post-column reaction detector), metal ions (and metal chelating agents) and substances derivatized with (iso)luminol. That catalyzing metal ions (and indirectly analytes that reduce the free metal ion concentration) can be detected is readily clarified. In chemiluminescence the number of emitted photons per second is measured which is proportional to the number of highly energetic indermediate compounds formed pro second, in other words to the reaction rate. The luminol CL reaction has been frequently applied in analytical chemistry especially in flow injection analysis, e.g. for various catalyzing metal cations (refs. 57-60), amino acids and proteins acting as metal chelating agents thus suppressing the CL intensity (refs. 61-63) and hydrogenperoxide produced by enzymes (refs. 64-69). For HPLC detection only few applications of luminol CL have been published. Metal ions have been determined after separation by ion exchange (ref. 70); here compatability is a serious problem since metals are usually separated under acidic conditions whereas the luminol CL requires high pH values. The CL detector has been combined with a photochemical reactor inducing photooxygenation of aliphatic alcohols, aldehydes, ethers and saccharides under releasing of H_2O_2 (ref. 71). Finally, the alkyl substituted isoluminol, N-(4-aminobutyl)-N-ethyl-isoluminol, has been successfully used for precolumn labeling of amines and caroboxylic acids (ref. 72).

The lucigenine CL reaction also requires strong basic solutions. Lucigenine reacts both with oxidants like sodium periodate or hydrogen peroxide as with reductants like ascorbic acid and glucose. In both situations the emitting species is N-methylacridone:

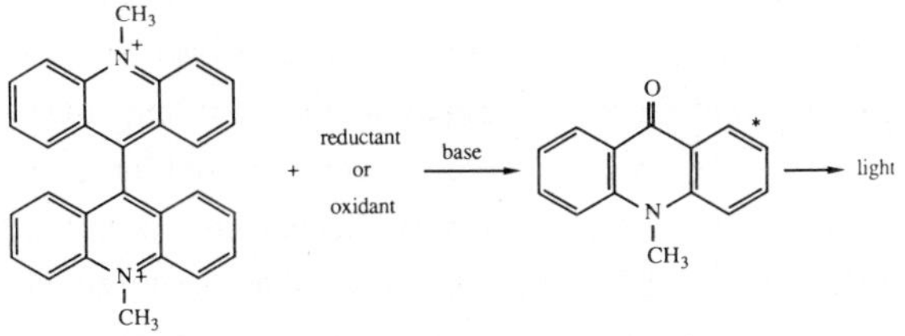

Hence it is possible to utilize lucigenine CL to measure both oxidants and reductants. For detection in HPLC it has been used for ascorbic and dehydroascorbic acid, for glucose, creatinine, heparin and steroids as cortisol (refs. 73-75). Furthermore, carboxylic acids have been measured indirectly after conversion to the p-nitrophenacyl esters (ref. 29).

Obviously the peroxyoxalate CL reaction is applied by far the most extensively for detection in HPLC. Traditionally the reaction is presented in the following way, although results of various groups unambiguously show that this presentation cannot account for all experimental data (refs. 76-79):

$$
Ar-O-\underset{O}{\overset{O}{\underset{\|}{C}}}-\underset{O}{\overset{O}{\underset{\|}{C}}}-O-Ar + H_2O_2 \longrightarrow
\begin{bmatrix}
\underset{O}{\overset{O}{\underset{\|}{C}}}-\underset{O}{\overset{O}{\underset{\|}{C}}} \\
\end{bmatrix}
+ 2\,ArOH
$$

$$
\begin{bmatrix}
\underset{O}{\overset{O}{\underset{\|}{C}}}-\underset{O}{\overset{O}{\underset{\|}{C}}} \\
\end{bmatrix}
+ \text{fluorophore} \longrightarrow \text{fluorophore}^* + 2CO_2
$$

$$\downarrow$$

light

Ar represents a substituted benzene nucleus: the most frequently used oxalate is TCPO, bis(2,4,6-trichlorophenyl) oxalate. Thusfar the peroxyoxalate CL reaction has been applied in HPLC for fluorescent or fluorescence-labeled compounds, for hydrogenperoxide (other oxidants are not appropriate) and for CL quenching analytes.

The CL intensities appear to vary strongly with the fluorophore applied in the reaction so that not in all cases CL detection is more favourable than fluorescence detection. Furthermore detectability is generally limited by background luminescence, i.e., chemiluminescence due to impurities or competitive reactions observed when no fluorophore is added to the eluent. Interesting results have for instance been reported for polycyclic aromatic hydrocarbons (refs. 80-82) but also for dansyl-labeled compounds as amino acids, amines, catecholamines and steroids (refs. 28, 29, 83-88). Furthermore, likewise dansylation, other labeling reactions originally developed for fluorescence detection have been examined as the reactions with OPA, orthophthalaldehyde, and NBD, 7-nitrobenzo-2-oxa-1,3-diazole (refs. 86, 88). An approach specifically directed on CL is the use of amino-substituted aromatics which have extremely high CL efficiencies for labeling carboxylic acids (ref. 89), aldehydes and ketones (ref. 90).

2.2 SOLID STATE REACTORS IN CL

Solid state reactors have been introduced in CL detection for various reasons. In the first place they can be used to extent the applicability of the CL reaction under consideration. An example is the Zn-reductor transferring nitro-PAHs to amino-PAHs; whereas the former have very low CL efficiencies in the peroxyoxalate reaction, the latter can be detected extremely selective and sensitive in complex samples (ref. 81). Also enzymes can be immobilized for this purpose, as for instance immobilized enzyme reactors (IMERS) producing H_2O_2 that in turn can be measured via CL (refs. 91, 92). Thus various important substrates can in principle be analyzed; successful reports have been published for glucose (refs. 33, 65) and choline/acetylcholine (refs. 34, 93). Alternatively, IMERS can be utilized to enhance the CL intensity via the reaction rate. As an example we mention immobilized oxidase catalyzing the luminol reaction (ref. 94). Obviously, a very favourable property of both reactor types is that they in principle can be used during long time periods.

Nevertheless as a third possibility of applying solid-state reactors also immobilized reagents that are consumed have been used, for instance immobilized luminol (ref. 95). Because of the large surface area available on small solid particles this offers a way to employ these reagents at much higher effective concentrations than allowed by solubilities. In our laboratory we have utilized a packed bed-type reactor for TCPO in the peroxyoxalate CL reaction to simplify the experimental set-up and to reduce instability problems (ref. 26). This compound being poorly soluble in the solvents usually applied for peroxyoxalate CL, is slowly dissolved in the carrier stream. In some cases immobilized CL reagents are used to quantitate species that quantitatively liberates the CL reagent from its support (refs. 96, 97). As an example a detection system for thiols has been developed: a thiol-modified luminol is bound on polysaccharide but released after thiol-disulfide interchange.

Finally, it is emphasized that in the peroxyoxalate CL the luminescing compound can be successfully bound on controlled pore glass or silica. It is not consumed during the reaction. In our laboratory we have immobilized 3-aminofluoranthene, one of the most efficient CL reagents in the peroxyoxalate system known at the moment (ref. 32). Immobilization is important since this compound has carcinogenic properties. Another favourable point is that the CL reaction can be localized in the detector cell which is packed with immobilized fluorophore.

2.3 H_2O_2 DETECTION BY PEROXYOXALATE CL

It is worthwhile to improve H_2O_2 detection methods in FIA and to develop detection methods in HPLC not only because H_2O_2 is an important analyte itself, but also because a coupling can be made with photochemical (refs. 98, 99) and enzymatic reactions (refs. 34, 92, 93). Development of such coupling techniques is not only important concerning sensitivity; in some cases the selectivity parameter is even more important. For example, it is the selectivity of the IMER CL combination that allows the quantitation of choline and acetylcholine in extremely complex mixtures (refs. 34). Furthermore, because of this selectivity the additional band broadening caused by the reactors is less important than usual in HPLC detection.

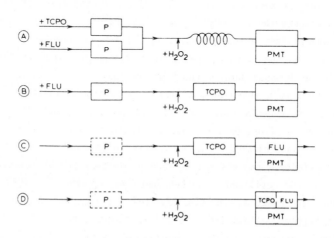

Fig. 2   Detector configurations investigated: (A) conventional system; (B) system with solid-state TCPO reactor; (C) system with separate TCPO and immobilized fluorophore reactors; (D) mixed reactor system (FLU = fluorophore, P = pump and PMT = photomultiplier tube).

2.3.1 THE TCPO REACTOR (refs. 30,31)

The significant reduction in complexity of H_2O_2 detection which can be achieved by applying solid-state reactors is readily conceived from Fig. 2: manifold A is the conventional one; in B a (hand packed) solid-state TCPO reactor is used, in C and D both a solid-state TCPO reactor and a reactor with 3-aminofluoranthene immobilized on controlled pore glass (CPG) are applied.

In B and C the TCPO reactor is the same. It is a precolumn of the type designed in our lab by Goewie et al (ref. 100) using a teflon coating cartridge of 4.6 mm I.D. and 22 mm length. Various experimental parameters for its use in FIA have been evaluated on the basis of configuration B, utilizing perylene as fluorophore and aqueous acetonitrile as eluent. Parameters as reagent purity and acetonitrile to water ratio play an important role. To achieve a background signal as low and as stable as possible, special attention has to be paid to reagent purity and to the acetonitrile to water ratio. Optimal results were obtained by passing acetonitrile over an alumina column prior to use and furthermore by adjusting the composition of H_2O_2 injection plug to the composition of the carrier stream as much as possible. Detection limits of 6×10^{-9} M H_2O_2 (0.2 $\mu g/l$) were achieved combined with a linear dynamic range up to 10^{-5} M and a good reproducibility (R.S.D. 2.8% at 2×10^{-7} M for 10 injections). Large injection volumes were required using the rather big TCPO reactor. In acetonitrile/water 80:20 this big reactor could be operated at least 8 hours. Unfortunately, large sample injection volumes are required which indicates that the reactor has a large dead volume so that it cannot be used in series with the analytical column in an HPLC system. To reach the maximum CL signal in FIA an injection volume of at least 650 μl was necessary.

Obviously for combining the H_2O_2 detection system with HPLC a smaller TCPO reactor is required to circumvent desastrous band broadening. From this point of view configuration D is more appropriate than configuration C. Here, a dual layer detector cell is applied of typically 3 mm internal diameter and 27 mm length; about 1/3 of the cell is filled with TCPO. Since TCPO is consumed during the reaction it is appropriate to place a frit between the TCPO and the luminophore layer. Also an inlet frit is necessary to spread the flow evenly over the whole reactor. The two layer cell can be used during ca. 3 h continuous operation at flow rates of about 1 ml min^{-1} (acetonitrile/water 80:20) without excessive additional band broadening caused by voids in the TCPO layer. Repacking with TCPO is extremely simple. In FIA experiments applying manifold D, maximum CL signals were reached at injection volumes not larger than 100 to 150 μl. Presumably this is not only due to the small size of the TCPO reactor, but also to the localization of the CL reaction (the light emission) in the detector cell. The latter implies that no band-broadening is caused by detection of CL occurring in the inlet and outlet capillaries of the flow cell.

Another important difference between the experimental configurations C and D is their dependence on flow rate (for 80% acetonitrile as a model eluent). In C there is a significant dependence, whereas in D it is negligible. Most probably this is due to the time lag in C between the reaction of hydrogen peroxide with TCPO in the first reactor and the actual excitation step in the second reactor. During this time interval some of the intermediate formed in the first reactor decomposes, so that the amount of intermediate that reaches the second reactor depends on the flow rate.

To test the solvent compatability of the TCPO reactor, in addition to acetonitrile, tetrahydrofuran (THF), acetone and methanol were studied all in mixtures of 20% aqueous Tris buffer and 80% organic solvent. THF is unsuitable; it gives a very high CL background, presumably because of peroxide formation. Acetone can be applied without any problem; alike acetonitrile the CL signal is not influenced by the flow rate. This is not true for methanol; in 80% aqueous methanol (even in manifold D) a strong flow-rate dependence is observed that varies from one TCPO batch to another. It has been shown that this dependence can be eliminated by addition of 2,4,6-trichlorophenol (TCP) which is a precursor in the synthesis of TCPO. It has been suggested that in methanol TCPO undergoes side reactions in competition with the $TCPO/H_2O_2$ reaction; if a reproducible TCP concentration is maintained, the side reactions are controlled and the flow dependence is suppressed. Concerning the optimal pH, for acetonitrile and acetone there is no significant effect on the signal between pH 7.5 and 10; in methanol (apart from the role of TCP) the flow rate dependence at pH 7.5 is significantly lower than at pH 9.5.

The flow independence of configuration D (apart from methanol) and the low dead volume of the two-layer cell implies that it can be appropriately applied in HPLC. On the other hand it has been successfully applied for H_2O_2 quantitation in rain and cloud water samples by FIA; manual injections can be readily performed because of its low back pressure. For this type of samples field monitoring is essential because very dilute aqueous peroxide samples are unstable. H_2O_2, which is believed to be one of the key intermediates in the atmospheric decomposition cycle of the sulphur oxides causing acid rain, is present in rain water in concentrations varying from about 1 μg l^{-1} (3×10^{-8} M) in polluted areas to more than about 1 μg ml^{-1} (3×10^{-5} M) in relatively clean areas. The present quantitation method based on the TCPO-CL system in Fig. 2D gives a detection limit of 1.5×10^{-8}M and a linearity over 6 orders of magnitude; the R.S.D. is 3% (at 17 μg l^{-1}) and by manual injection 40 samples can be analyzed per hour.

To sum up the results for the TCPO reactor described in this section, it will be obvious that the two layer cell can be combined easily with HPLC. Unfortunately, it can only be used during a limited time period, typically 3 hours, but repacking is quite easy. It is noted, however, that also the bigger reactor (as used in B and C) can be fruitfully applied in HPLC if it is placed in a separate flow line, not in line with the analytical column. In a set-up of this type, formation of voids in the

TCPO reactor is not so critical. Frequently the solid finely ground TCPO is mixed with glass beads of 40-80 μm diameters to reduce backpressure.

2.3.2 IMMOBILIZED FLUOROPHORE (ref. 32)

It is obvious from the simplified reaction scheme for peroxyoxalate CL (see section 2.1) that the fluorophore is not consumed during the reaction. From this point of view, immobilization is quite appropriate. With respect to the choice of the fluorophore, it is well-known from CL measurements on fluids solutions that 3-aminofluoranthene (3-AF) is one of the most efficient fluorophoric compounds (ref. 81). Immobilization of this compound is not only useful in order to extent the freedom of solvent choice (the solubility of the fluorophore is no longer a limiting factor), but especially because of its toxic properties.

Various factors determine the CL signal to noise ratio that can be reached in a peroxyoxalate system utilizing immobilized 3-AF. First of all, after coupling to the solid support the electronic structure of the fluorophore should not have changed basically. Only if this condition is fulfilled a CL efficiency approaching that of liquid state 3-AF will be attainable. Secondly, the coupling to the support must be realized by a spacer, thus creating a pseudo-liquid solution for 3-AF. Thirdly, the solid support itself should be applicable under HPLC conditions without any problems. Cellulose for instance is not suitable, because it swells in aqueous acetonitrile and methanol thus causing back pressure problems in a HPLC set-up. Finally, on the one hand a high surface coverage should be realized to enhance the CL signal, while on the other hand the transparancy of the system for the outcoming luminescence light should not be reduced.

A successful immobilization procedure for 3-AF on controlled pore glass (CPG) and silicagel developed by Gübitz et al. (ref. 32) is schematically given in Fig. 3. In the first step a reaction with 3-glycidoxypropyltrimethoxysilane in toluene is performed producing a support carrying long chains ending with an epoxide group. In the second step the epoxide group reacts with the amino group of 3-AF. Although for silica higher surface coverages have been found, the highest CL signals have been obtained for CPG. Using CPG glass beads of 200-400 mesh, good stability over long periods was obtained.

Poulson et al. have compared the CL efficiencies of immobilized 3-AF (on glass beads) and liquid state rubrene (ref. 98). From results of Sigvardson et al. (ref. 81) on liquid state efficiencies of both fluorophores one would expect a significant higher efficiency for 3-AF.

However, for immobilized 3-AF the difference is only 20 per cent, indicating that immobilization affects the role of 3-AF in peroxyoxalate CL in a negative way. Nevertheless, despite of this reduction the efficiency of immobilized 3-AF is very favourable in comparison to other fluorophores.

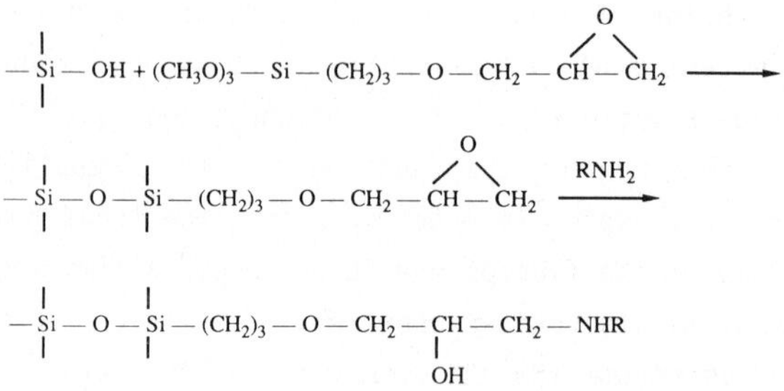

$$-\overset{|}{\underset{|}{Si}} - OH + (CH_3O)_3 - Si - (CH_2)_3 - O - CH_2 - \overset{\overset{O}{\diagup \diagdown}}{CH} - CH_2 \longrightarrow$$

$$-\overset{|}{\underset{|}{Si}} - O - \overset{|}{\underset{|}{Si}} - (CH_2)_3 - O - CH_2 - \overset{\overset{O}{\diagup \diagdown}}{CH} - CH_2 \quad \overset{RNH_2}{\longrightarrow}$$

$$-\overset{|}{\underset{|}{Si}} - O - \overset{|}{\underset{|}{Si}} - (CH_2)_3 - O - CH_2 - \underset{\underset{OH}{|}}{CH} - CH_2 - NHR$$

Fig. 3   Immobilization of 3-aminofluoranthene (R-NH$_2$) on silicagel and glass beads, derivatized with 3-glycidoxypropyltrimethoxysilane.

2.3.3 COUPLING WITH PHOTOCHEMICAL REACTORS (refs. 98, 99)

Poulson et al. have coupled the peroxyoxalate CL system with a post-column photochemical reactor producing hydrogenperoxide for detection in HPLC. It is emphasized that the photochemical reactor requires methanol, a quite interesting point in view of the little stability of TCPO in this solvent (see section 2.3.1). The analytes were quinones, commonly used in the wood pulp industry. The reaction is initiated by a protontransfer from methanol to the excited quinone. Then, in presence of oxygen various reaction pathways can be followed (as is visualized in Fig. 4) all eventually producing H$_2$O$_2$, while quinone is not consumed. In that sense the reaction is a photocatalytic one; in practice up to 100 H$_2$O$_2$, molecules are produced for any analyte molecule. Regarding selectivity it is noted that only a limited class of compounds undergoes the above type of photooxygenation. Thus photochemical reaction detection for quinones has an inherent selectivity, a clear advantage over conventional UV detection.

In a first study the two-layer TCPO reactor (see section 2.3.1) was simply placed in the main stream of the analytical system. As a photochemical reactor PTFE (polytetrafluorethylene "teflon") tubing crocketed into cylinders that fit over a fluorescent poster ("black") lamp was applied; tubing of different lengths provides different residence times for the analytes.

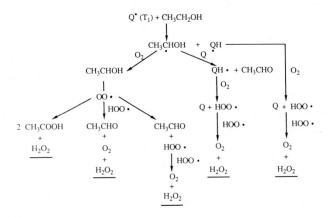

Fig. 4   Possible photochemical reaction pathways for quinones (Q) in the presence of alcohols and oxygen leading to H_2O_2 (refs. 98, 99).

To discuss the performance of this detection system, one must consider that the peaks are already broadened by the post-column photochemical reactor so that addition of the CL detector cell causes relatively little extra reduction in chromatographic resolution. Furthermore, it should be remembered that for real samples some loss of separation efficiency due to band broadening caused by the detector is less important if the detection system is more selective; if the detector shows only a response for a few compounds, separation problems can more easily be solved. Of course the band broadening caused by the photochemical reactor depends on the residence time. Poulson et al. applied a flow rate of 0.76 ml min^{-1} methanol (95%); then for a reactor of 9.8 m long the residence time is 69 s and a band broadening of 600 μl^2 was observed, while for a 29 m loop reactor with a residence time of 187 s the broadening was 1400 μl^2.

Under these conditions the contribution of the TCPO/3-AF detector cell (see Fig. 2D) to the band broadening is 4000 μl^2. Though such a contribution is fairly high, good chromatograms could be obtained. In fact, Poulson et al. only lost a factor of 5 in sensitivity compared to the complicated three pump liquid-phase system (see Fig. 2A).

Unfortunately, in the set-up under consideration, after about 2.5 hours of use the broadening caused by the dual-layer detector cell becomes more serious. Since the photochemical reactor requires high methanol concentrations, after that time void formation in the TCPO layer occurs and the dead volume and thus the peak variance increases significantly with use. Longer lifetimes are observed at lower methanol contents, but this is associated with a strong reduction of the CL signal.

To avoid these band broadening problems, in a subsequent study the TCPO reactor was placed outside the path of the analytical column effluent (see Fig. 5). Obviously in this set-up the constant contribution of the immobilized fluorophore cell to the peak variance remains: in a 3 cm x 1.5 mm I.D. cell packed with 40-80 μm CPG glass beads (derivatized or underivatized) the peak variance is 1200 μl^2. Besides it is noted that the variance caused by the fluorophore cell would be reduced strongly if smaller glass beads were applied.

Since the analyte does not pass through the TCPO reagent addition bed, the lifetime of the bed can be extended by increasing its capacity without causing an increase in peak variance. In practice a bed of 4.6 mm x 4 cm was used, packed each day with a mixture of finely ground TCPO and 40-80 μm glass beads, 70:30 by weight. Depending on the desired lifetime, the TCPO charge varied from 50-250 mg; the remaining space at the inlet side of the reactor was simply filled with glass beads. The glass beads improve the flow characteristics, reduce back pressure, and siplify packing by decreasing the static electrical charge on the solid TCPO particles.

Two configurations were developed to investigate the possibilities of TCPO addition from off-path solid reagent beds. In the first, a dual-pump design (Fig. 5a), choice of the solvent delivered to the reagent bed does not affect the HPLC separation. With this system the solvent dependences of the photochemical reaction and chromatography are isolated from the chemiluminescent response, facilitating the optimization of the reagent addition conditions. In the second, TCPO is solubilized from the reagent bed without any additional reagent pump. To accomplish this, the flow from the HPLC mobile phase pump is split into a reagent addition- and a chromatographic stream (Fig. 5b). Flow splitting reduces equipment demands at the expense of flexibility in choice of reagent addition solvent. The flow rate through the reagent bed is an important parameter in the optimization of these systems. Concentration of TCPO in the detector flow cell is determined by its solubility in the reagent bed solvent, the relative flow rates of the two solvent streams, and the rate of reagent decomposition in the solvent. Residence time of the reagent and the analyte in the detector cell decreases as the total flow rate increases. Peroxyoxalate chemiluminescence is long lived relative to the residence time in the HPLC detector cell so that the efficiency of light collection will be reduced by high total flow rates.

In Table I some detection limits obtained with the dual pump system are compared with those obtained with liquid-phase TCPO addition, combined with the immobilized 3-AF detector cell or with liquid-phase rubrene as fluorophore. It is obvious that the results obtained with solid-state TCPO addition combined with immobilized AF compare favourably with the other data.

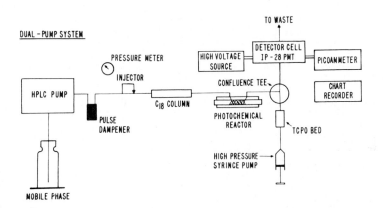

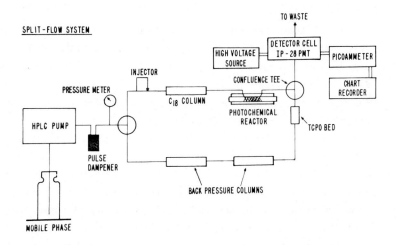

Fig. 5   Schematic diagram of the dual-pump system (a) and the split-flow system (b) (ref. 99). Further details see text.

TABLE I Dual-pump system detection limits (S/N = 3) in picomoles on
column. Solvent dependence and comparison to liquid phase
reagent addition (refs. 98, 99).

	Solid state TCPO addition with immobilized fluorophore			Liquid phase addition of TCPO	
Compound phase	Reagent addition solvent			Immobilized fluorophore (3-AF)	Liquid fluorophore (rubrene)
	100% CH$_3$OH	80% CH$_3$OH	100% CH$_3$CN		
2-methyl-1,4-naphthoquinone (vitamin K-3)	0.29	1.8	0.31	2.0	0.84
9,10-anthraquinone	0.29	1.6	0.30	1.8	0.68
2-t-butyl-anthraquinone	0.24	1.3	0.24	1.5	0.53

The HPLC mobile phase is 95% methanol at a flow rate of 0.76 ml/min in all
cases. In the dual-pump system, the TCPO bed flow rate is 0.42 ml/min. For
liquid phase addition of TCPO it is delivered in acetone at a concentration of 0.92 g/l with a flow rate of 0.28 ml/min. Rubrene, 45 mg/l is
the liquid phase fluorophore and is delivered in 99:1 acetone: TRIS buffer
(pH = 8.0) at 0.15 ml/min. The final buffer concentration is 0.5 mM.

In Fig. 6 some chromatograms are shown. The detection limits
attainable with the split-flow system are only little higher than for the
dual-pump system. Of course the percentage of water in the eluent in the
split-flow system plays an important role: on the one hand it influences
the retention times, on the other hand the peak height since the TCPO
reaction gives lower CL signals at higher water contents.

We conclude that addition of TCPO from solid-state reagent beds has
clear advantages over liquid-phase delivery. It widens the range of
reactions applicable to HPLC detection by relaxing the stability
requirements of the reagent in the delivery solvent. Even in the case of
partial decomposition as in methanol, the decreasing response over time
associated with reagent breakdown during liquid-phase addition is not
observed. Since the time spent in the liquid phase remains constant, the
levels of reagent and reagent breakdown products introduced to the cell
also remain constant over the life of the reagent bed. Thus any reagent of
limited solubility which exhibits a reasonable degree of stability in
solution is amenable to this method of addition.

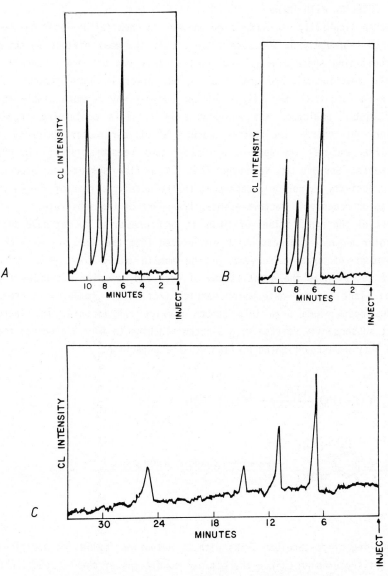

Fig. 6   Liquid chromatograms of quinones detected via the post-column
photochemical reactor/peroxyoxalate CL detection system (ref.
99). The four peaks belong to menadione (5.9 pmol), anthraquinone (4.5 pmol), 2-methylanthraquinone (3.6 pmol) and
2-t-butyl anthraquinone (4.7 pmol). For chromatogram "C" the
amounts injected are 2.5 times higher.
a) Dual-pump system; 94% CH$_3$OH HPLC flow of 0.72 ml/min. TCPO is
   added with CH$_3$OH + TRIS Buffer 99:1 at 0.42 ml/min.
b) Split-flow system; 94% CH$_3$OH HPLC flow of 0.73 ml/min and the
   reagent bed flow is 0.32 ml/min.
c) Split-flow system; 80% CH$_3$OH HPLC flow of 0.73 ml/min and the
   reagent bed flow is 0.32 ml/min.

2.3.4 COUPLING WITH IMERs

The applicability of immobilized enzymes in chemical analysis has been discussed thoroughly by L.D.Bowers (ref. 91). The most obvious advantage of immobilizing these biocatalysts is that they can be readily separated from the reaction mixture and thus reused. Other important aspects are that, as a result of immobilization, the enzymes may be more stable than there soluble analogues and applicable in solvents containing organic modifier. Of course the latter point is of particular interest if immobilized enzymes are applied as post-column reactors (IMERs) in HPLC experiments. Besides it is noted that in a flow system the apparent enzymic activity is not only dependent on the catalytic rate of the enzyme but also on nonenzymic factors as mass transport of the substrate.

Most of the applications of IMERs in HPLC presented so far have dealt with polar natural products which are eluted from the HPLC system with a low content of organic solvent in the mobile phase (ref. 92). As an example we refer to the quantitation of urea and ammonia in samples from an urea plant and in waste water samples (ref. 101). Immobilized urease degrades post-column urea into carbon dioxide and ammonia. The latter product subsequently reacted with o-phthalaldehyde to form a compound that can be very well quantitated by fluorescence detection:

$$(NH_2)_2CO + H_2O \xrightarrow{\text{urease}} CO_2 + 2NH_3$$

$$NH_3 + \underset{\text{CHO}}{\overset{\text{CHO}}{\bigcirc}} \longrightarrow \text{fluorescent compound}$$

Coupling of post-column IMERs with CL detection implies an additional compatibility problem, since the optimal conditions of the CL reaction do not match with those of the enzymatic reaction. For example, the luminol reaction requires strongly alkaline conditions (pH about 12), while for peroxyoxalate CL in highly aqueous media the CL efficiency is extremely low so that high organic modifier concentrations are needed (e.g., 80% acetonitrile). Coupling of the luminol reaction to enzymatic reactions has been reported (refs. 65, 102). Scott et al. have shown the potential of the peroxyoxalate CL reaction in combination with immobilized uricase for the determination of uric acid (ref. 103).

One of the ultimate goals of the combination of IMERs and HPLC is the application of group-specific and/or stereoselective enzymes. If the eluate from the analytical column flows through the IMER, the enzyme causes a reaction selective for the substrate molecules leading to the formation of products which can be detected by suitable methods. It is emphasized that, despite of the selectivity of the IMER in complex (natural) samples as urine or serum, the eventual detection of the products formed in the IMER may be hindered by interferences. That is the reason why combination with the highly selective CL detection techniques is interesting.

In order to investigate the compatibility of immobilized oxidases with the solid-state peroxyoxalate CL detection system, Van Zoonen et al. have tested (the low cost enzyme) glucose oxidase as a model system (ref. 33). Two immobilization procedures for glucose oxidase were examined, i.e., immobilization on an ion-exchanger simply by electrostatic interaction according to Meek et al. (ref. 104) and immobilization via chemical bonding on glass beads following the glutaraldehyde method according to Weetall (ref. 105). In the latter method the glass matrix, after activation, is coated with an amino functional group and subsequently the following steps are carried out:

$$\sim NH_2 + H-\overset{\overset{\displaystyle O}{\|}}{C}-(CH_2)_3-\overset{\overset{\displaystyle O}{\|}}{C}-H \longrightarrow \sim N=\overset{\overset{\displaystyle H}{|}}{C}-(CH_2)_3-\overset{\overset{\displaystyle O}{\|}}{C}-H$$

$$\xrightarrow{H_2N-R} \sim N=\overset{\overset{\displaystyle H}{|}}{C}-(CH_2)_3-\overset{\overset{\displaystyle H}{|}}{C}=N-R$$

The ion-exchanger support appeared to be unsuitable for immobilization of oxidases: it strongly retains the formed hydrogenperoxide. Such problems are not encountered for the IMER based on glass beads.

Two FIA experimental set-ups were compared. In the former a single flow line was applied and the flow composition was simply optimized for the CL reaction, i.e., 80% aqueous acetonitrile containing a small amount of Tris buffer. Even for such a high modifier concentration in an IMER of 6 cm length and 3.0 mm I.D., at pH = 8.0 a glucose conversion as high as 15% was achieved. Rather surprisingly the conversions are almost independent of flow rate in the range between 0.3 and 1.5 ml min^{-1}. The limit of detection for glucose was 8×10^{-8} M.

To avoid the entrance of high acetonitrile concentrations in the IMER, in the second set-up acetonitrile was added according to the make-up flow principle. The aqueous flow containing Tris buffer passes trough the IMER at a rate of 0.3 ml min^{-1} and combines with an acetonitrile flow of 1.0 ml min^{-1} before entering the solid-state CL reactor cell. Under these conditions the maximum conversion of about 50% was found (only ß-D glucose is converted) and with a smaller IMER (length 0.4 cm) for glucose a LOD of 5 x 10^{-8} M was achieved and linear range up to 10^{-5} M. These encouraging results indicate the feasibility of the (oxidase) IMER - solid-state peroxyoxalate CL combination for HPLC.

Honda et al. have applied the combination of IMERs and liquid state peroxyoxalate CL in HPLC for the simultaneous determination of acetylcholine (ACh) and choline (Ch) (ref. 89). A mixed bed reactor of immobilized acetylcholine esterase and cholineoxidase was applied enabling the following reaction pathway for acetylcholine:

$$(CH_3)_3\overset{\oplus}{N}CH_2CH_2O\underset{\underset{O}{\|}}{C}CH_3 + H_2O \xrightarrow{\text{esterase}} (CH_3)_3\overset{\oplus}{N}CH_2CH_2OH + CH_3COOH$$

$$\text{acetylcholine} \qquad\qquad\qquad\qquad\qquad\qquad \text{choline}$$

$$(CH_3)_3\overset{\oplus}{N}CH_2CH_2OH + O_2 + H^+ \xrightarrow{\text{oxidase}} (CH_3)_3\overset{\oplus}{N}CH_2CHO + H_2O_2$$

The optimum pH for these enzymes is from 8.1 to 8.5 so that there is no pH problem for the application of TCPO CL. Nevertheless, the HPLC set-up requires three pumps as shown in Fig. 7. The separation of choline and acetylcholine is based on paired-ion chromatography: the column is an RP-18 column and the eluent (10 mM phthalic acid, 1.2 mM triethylamine and 76 mM sodiumoctanesulfonate pH adjusted to 5.0 with KOH) has a pH too low to be applicable to the IMER. Hence, after the separation column a flow of Tris buffer (pH = 8.5) was provided; nitrate was used instead of chloride as a counterion because chloride is known to quench the CL reaction (see section 2.5). The third pump delivers TCPO and the fluorophore perylene in a mixture of ethylacetate and acetone. Flow rates were chosen so that in the CL reaction medium the flow has the composition ethylacetate:acetone:buffer:eluate in 15:45:4:8. Due to the low solubility of TCPO in this medium its concentration in the second addition line could not be higher than 1.2 mM in order to prevent precipitation effects. Good results were obtained for standard solutions: detection limits for Ch and ACh of about 1 pmol with linear ranges from 10 pmol to 10 nmol.

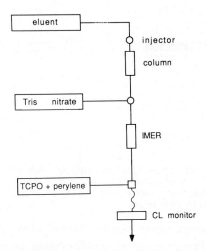

Fig. 7   Schematical representation of the experimental set-up for the
           simultaneous determination of acetylcholine and choline via
           IMER-peroxyoxalate CL detection, according to Honda et al. (ref.
           89). Further details, see text.

It is interesting to compare this approach, based on liquid-state peroxyoxalate CL, with the solid-state principle applied to the same problem,
i.e. the simultaneous determination of ACh and Ch (ref. 34). A detailed
block diagram of the experimental set-up is presented in Fig. 8. Another
separation principle has been applied, (derived from Damsma et al. (ref.
106) based on a (home-packed) cation-exchange column for the separation of
ACh and Ch: the mobile phase is aqueous 0.05 M potassium phosphate (pH =
7.4) containing tetramethylammoniumnitrate (again nitrate instead of
chloride to prevent CL quenching). A precolumn was placed before the
injector as a guard column and pulse dampener. The IMER (ACh esterase and
Ch oxidase covalently bonded to sepharose, dimension 75 x 2.1 mm) was
directly coupled to the analytical column by means of a valco union. An
acetonitrile make-up flow was applied containing 18-crown-6, a crown ether
that efficiently forms complexes with potassium ions. Under these
conditions even relatively high concentrations of potassium phosphate
buffers can be mixed with acetonitrile without precipitation problems.
Furthermore the acetonitrile flow contains triethylamine (TEA), causing a
10-fold improvement of S/N ratio. Optimal flow rates were 0.5 ml min^{-1} for
the chromatographic and 1.5 ml min^{-1} for the make-up flow. Essential is
the use of an efficient vortex mixer because mixing-noise is the main
factor determining the detection limits that can be reached. For the CL
detection the two-layer bed reactor of solid TCPO and 3-AF immobilized on
glass beads was used.

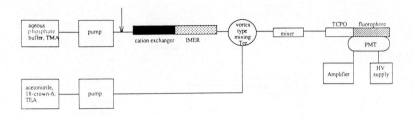

Fig. 8   Block diagram of the experimental set-up for the simultaneous
         determination of acetylcholine and choline via IMER-per-
         oxyoxalate CL detection, according to Van Zoonen et al. (ref.
         34). Further details see text.

     Chromatograms of untreated urine samples spiked with ACh and Ch are
presented in Fig. 9; similar pictures on deproteinated serum samples are
shown in Fig. 10. The detection limits are comparable with those reported
by Honda et al. (ref. 89). However, the solid state set-up is easier to
handle since only one post-column pump is utilized. The TCPO layer has a
lifetime of about 4 hours, but repacking is quite easy and can be
performed with a microspatula. The IMER can be applied for several hundred
samples; if it is used continuously for two weeks, the sensitivity is
decreased by about 50%.

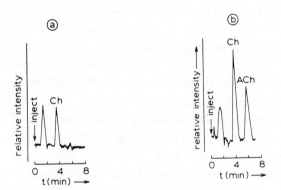

Fig. 9   (a) Chromatogram of an undiluted urine sample, (b) chromatogram
         of a urine sample spiked with 20 pmol of Ch and ACh detected
         with the set-up depicted in Fig. 8 (ref. 34). Further details,
         see text.

     Summarizing this section, it is concluded that the combination of
IMERs and solid state peroxyoxalate reactors have potential for analysis

of complex samples especially because of the high selectivity inherent to the combination of rather specific enzymatic and chemiluminescence reactions. Developments along these lines have to be anticipated.

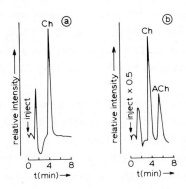

Fig. 10   (a) Chromatogram of a deproteinated pooled serum sample, (b) serum sample spiked with 200 pmol of Ch and ACh detected with the set-up depicted in Fig. 8 (ref. 34). Further details, see text.

2.4   USE OF THE SOLID TCPO REACTOR FOR DETECTION OF FLUOROPHORES (ref. 107)

Although in this chapter we are primarily concerned with non-fluorescent compounds, it is appropriate to point out that the solid-state TCPO reactor has been invoked successfully for the detection of fluorophoric compounds. Compared to the conventional peroxyoxalate CL detection system, a significant simplification has been reached without a substantial loss in sensitivity. As noted above, in addition to the reduced experimental complexity, reagent addition from a solid bed at least partially circumvents the chemical decomposition problems encountered in liquid-phase addition of oxalates. In this way a larger range of solvents can be utilized, because TCPO is dissolved very shortly before it is actually used in the CL reaction.

Two experimental set-ups have been examined with the solid TCPO reagent bed situated parallel to the analytical column in order to reduce band broadening (see Fig. 11). The split-flow system, utilizing only the mobile phase pump, can be employed if the chromatographic separation can be achieved under conditions matching those of the CL reaction (in the model system presented more than 80% acetonitrile). A highly sophisticated system to regulate the split ratio was not necessary; retention times were reproducible within 2% (over 14 chromatograms). The two pump system could be used very well for mobile phase compositions with at least 50%

284

acetonitrile. In more aqueous media precipitation of TCPO in the mixing tee-piece was encountered.

The CL signal is linearly dependent on the H_2O_2 concentration in the eluent. However, if the concentrations become too high also background is increased. For practical reasons such as corrosion of pumps and stainless steel parts of the system, concentrations higher than 10^{-1} M H_2O_2 were not utilized. Under these conditions the performance of the analytical column (Spherisorb ODS-5 in the present example) remained constant for at least a month.

The TCPO reactor (in the present example a 6 cm, 3 mm I.D. stainless steel column with a mixture of solid finely ground TCPO and 40-80 μm glass beads) could be employed at least 8 hours without any drifts; each day TCPO was directly added to the reactor. After about 2 weeks the reactor should be repacked completely since after a prolonged time of use impurities can accumulate in the reactor.

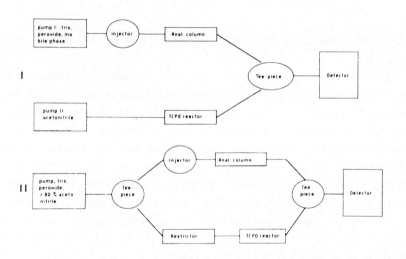

Fig. 11   Schematical representation of the two-pump system I and the single-pump system II for the detection of fluorescers based on peroxyoxalate chemiluminescence. As restrictor a 25 cm Spherisorb ODS-5 column was applied. The luminescence was measured with a Kratos FS 970 fluorescence detector (ref. 107).

TABLE II   Comparison of detection limits (pg) obtained by Sigvardson et al. (ref. 80) and van Zoonen et al. (ref. 107).

Analyte	Sigvardson et al. S/N = 2	Van Zooner et al. 2 pump system S/N = 3	split system S/N = 3
perylene	0.77	1.6	2.0
3-aminofluoranthene	0.30*	0.6	0.7
9,10-dephenylanthracene	20	30	35
anthracene	130	150	200
tetracene	735	860	980
benz(a)pyrene	45	80	120
1,2-benzanthracene	20.5	53	65

*Calculated from the text

In Fig. 12a a chromatogram obtained by the split-flow method is shown. In Table II detection limits of the both solid TCPO addition systems (see Fig. 11) with those of the conventional CL system determined by Sigvardson et al. (ref. 80) are compared. It is obvious that the simplified system provides no substantial loss of sensitivity. Of course the variation in detection limits for the fluorophores under investigation reflects their CL efficiency in the peroxyoxalate reaction, which is not only determined by the energy of the lowest excited electronic state but also by the oxidation potential. As such, 3-aminofluoranthene, the compound applied for immobilization on glass beads (see section 2.3.2) is the most efficient CL fluorophore.

2.5   QUENCHED PEROXYOXALATE CHEMILUMINESCENCE DETECTION

Various compounds, also if present at low concentrations, are able to quench peroxyoxalate chemiluminescence (refs. 108-110). Obviously it is important to acquire some knowledge about the background of this phenomenon in order to be aware of possible and potential interferences when peroxyoxalate CL is applied for detection purposes in HPLC and FIA. On the other hand it is interesting to examine the potential of peroxyoxalate CL for detection of these quenching compounds, especially if other existing detection techniques have some disadvantages.

In this section first some points regarding new insights into the mechanism of the peroxyoxalate CL reaction are discussed. Such a discussion is needed since in the mechanistic studies reported thus far no attention has been paid to quenching phenomena at very low concentration levels. Subsequently, the applicability of Quenched CL detection is considered and finally an interpretation of the quenching process is presented.

286

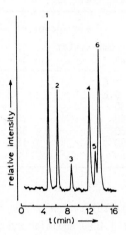

Fig. 12   Chromatogram obtained by the split-flow method for the determination of fluorophores via solid-state TCPO addition (ref.
107). Mobile phase composition: 90% acetonitrile, 10% aqueous
Tris (5 mM, pH = 8.0), 10^{-1} M H_2O_2. Injection volume: 20 μl.
1. 25 pg 3-aminofluoranthene
2. 2 ng anthracene
3. 300 pg 1,2-benzanthracene
4. 30 pg perylene
5. 149 pg benz(a)pyrene
6. 350 pg 9,10-diphenylanthracene.

2.5.1 ON THE MECHANISM OF PEROXYOXALATE CL

By now it is unambiguous that the reaction scheme of the peroxyoxalate
CL reaction presented in section 2.1 does not account for all experimental
data available in the literature. There are two obvious reasons why a
direct energy transfer between the 1,2-dioxetanedione and the fluorophore
is extremely unlikely. First of all, the intensity of CL is strongly
dependent upon the electronegativity of the aryl group of the oxalate
esters which excludes a common intermediate. Secondly, it is not only the
energy of the lowest excited electronic state of the fluorophore that
determines the CL intensity; also its ionization potential plays a role.

Catherall, Palmer and Cundall have published a detailed kinetic study
using bis(pentachlorophenyl)oxalate (PCPO) as oxalate, 9,10-diphenyl-
anthracene (DPA) as fluorophore and sodiumsalicylate as base (ref. 76).
They observed that the decay of CL is independent of the DPA concentration
while the quantum yield increases linearly with (DPA) and eventually
reaches a maximum. The rate-determining step is probably a reaction
between PCPO and the hydrogenperoxide anion OOH⁻(which explains the role

of base catalysts) and does not involve the fluorophore. The lifetime of the intermediate formed in this step was found to be about 5×10^{-7}s, which is an indication for an unstable compound and is not consistent with attempts to identify a relative stable dioxetane. The same authors compared the efficiencies of a number of fluorophores. They were able to show the existence of a relationship between the normalized CL quantum yields (Φ_{CL}/Φ_F where Φ_{CL} and Φ_F are the chemiluminescence and fluorescence quantum yields, respectively) and the oxidation potentials of the fluorophores pointing to an electrontransfer mechanism. Formally, the mechanism according to Catherall et al. can be represented as follows:

$$PCPO + OOH^- \rightarrow X \tag{1a}$$

$$PCPO + OOH^- \rightarrow \text{non CL products} \tag{1b}$$

$$X \rightarrow \text{non CL decay} \tag{2a}$$

$$X + F \rightarrow XF \tag{2b}$$

$$XF \rightarrow F^* + \text{products} \tag{3a}$$

$$XF \rightarrow F + \text{products} \tag{3b}$$

$$F^* \rightarrow F + h\nu_F \tag{4a}$$

$$F^* \rightarrow F + \text{heat} \tag{4b}$$

where (OOH^-) is proportional to the initial concentrations of both H_2O_2 and PCPO and X is a reactive intermediate (ref. 76). Conventional kinetic treatment, assuming that the lifetimes of X, XF and F* are short, gives for the intensity of chemiluminescence at any time, t, defined as

$$I_t = \frac{d}{dt}(h\nu) = k_{4a}[F^*] \tag{6}$$

the following expression

$$I_t = \Phi_{4a} \cdot \Phi_{3a} \cdot \Phi_{2b} \cdot k_{1a}[PCPO]_0[H_2O_2]_t \tag{7}$$

if oxalate is in excess. $[PCPO]_0$ is the oxalate ester concentration which is effectively constant with time and $[H_2O_2]_t$ is the concentration of hydrogenperoxide at any time t. The Φ's are the efficiencies of the reaction steps (4a), (3a) and (2b), respectively; Φ_{4a} of course being equivalent to the fluorescence quantum yield Φ_F. The time dependence of the hydrogenperoxide concentration can be written as

$$[H_2O_2]_t = [H_2O_2]_o \ \exp\{-(k_{1a}+k_{1b})\,[PCPO]_o\,t\} \qquad (8)$$

The number of quanta emitted (QE) is given by

$$QE = \int_0^\infty I_t\,dt \qquad (9)$$

so that QE is equal to

$$QE = \Phi_{4a}\,\Phi_{3a}\,\Phi_{2b}\,\Phi_{1a}\cdot[H_2O_2]_o = \Phi_{CL}\cdot[H_2O_2]_o \qquad (10)$$

wherein Φ_{1a} is the efficiency of reaction step (1a).

Thus, in excess of oxalate, Φ_{CL} is equal to the product of four efficiencies (i.e., those of steps (1a), (2b), (3a) and 4a)) and the initial H_2O_2 concentration. The role of F is readily visualised via Φ_{2b}, i.e.,

$$\Phi_{2b} = \frac{k_{2b}\,[F]}{k_{2b}\,[F] + k_{2a}} \qquad (11)$$

which approximates $(k_{2b}/k_{2a})\,[F]$ under conditions where $k_{2a} \gg k_{2b}\,[F]$. This is generally met for fluorophore concentrations below 10^{-4} M since $k_{2b}/k_{2a} \approx 5 \times 10^3$ M^{-1}; as noted before k_{2a}^{-1} is about 5×10^{-7} s (ref. 76). Therefore, at low fluorophore concentrations QE (the CL signal) is proportional to [F]. Furthermore Eq. (ref. 10) shows the linear dependence between QE and $[H_2O_2]_o$. These two relationships underline the applicability of the peroxyoxalate CL reaction for fluorophore and H_2O_2 detection purposes.

As the key intermediate X, Catherall et al. assume 3-pentachlorophenoxy-3-hydroxy-1,2-dioxetanone, denoted as which reacts with F to a radical ion pair

$$\left[\begin{array}{c} R - O - \underset{OH}{\overset{\overset{O-O}{|\quad|}}{C}} - C \diagdown_{O} \quad F^{\cdot+} \end{array} \right]^{\cdot -}$$

that subsequently undergoes the following sequence

$$\longrightarrow \left[\begin{array}{c} R - O - \underset{OH}{\overset{\overset{O}{\|}}{C}} \quad F^{\cdot+} \end{array} \right]^{\cdot -} + CO_2$$

$$ROH + CO_2 + F^* \qquad\qquad ROH + CO_2 + F$$

More recently Alvarez et al. (ref. 77) published a detailed kinetic study on the same reaction utilizing TCPO as the oxalate, triethylamine as the base, DPA as the fluorophore in ethyl acetate as the solvent. Their results strongly suggest that instead of a single intermediate at least two intermediate compounds X_1 and X_2 producing the same singlet excited state of DPA, play a role, a conclusion based on a two-pulse intensity/ time profile occuring at lower concentrations of triethylamine. A possible structure for X_2 is the intermediate X of Catherall et al. given above. A possible sturcture for X_1 is

$$HO - O - \underset{O}{\overset{\|}{C}} - \underset{O}{\overset{\|}{C}} - O - R$$

which in Catherall's work is a precursor for X. Thus an important consequence of Alvarez' study is the existence of an additional reaction pathway for the fluorophore.

Finally, it should be realized that in HPLC and FIA frequently solvents partially composed of water are applied while furthermore eluents as methanol are very important. In these solvent compositions additional reaction pathways such as a direct reaction of the oxalate with H_2O or methanol may complicate the establishment of a reaction mechanism. Thus, for practical reasons it is relevant to evaluate aryl oxalates in terms of maximum chemiluminescence intensity (maximum of I_t), decay rate, solubility in different solvents, stability in presence of hydrogenperoxide and pH working range as has been done by Honda, Miyaguchi and Imai (ref. 111). They have shown that among others bis(2-nitrophenyl) oxalate, 2-NPO, has favourable properties: it is six times more soluble in acetonitrile than TCPO, has a reasonable stability in presence of H_2O_2 and is optimally applicable in the pH range 4-6.

At this point we emphasize that it is not only the solubility of the oxalate that hampers the success of the peroxyoxalate CL detection in aqueous solvents. Apparently, the CL efficiency is very low in water, which is readily conceived in view of the mechanism discussed above: an immediate dissociation of the radical ion pair as probably occurs in water obviously prevents the formation of F*.

2.5.2 APPLICABILITY OF QUENCHED CL DETECTION (refs. 109, 110)

Since electron transfer from F to X is initiating the formation of F*, it is appropriate to examine whether other easily oxidizable (but non-fluorescent) compounds are able to consume X thus reducing the overall CL quantum yield.

Van Zoonen et al. have shown that various compounds are able to induce quenching of peroxyoxalate CL, even if present at low concentrations (refs. 109, 110). Some examples are presented in Table III. They do obey a Stern-Volmer type relationship, i.e.

$$I_0/I_Q = 1 + k_Q[Q] \qquad (12)$$

wherein k_Q is a constant of quenching (in M^{-1}), $[Q]$ is the quencher concentration (in M) and I_0/I_Q is the ratio of the CL signals in absence and presence of Q, respectively. Obviously k_Q determines the sensitivity of Quenched CL detection (QCL) for a particular (quenching) analyte, so that the method has an inherent selectivity. On the other hand it should be realized that QCL is based on a decrease of luminescence. Hence the noise of the luminescence signal in absence of quencher should be reduced as much as possible to achieve a favourable signal to noise ratio for QCL. This can be reached most easily under experimental conditions where I_0 is high.

TABLE III   Detection limits (S/N = 3) for quenched CL using 2-NPO as
            oxalate, derived from chromatograms

Analyte	l.o.d. (ng)
bromide	1.5
iodide	0.3
sulphite	1.1
nitrite	0.3
p-isopropylaniline	1.4
N,N-dimethylaniline	0.6
N-ethyl-m-toluidine	1.0
N,N-dipropylaniline	8.0
thiourea	1.0
N-allylthiourea	1.6
ethynyl thiourea	2.0
methimazole	0.4

The influence of the H_2O_2 concentration and the nature and concentration of fluorophore on k_Q have been examined. $[H_2O_2]$ does not influence k_Q for concentrations between 10^{-2} and 10^{-6} M. This implies that in hydrogenperoxide determinations straight calibration lines will be observed in the presence of quenchers. Of course the associated slopes will depend on the concentrations of quenchers, so that standard addition procedures should be applied to circumvent systematic errors. For QCL, the applied H_2O_2 concentration should be as high as possible in order to reach a high I_o; in practice concentrations from 10^{-3} to 10^{-2} M are appropriate.

In fluid samples both concentration and nature of F (i.e., its oxidation potential) hardly influence k_Q. As an example the results for variable perylene concentrations are given in Table IV. In view of the achievable signal to noise ratio this means that a fluorophore with a high CL efficiency is appropriate for QCL, so that (immobilized) 3-aminofluoranthene should be a good choice. Unfortunately, the quenching constant appears to be affected by the immobilization procedure of 3-aminofluoranthene.

The reproducibility between different batches for the synthesis of immobilized fluorophore with respect to the quenching constants is rather poor. As a general trend it might be concluded that silianization in dry toluene (see section 2.3.2) gives the most suitable product for QCL, since it yields the highest quenching constants at intermediate CL intensity.

Nevertheless, the poor reproducibility is not a very serious problem, since a single batch of immobilized 3-AF can be used over long periods of time.

TABLE IV   Effect of the concentration of the fluorophore (perylene) on the quenching constant of peroxyoxalate CL (applying solid state TCPO addition) measured with methimazole (5.2×10^{-5} M)

Concentration	I_o in relative units	k_Q in $10^3 M^{-1}$
4×10^{-5}	40500	44
4×10^{-6}	10000	48
4×10^{-7}	1200	42

For the development of QCL detection both solid TCPO (in a dual-cell configuration or in a separate make-up flow line) and liquid-state 2-NPO were applied. The latter can be mixed with H_2O_2 in acetonitrile without decomposition problems. It was found that for 2-NPO the QCL peak heights (and thus the k_Q-values) were about 10 times higher than for TCPO.

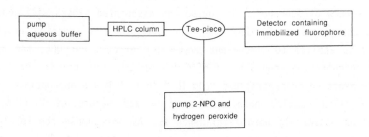

Fig. 13   Experimental set-up for HPLC with Quenched Chemiluminescence detection (ref. 110).

This explains the experimental set-up for QCL detection, see Fig. 13. For the analytes under study aqueous mobile phases were applied (for instance the chromatograms of iodide, bromide, sulphite and nitrite were obtained with the mobile phase aqueous ammoniumbenzoate, 10 mM, pH 5 and PRPX-100 column) as the mobile phase at a flow rate of 1 ml min^{-1} and a reagent flow of 0.05 M H_2O_2 and 8mM 2-NPO in acetonitrile at a rate of 1.2 ml min^{-1}. Thus, CL detection method is fully compatible with aqueous separation systems as for instance commonly used in ion chromatography. As such, the limits of detection presented in Table III are quite interesting, indicating the potential of QCL detection. Its selectivity is satisfactory as can be seen from Fig. 14 where a chromatogram of spiked urine is shown; no pretreatment or dilution of the samples was necessary to detect methimazole and N-allyl thiorurea in this matrix.

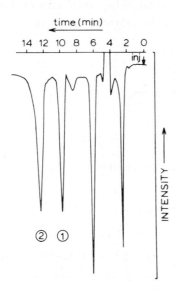

Fig. 14   Chromatogram of a urine sample spiked with 100 ng of N-allylthiourea (peak 1) and 25 ng methimazole (peak 2) with Quenched CL detection; injection volume 20 μl; chromatographic conditions: RP-18 column; mobile phase aqueous ammoniumbenzoate (10 mM, pH 5) at flow-rate 0.8 ml min^{-1}; reagent flow, 0.05 M H$_2$O$_2$ and 8 mM 2NPO in acetonitrile at 1.2 ml min^{-1} (ref. 110).

2.5.3   THE QUENCHING ACTION (ref. 112)

The results described in section 2.5.2 obtained during development of the QCL detection method have to be discussed within the framework of the peroxyoxalate mechanism outlined in section 2.5.1. Summarizing, the following points have to be considered:

(1) readily oxidizable compounds are efficient quenchers;
(2) the quenching constant k_Q depends on the character of the arylgroup in the oxalate (for 2-NPO k_Q is about 10 times higher than for TCPO);
(3) the quenching constant k_Q is independent of the nature and the concentration of the fluorophore in the liquid state;
(4) for immobilized fluorophore, the immobilization procedure affects k_Q;
(5) the quenching constant k_Q is independent of the concentration of H$_2$O$_2$.

Point (5) rules out that the action of Q is connected with the reaction between oxalate and H_2O_2 (or OOH^-), i.e., step (1) in the reaction scheme. At a first sight, the combination of points (1) and (3) is rather puzzling; as F and Q have similar properties concerning electrontransfer, why is the quenching effect of Q not attributable to a competition between F and Q? Fortunately, this paradox can be readily solved on the basis of Eq. (11). In presence of Q, the magnitude of Φ_{2b} will be reduced to

$$\Phi_{2b,Q} = \frac{k_{2b}\,[F]}{k_{2b}\,[F] + k_{2a} + k_{2c}\,[Q]} \tag{13}$$

where k_{2c} is the rate constant of the reaction

$$X + Q \rightarrow X^{-\cdot}Q^{+\cdot} \rightarrow \text{non CL decay}$$

However, since k_{2b} and k_{2c} will have a comparable magnitude, also under these conditions k_{2a} is commonly by far the largest term in the denumerator of Eq. (13) and thus effectively $\Phi_{2b,\,Q} = \Phi_{2b}$.

Summarizing the previous paragraph in terms of Eq. (10), Φ_{1a} and Φ_{2b} are not affected by [Q]. The same holds for Φ_{4a}, the quantum yield of fluorescence; this is evident since fluorescence is not quenched under the experimental conditions under consideration. Thus, in terms of the Catherall reaction scheme only one possible explanation remains: Q is particularly influencing efficiency Φ_{3a}, i.e., the fraction of XF that leads to F*. This is not unreasonable in view of Catherall mechanism; before F* is produced, $X^{-\cdot}$ undergoes rearrangement while $F^{+\cdot}$ is not changing. Collision with an electron donating compound would immediately destroy $F^{+\cdot}$ which explains that readily oxidizable compounds are good quenchers, point (1). Furthermore the lifetime of the radical ion pair determines the efficiency of the collisional quenching; this is in-line with point (2) since the nature of X determines the lifetime of XF. Finally, it will be obvious that this interpretation can also account for the fact that the fluorophore immobilization procedure influences k_Q: quenching requires a collision between Q and the fluorophore radical cation in the ion pair. In other words, k_Q will be influenced by the ability of Q to approach immobilized 3-AF, which of course depends on the detailed structure of the surface layer on the solid substrate.

2.6   CONCLUDING REMARKS

The recent results obtained for peroxyoxalate chemiluminescence as a chromatographic detection method for non-fluorescent analytes indicate a significant progress. The main line is the selective and sensitive detection of hydrogenperoxide produced in a post-column photochemical or immobilized enzyme reactor.

The peroxyoxalate CL reaction does not only provide favourable detection limits for H_2O_2 compared to other CL reactions, but even more important, also a high degree of selectivity. Only a minor amount of compounds do affect the CL efficiency of the reaction and furthermore, if quenchers are present, the signal remains linearly related to the hydrogenperoxide concentrations. Also post-column systems as IMERS and photochemical reactors producing H_2O_2 generally have a good selectivity. This explains why the combination of such reactors and peroxyoxalate CL detection allows the quantitation of analytes in very complex matrices without elaborate sample pretreatment as has been shown for choline and acetyl choline in urine and serum. Thus it is expected that further interesting applications will be realized for instance by applying group-specific enzymes.

Of course an important aspect of the applicability of the described system is the reduction in the complexity of the experimental set-up that has been realized following the solid-state approach. On the one hand this is achieved by the development of the immobilized fluorophore reactor. Since the fluorophore is not consumed during the reaction one of the most efficient fluorophores, i.e., 3-aminofluoranthene, can be used despite of its toxic properties. Furthermore, the emission of luminescence is localized in the detector cell. On the other hand the application of a solid state oxalate (TCPO) reactor simplifies the system. A disadvantage of such a reactor is its limited lifetime since the oxalate is consumed. Nevertheless, the system is easy to handle and in practice it can be used over 8 hours without any reduction of chromatographic integrity provided that it is positioned not in line with the analytical column but in a flow addition line. An important positive aspect of solid state reactors is that instability of reagents is of minor importance. This is illustrated by the fact that in methanol/water eluents good results have been obtained.

Of course for any particular application the compatability of the CL reaction conditions and the chromatographic separation conditions including the conditions required by the post-column reactor had to be considered. As such it is interesting that the examples presented indicate that aqueous eluents do not exclude the applicability of peroxyoxalate CL. In fact the most interesting applications reported thus far are those requiring aqueous mobile phases.

3. LIQUID PHASE PHOSPHORESCENCE DETECTION

3.1 FUNDAMENTAL ASPECTS OF PHOSPHORESCENCE

Phosphorescence was identified in 1944 by Lewis and Kasha as the emission of radiation from the lowest triplet state of a molecule to the singlet ground state (ref. 113). Fig. 15 shows a simplified Jablonski energy diagram of an analyte molecule (denoted as An).

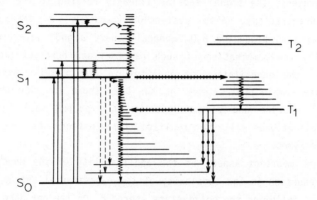

Fig. 15 Jablonski energy diagram ⟶ absorption (I)
- - - -> fluorescence (f)
∿∿∿> internal conversion (ic)

(nr = non radiative)

⋀⋀⋀⋀⋀⋀> vibrational relaxation (vr)
⊦⊦⊦⊦⊦⊦> intersystem crossing (isc)
•—•—•> phosphorescence (p)

Irradiation with light of a suitable wavelength transfers An from its electronic ground state S_0 to an excited electronic state S_n (Eq. 14), while the singlet spin state is conserved. Then a non-radiative decay process takes place, eventually ending in the lowest vibrational level of the first excited singlet state S_1 (Eq 15). The efficiency of this decay process, comprising internal conversion of energy and vibrational relaxation via collisions with the solvent molecules surrounding An, is generally hundred per cent. Transition from the S_1 state back to the S_0 state can occur either radiationless via internal conversion and subsequent vibrational relaxation (Eq. 16) or by emission of radiation

(Eq. 17). Furthermore, bimolecular quenching via reactions with a quencher q is a possible decay pathway (Eq 18). And finally, intersystem crossing from the S_1 state to the lowest excited triplet state T_1 is another non-radiative decay of the S_1 state (Eq. 19). Once the analyte has reached the T_1 state transition to the S_0 state can take place via three competitive pathways, i.e., radiationless (Eq. 20), by emission of radiation (Eq. 21), called phosphorescence and by a bimolecular quenching reaction with a quenching compound Q (Eq. 22); the capital Q is utilized to emphasize that compounds quenching efficiently a T_1 state of the analyte do not necessarily quench the analyte in its S_1 state and vice versa.

$$An\,(S_0) + h\nu_{ex} \quad \rightarrow \quad An\,(S_x) \tag{14}$$

$$An\,(S_x) \quad \overset{k_{nz}}{\rightarrow} \quad An\,(S_1) \tag{15}$$

$$An\,(S_1) \quad \overset{k_{nf}}{\rightarrow} \quad An\,(S_0) \tag{16}$$

$$An\,(S_1) \quad \overset{k_f}{\rightarrow} \quad An\,(S_0) + h\nu_f \tag{17}$$

$$An\,(S_1) + q \quad \overset{k_q}{\rightarrow} \quad An\,(S_0) + q \tag{18}$$

$$An\,(S_1) \quad \overset{k_{isc}}{\rightarrow} \quad An\,(T_1) \tag{19}$$

$$An\,(T_1) \quad \overset{k_{np}}{\rightarrow} \quad An\,(S_0) \tag{20}$$

$$An\,(T_1) \quad \overset{k_p}{\rightarrow} \quad An\,(S_0) + h\nu_p \tag{21}$$

$$An\,(T_1) + Q \quad \overset{k_Q}{\rightarrow} \quad An\,(S_0) + Q \tag{22}$$

The rate constants $k_{nf}(S_1 \rightarrow S_0)$ and $k_{isc}(S_1 \rightarrow T_1)$ usually range from 10^5-10^7 and 10^6-10^9 s^{-1}, respectively (ref. 114). The rate constant of the fluorescence process k_f is in the order of 10^7-10^9 s (ref. 114). According to the rules of quantum mechanics the phosphorescence process is strictly forbidden as two electronic states of different spin multiplicities are involved. Nevertheless this process can be observed for certain molecules, as a result of spin-orbit coupling. When this mechanism occurs a triplet state is not pure but has some singlet character and a singlet state has some triplet character. The result is a triplet-singlet transition probability unequal zero, which means that the rate constant of the phosphorescence process k_p is 10^{-1}-10^2 s^{-1}. The efficiencies of fluorescence (θ_f), intersystem crossing (θ_{isc}) and phosphorescence (θ_p) can be readily expressed in terms of the rate constants of the reactions Eqs. (16) to (22).

The fluorescence efficiency is given by

$$\theta_f = \frac{k_f}{k_{isc} + k_f + k_{nf} + \sum_q k_q [q]} = \Phi_f \tag{23}$$

It is equal to the fluorescence quantum yield Φ_f, since it gives the probability that An after absorption of radiation emits fluorescence. On the contrary, the quantum yield of phosphorescence Φ_p, is the product of the efficiency of intersystem crossing (the probability for An to reach the T_1-state) and the efficiency of phosphorescence, i.e.,

$$\Phi_p = \theta_{isc} \cdot \theta_p \tag{24}$$

wherein

$$\theta_{isc} = \frac{k_{isc}}{k_{isc} + k_f + k_{nf} + \sum_q k_q [q]} \tag{25}$$

and

$$\theta_p = \frac{k_p}{k_p + k_{np} + \sum_Q k_Q [Q]} \tag{26}$$

It is appropriate to invoke lifetimes in Eqs. (23) and (24). The lifetime of fluorescence τ_f is equal to the lifetime of An(S), so that

$$\theta_f = k_f \tau_f \tag{27}$$

This lifetime should be distinguished from the radiative lifetime of An(S_1) which is defined as k_f^{-1}; they are only equal if radiation is the only decay path for An(S_1), in other words if Φ_f is hundred per cent.

Similarly the phosphorescence efficiency can be written as

$$\theta_p = k_p \tau_p \tag{28}$$

in which τ_p, the lifetime of phosphorescence, is equal to the lifetime of An(T_1). Analogously with the fluorescence lifetime, $\tau_p < k_p^{-1}$ where k_p^{-1} is the radiative lifetime of phosphorescence.

Generally, bimolecular quenching of An(S_1) is negligible, implying that molecular fluorescence in fluid solutions is a quite common phenomenon. This is due to the high values of the intramolecular decay rate constants so that usually

$$(k_{isc} + k_f + k_{nf}) \gg \sum_q k_q [q] \tag{29}$$

despite of the fact that k_q can be as high as the diffusional-controlled constant. If both $\sum_q k_q[q]$ and k_{nf} play a minor role for the decay of An $[S_1]$ combination of Eqs. (23) and (25) reveals that

$$\theta_{isc} = 1 - \Phi_f \tag{30}$$

This clearly shows the complementary character of fluorescence and phosphorescence.

For phosphorescent compounds θ_{isc} must be relatively high. This efficiency depends on the amount of spin-orbit coupling which increases with decreasing difference between the energies of the T_1 and S_0 states. Moreover it can be enhanced by the introduction of heavy atoms into the phosphorescent compound itself, but also into the solvent molecules (internal, respectively, external heavy atom effect). Besides it is noted that no effect is encountered if in absence of heavy atoms θ_{isc} already approximates hundred per cent.

In contrast with fluorescence, molecular phosphorescence in fluid solutions requires special experimental circumstances. Only under the prerequisite that the competitive intra- and intermolecular deactivation processes of An(T_1) are diminished as much as possible there is a substantial probability for a radiative transition. Confining our attention to bimolecular deactivation it is readily seen that commonly

$$k_p \ll \sum_Q k_Q [Q] \tag{31}$$

and θ_p is negligible. If for example $k_p = 10 \text{ s}^{-1}$ and $k_Q = 10^{10} \text{M}^{-1}\text{s}^{-1}$ a quencher concentration as low as 10^{-8} M already fulfils Eq. (31). As a result phosphorescence in homogeneous solutions will be quite exceptional. Only for compounds with a high k_p value the radiative transition may be able to compete successfully with bimolecular quenching provided that the solvent is thoroughly purified and deoxygenated. A second requirement is that the intramolecular radiationless decay process does not dominate. This explains why frozen glassy samples have been applied extensively in phosphorimetry. The rate constant k_{np}, especially for compounds with flexible structures, can be much higher in a fluid solution than in a frozen solution. The incorporation of heavy atoms enhances both k_p and k_{np} leading to a reduction of τ_p, while the net effect on θ_p is unpredictable, see Eq. (26).

3.2   NEW DEVELOPMENTS IN PHOSPHORIMETRY

In the last decade interesting new developments in phosphorimetry have been realized, directed on the possibility to circumvent the need of freezing samples. They include solid-surface-, micelle-stabilized-, solution-sensitized and solution-quenched room temperature phosphorescence.

3.2.1   SOLID-SURFACE RTP

The observation of room temperature phosphorescence (RTP) from organic molecules adsorbed on a solid matrix has been done several times in the sixties (refs. 115-117). The analytical potential of this technique was shown by Schulman and Walling in 1982 (refs. 118-119). They studied a number of organic compounds adsorbed on silica, alumina and filter paper. Important research in this field is carried out in the groups of Winefordner and Hurtubise.

A considerable number of solid substrates has been used to induce phosphorescence from adsorbed compounds; the most promising seems to be filter paper (refs. 120, 121). Beside several qualities of filter paper (refs. 119, 122-133), also silica gel (refs. 134, 135), alumina (ref. 136), sodium acetate (refs. 137-141), polyacrylic acid - sodium chloride mixtures (refs. 142-144) and cellulose (ref. 145) have been tried. RTP has also been observed from compounds adsorbed on streched polymer films (ref. 146).All substrates give rise to a broad band background emission (400-600 nm) which often interferes with quantitative and qualitative measurements (refs. 147-148). The influences of phosphorophore/substrate combination, sample preparation, amount of moisture and oxygen present (refs. 149, 150), optimum pH value and presence of heavy atoms (refs. 151, 155) on RTP have been studied extensively. Furtheron the nature of the phosphorophore-substrate interaction has been investigated (refs. 156-160).

RTP spectra are very similar in shape to LTP (low temperature phosphorescence) spectra although intensities and lifetimes can be significantly affected by the actual experimental conditions. RTP emission has been observed from ionic organic compounds (refs. 119, 134, 137-140, 157), non-polar polynuclear aromatic hydrocarbons (PAH's) (refs. 121, 130, 132, 135, 153, 156, 159, 160) and compounds of pharmaceutical (refs. 127, 133, 153, 161, 162) and biological (refs. 120, 128, 154, 161) interest. The rotating hollow drum developed by Miller (refs. 163) and the rotating-mirror phosphorescence as developed by Vo-Dinh et al. (ref. 164) made it possible to scan thin layer chromatograms for phosphorescent compounds. Lloyd (ref. 165) described a flow cell packed with a mixture of crushed quartz and paper-derived lint to detect analytes with RTP after liquid chromatographic separation.

Although RTP has a poorer sensitivity than LTP the analytical procedure is very simple. Additionally, a chromatographic separation can be performed on the solid substrate before the analysis.

A considerable gain in sensitivity in phosphorimetry was obtained by the introduction of the pulsed source-time resolved detection technique (ref. 166). With this technique it is possible to analyze mixtures of phosphorphores. After a short excitation source flash, the phosphorescence emission is measured, after a certain delay time t_d, during a gating time t_g. By this approach, the phosphorescence signal can be temporally discriminated from rapidly decaying species (e.g., fluorescent impurities) and source light scatter. Because of their excellent temporal characteristics the use of pulsed lasers instead of the normally applied pulsed Xenon sources is a promising development (refs. 167-169).

Another instrumental technique to increase the selectivity of phosphorimetry is synchronous scanning as proposed by Vo-Dinh (ref. 170). The excitation and emission monochromators of a phosphorimeter are set with a constant wavelength difference of $\Delta\lambda = \lambda_{em} - \lambda_{exc}$ and both monochromators are scanned at the same rate. A phosphorescence peak only occurs when both

λ_{xc} and λ_{em} correspond simultaneously to wavelengths at which excitation and emission of a particular compound occurs. In this way, sharper peaks are obtained.

Second derivative phosphorimetry has also been used by Vo-Dinh and co-workers (ref. 171). By means of taking the second derivative of an RTP emission spectrum, overlap between phosphorescence bands could be reduced and the phosphorescence background decreased. Both the synchronous scanning and the second derivative technique have been applied to the analysis of PAH mixtures (refs. 172, 173).

3.2.2 MICELLE-STABILIZED RTP

The use of organic media such as micellar solutions and cyclodextrins to induce RTP in certain compounds has been introduced by Cline Love et al. (refs. 174, 175). Reviews of the analytical implication of micelle chemistry in phosphorimetry have appeared recently (refs. 176, 177). The advantages of this method are: 1) an increase in sensitivity because the organized environment reduces intramolecular processes competing with photoemission; 2) better solubility of non-polar compounds with respect to aqueous solutions; 3) the possibility to bring analytes and heavy atoms together very efficiently to create a heavy atom effect. A disadvantage is that oxygen still has to be removed from micellar solutions because the micelles do not protect the phosphorophores against quenching species. The application of the principle of micelle enhanced phosphorescence as a detection method in liquid chromatography has been proposed by Weinberger et al. (ref. 178) in two ways. First by using a micellar solution as the mobile phase and secondly by post-column addition of the micellar solution to the column effluent. Unfortunately, the use of micellar solutions is not always easily compatible with liquid chromatography conditions. DeLuccia and Cline Love studied the sensitized phosphorescence of biacetyl in organized media (refs. 179, 180). The potential of synchronous scan and second derivative techniques in micellar RTP was examined by Femia and Cline Love (ref. 181).

3.2.3 RTP IN NORMAL FLUIDS

It is generally accepted that in normal fluid solutions, where factors suppressing the diffusion of triplet quenchers are commonly absent, phosphorescence intensities are too low to be used for analytical purposes. Several fundamental studies in this field have been published during the last 20 years (refs. 182 - 185). The highest phosphorescence emission intensities at room temperature have been reported by Almgren (ref. 183) for biacetyl in benzene with a phosphorescence quantum yield Φ_p of 0.08 and by Parker and Joyce (ref. 184) for acetophenone in per-fluorormethylcyclohexane with a Φ_p of 0.0581. For benzophenon, with a quantum yield of 1.0 and a triplet lifetime of around 7 msec at 77 K no RTP could be observed in hexane; Turro (ref. 186) reported a Φ_p of only 9.1×10^{-3} in water, and Joyce (ref. 184) a Φ_p of 0.097 in perfluoro-methylcyclohexane. Turro et al. (ref. 185) showed that in acetonitrile, a solvent widely used in reversed phase liquid chromatography, a phospor-escence emission could be achieved for 1,4-dibromonaphthalene, with a quantum yield in the order of 10^{-3}. Table V includes a number of exceptional compounds that emit "strong" phosphorescence in normal fluid solutions. From an analytical point of view, this phenomenon can only be utilized in an indirect way. The phosphorophore is present as a solute and the analyte acts either as a sensitizer or as a quencher of phosphor-escence. Both techniques have been successfully applied as detection methods in HPLC.

TABLE V Phosphorescence data for naphthalene (N), 1-bromonaphthalene (1-BrN), 2-bromonaphthalene (2-BrN), 1,4-bromonaphthalene (1,4-BrN), 4,4'-dibromobiphenyl (4,4'-Br$_2$B), 2-bromobiphenyl (2-BrB) and 4-bromobiphenyl (4-BrB) in 2-methyltetrahydrofuran at 77 K and in n-hexane at room temperature (295 K); from (ref. 187).

Compound	77 K		295 K	
	Φ_p	τ_p,msec	Φ_p	τ_p,msec
N	0.03	2.1×10^3		
1-BrN	0.27	15.0	0.10	1.9
2-BrN	0.38	16.8	0.14	2.8
1,4-Br$_2$N	0.27	5.3	0.18	1.7
4,4'-Br$_2$B	0.49	12.5	0.08	0.86
2-BrB	0.15	-	0	
4-BrB	0.65	22.5	0.012	-

The state of the art in 1983 has nicely been overviewed by Hurtubise (ref. 166) and by Vo-Dinh (ref. 188). At that date the analytical potential of solution quenched phosphorescence was not yet known; this method has been introduced quite recently and its potential is subject of current research.

3.2.4 WHY RTP?

The reason that much effort has been devoted to the extension of phosphorimetry undoubtly is that phosphorescence, as explained above, can be considered as complementary to fluorescence. Furthermore for many purposes no additional instrumentation is required. Of course, compounds detectable by direct or by sensitized phosphorescence are also measurable by UV-VIS absorption spectroscopy. Nevertheless, in many applications luminescence measurements are essential not only because they are more selective, but especially because for trace analysis of real samples frequently lower detection limits are required than attainable by absorption measurements.

Solution quenched phosphorescence is applicable to compounds that rapidly react with the excited phosphorophore, their own absorption characteristics are not relevant. Therefore this technique is especially of interest for analytes badly detectable by direct UV-VIS absorption spectroscopy, as for instance inorganic ions (refs. 35-37).

3.3   EXPERIMENTAL ASPECTS
3.3.1 REMOVAL OF OXYGEN

Essential for RTPL detection is the long triplet state lifetime of the phosphorophore under consideration. This implies that special experimental requirements have to be met to make phosphorimetry in fluid solutions a useful analytical method. To this end the solutions have to be deoxygenated as much as possible since oxygen acts as a very efficient quencher and the solvents have to be purified carefully to avoid impurity quenching. Moreover, the experimental set-up has to be cleaned thoroughly and direct contact between solutions and synthetic materials as teflon should be minimized.

In practice, these conditions can be fulfilled relatively easy. For all types of experiments, in batch, in flow injection analysis and in liquid chromatography purging of solvents with nitrogen gas reveals a sufficient reduction in oxygen concentration (ref. 39). Commercially available nitrogen gas (containing about 5 ppm of oxygen) is passed through a column filled with a heterogeneous reduction catalyst (i.e., pyrophorous copper) and kept at a constant temperature of 100 $^{\circ}$C. In this way the oxygen content of the N_2 gas is reduced to less than 0.2 ppm. The purified N_2 flow is led through a washing bottle and (in batch experiments) subsequently through the sample solution. After 5-10 minutes of purging the deoxygenation is completed and a stable phosphorescence signal is obtained. During the measurements a constant N_2 flow is maintained over the sample in order to prevent re-entrance of oxygen.

In flow injection analysis and liquid chromatography the deoxygenation of the solutions occurs in the eluent vessel. In Fig. 16 a specially constructed vessel as described in ref. 43 is depicted; the crucial point is that the use of synthetic materials has been avoided so that only glass, quartz and/or stainless-steel have been applied. A schematic diagram of a HPLC system is given in Fig. 17. The interconnections between eluent vessel, pump, injection valve, column and detection are all stainless steel capillaries. The overall system is closed by leading the output capillary back to the eluent vessel, whereas during elution, the valve to waste is opened to avoid contamination of the eluent. In this way the eluent can be used continuously over weeks without any loss of phosphorescence sensitivity.

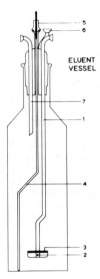

Fig. 16   Details of the eluent vessel, consisting of a 3 l glass bottle and glass stopper B55 which fits well in a ground glass joint. The nitrogen gas used for deoxygenation the eluent enters via a glass tube (1), with special glass joint (cup size 13/5, Rotulex^R), via an opening (2) and a glass filter (3); the outlet is via a glass tube (7). The deoxygenated eluent is pumped into the flow system via a stainless-steel capillary (4), which forms one unit with a stainless-steel ball part (5) for the outlet of deoxygenated eluent; (6) represents a construction identical to 4 and 5 for the inlet of eluent (ref. 43).

In order to be sure that the quality of the solution is constant over a longer period of time, this has to be checked. If biacetyl is applied as phosphorophore, to date by far the most indirect phosphorescence detection measurements are based on this compound, the phosphorescence to fluorescence signal ratio (see Fig. 18) is an indication for the quality of the

306

system regarding O_2 and impurities and thus for the sensitivity that can be obtained in the measurements (ref. 38).

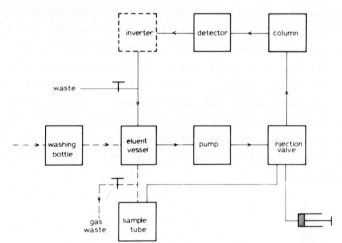

Fig. 17   Schematic representation of the dynamic system for liquid chromatography with phosphorescence detection. The broken lines represent stainless-steel capillaries for the nitrogen gas stream that after being washed in a washing bottle containing some eluent is led into the eluent vessel (depicted in Fig. 16) and goes eventually to waste or is used to deoxygenate the sample solution. The solid lines represent stainless-steel capillaries for the eluent stream, connecting eluent vessel, injection valve, analytical column and luminescence detector. The system is closed under normal conditions, to prevent entrance of oxygen or impurities; during the recording of the chromatograms the valve to waste is open. The inverter is not strictly necessary; it serves to record the inverted phosphorescence signal, which is useful in quenched phosphorescence (ref. 35)

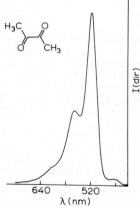

Fig. 18   Room temperature phosphorescence spectrum of 10^{-4} M biacetyl in deoxgenated acetonitrile/water (80%). λ_{exc} = 420 nm. The small peak at the low wavelength side is due to fluorescence (ref. 35).

3.3.2 INSTRUMENTATION

The detection devices that can be used for the measurements of phosphorescence in fluid solutions are standard commercially available fluorescence detectors, though some simplifications are possible. First, use can be made of a less expensive detector, as only a restricted number of wavelengths are important. In sensitized phosphorescence the choice of the excitation wavelength λ_{exc} depends on the absorption characteristics of the analyte, whereas the emission wavelength λ_{em} can be fixed; in quenched RTPL both λ_{exc} and λ_{em} are fixed. Secondly, for phosphorescence the difference between λ_{exc} and λ_{em} is larger than for fluorescence with the result that background radiation due to scattering and Raman effects is more readily reduced. For that reason the emission grating mono-chromator can be replaced by simple cut-off filters, which are not only less expensive but also allow a higher light throughput thus revealing higher sensitivities. As already mentioned a gain in sensitivity can be obtained with the pulsed source-time resolved detection technique. With a short excitation source flash the molecules are excited. In order to eliminate background due to rapidly decaying emission (impurity fluorescence and scatter) the phosphorescence signal is recorded during a time interval t_g (gating time) which starts a time interval t_d (delay time) after the flash (see Fig. 19). By choosing the suitable light pulse, delay - and gating time, background luminescence and scattering of the light source can be suppressed considerably.

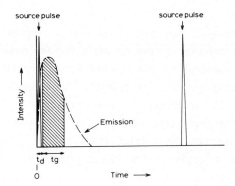

Fig. 19   Excitation and emission signal dependence of time after source flash at t = 0 operating in the phosphorescence mode: t_d, delay time; t_g, gate width of detector. The emission signal is gated after the source flash by a delay time t_d. By choice of an appropriate value for t_d, background emission caused by scattering and fluorescence impurities, which have lifetimes shorter than 10^{-8} s, is not detected. This results in a reduction in the noise of the system.

For liquid state phosphorescence with a lifetime of 1 μs to 10 ms the additional instrumental requirements needed for time resolution measurements can be met more easily than for fluorescence with a liftime of 1-100 ns (refs. 40, 48). Nowadays luminescence detectors with a pulsed Xe-lamp (pulses in the 50 Hz range, with a width of about 50 μs) and a gated photomultiplier are commercially available.

3.4   INDIRECT PHOSPHORESCENCE DETECTION
3.4.1   SENSITIZED PHOSPHORESCENCE
3.4.1.1   INTRODUCTION

In sensitized phosphorescence after exciting a donor molecule energy transfer to an acceptor molecule takes place and the phosphorescence of the acceptor is monitored. In general this indirect method is applied for non-fluorescent analytes with a high θ_{isc} which do not emit direct phosphorescence in liquid solutions (θ_p is negligible). This means that for these compounds the radiative phosphorescence transition is too slow to compete sucessfully with non-radiative decay. It is the aim of sensitized phosphorescence to circumvent this decay and to realize energy transfer to an acceptor, a compound with an exceptional high phosphorescence efficiency in liquid solutions. This method has been applied both in homogeneous and in micellar solutions. In most applications the analyte is the donor compound, however, also an interesting example has been reported in which the analyte acts as the acceptor.

3.4.1.2   THEORETICAL ASPECTS

The sensitized RTPL pathway is depicted in the energy diagram of Fig. 20 and the simplified reaction scheme in Table VI. The donor (D) is excited by means of light absorption and reaches eventually the lowest singlet excited state denoted as $^1D*$. Subsequently the molecule crosses over to the triplet state $^3D*$. In absence of the acceptor, the following step is the return to the ground state. However, in the presence of an acceptor (A), energy transfer to the triplet state of the acceptor may occur, so that an acceptor molecule in its lowest triplet excited state $^3A*$ is produced:

$$^3D* + {}^1A \xrightarrow{k_t} {}^1D + {}^3A*$$

$$(32)$$

k_t is the bimolecular rate constant of this energy transfer reaction, expressed in $M^{-1} s^{-1}$. The final step is the phosphorescence emission from $^3A*$.

It is readily seen that the intensities of sensitized phosphorescence I_p (sens) can be expressed as a product of four independent factors (ref. 39), i.e.,

$$I_p (\text{sens}) = I_{abs}^D \cdot \theta_{isc}^D \cdot \theta_t^{DA} \cdot \theta_p^A \tag{33}$$

I_{abs}^D is the rate of light absorption by D, θ_{isc}^D is the intersystem crossing efficiency of D and thus the efficiency of triplet formation of the donor, θ_t^{DA} is the efficiency of energy transfer from D to A and θ_p^A is the phosphorescence efficiency of A.

In absence of inner filter effects I_{abs}^D is proportional to the concentration of D:

$$I_{abs}^D = 2.3\ I_{o,exc} \cdot \epsilon_{exc}^D \cdot [D]\, l \tag{34}$$

wherein $I_{o,exc}$ the intensity of the light source at λ_{exc}^D, ϵ_{exc}^D the molar absorptivity of D at λ_{exc} and l the optical pathlength. If the analyte acts as a donor the sensitized phosphorescence signal I_p (sens) is linearly dependent on the analyte concentration, as is obvious from Eqs. (33) and (34).

Obviously, the crucial factor in sensitized phosphorescence detection is θ_t^{DA}. This efficiency depends on the rate constant k_t, the acceptor concentration [A] and the lifetime of $^3D^*$ in absence of the acceptor, denoted as τ_o^D:

$$\theta_t^{DA} = \frac{k_t [A]}{1/\tau_o^D + k_t [A]} \tag{35}$$

To approach a transfer of 100 per cent, the conditon

$$k_t [A] \gg \frac{1}{\tau_o^D} \tag{36}$$

should be fulfilled. This means that energy transfer should be much faster than the overall decay rate of the triplet state of the donor. In general, provided that the energy transfer reaction is exothermic, k_t approximates the diffusional-controlled rate constant. Nevertheless some caution should be taken. The transfer reaction involves a change of spin state, which is only allowed in the electron exchange mechanism, requiring collisional interaction between $^3D^*$ and ^1A. In this mechanism k_t is proportional to the overlap between the normalized emission spectrum of D, i.e., $^3D^* \to$ ^1D, and the normalized singlet-triplet absorption spectrum of A, $^1A \to {}^3A^*$. In other words the intensities of the emission and the absorption band play no rule but nevertheless their shapes are important (ref.189). The problem is that S $\to$ T absorption spectra are not simply accessible. So it might be possible that for exothermic energy transfer k_t is much smaller than expected due to unfavourable overlap. Unidirectional energy transfer from $^3D^*$ to ^1A takes place if the energy of $^3D^*$ is at least 20 J/mol higher than the energy of $^3A^*$ (ref. 190); if the difference is smaller back transfer of energy plays a role (ref. 191).

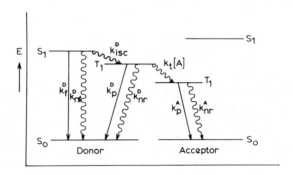

Fig. 20   Decay pathways in sensitized phosphorescence. k_f^D, k_{nf}^D and k_{isc}^D are the rate constants in s^{-1} of the intramolecular deactivation of the donor via fluorescence, internal conversion and intersystem crossing, respectively. k_p^D and k_p^A are the phosphorescence rate constants of the donor and the acceptor, while k_{nr}^D and k_{nr}^A are the overall rate constants of intra- and intermolecular nonradiative deactivation in s^{-1} of the T_1 state of the donor and acceptor. $k_t[A]$ is the apparent rate constant of the energy transfer reaction in s^{-1} (ref. 39).

Furtheron it is obvious from Eq. (36) that for a high sensitized RTPL signal the concentration of the acceptor should be taken as high as possible. It should be realized, however, that direct excitation of the acceptor must be avoided, because it would lead to a background phosphorescence signal. Fortunately, biacetyl, a compound with a high θ_p in a variety of liquid solutions, has very low molar absorption coefficients over a wide range of excitation wavelegths. Nevertheless, its concentration must not be chosen too high, i.e., about $10^{-4}M$.

Finally, it is emphasized that obviously the phosporescence efficiency of the acceptor is unfavourable influenced by impurity quenching. This means that quenchers, including oxygen should be removed from the solution, as much as possible.

TABLE VI   Reaction scheme of sensitized RTPL by triplet-triplet energy transfer.

---

1. Exitation of the donor (D)

$$^1D + h\gamma^D_{ex} \longrightarrow {}^1D*$$

2. Intersystem crossing to the triplet state

$$^1D* \longrightarrow {}^3D*$$

3. Energy transfer to the acceptor (A)

$$^3D* + {}^1A \xrightarrow{k_t} {}^1D + {}^3A*$$

4. Phosphorescence of the acceptor

$$^3A* \longrightarrow {}^1A + h\gamma^A_p$$

3.4.1.3   APPLICATIONS

Various organic compounds with low native fluorescence can be sensitively detected by sensitized RTPL; especially those compounds that undergo an efficient non-radiative decay via intersystem crossing. In principle sensitized RTPL is generally applicable to all compounds meeting the following two requirements: their T_1 state energy must be higher than that of the acceptor and they must show phosphorescence in rigid solutions at 77 K, since the main condition to be fulfilled is an efficient triplet formation. As sensitized RTPL is an indirect emission method, only the excitation properties of the analyte (acting as donor) play a role. All compounds are detected at a single wavelength, i.e., the emission wavelength of the acceptor which is relatively long. Monitoring of the same phosphorescence signal for all analytes implies that usually chromatographic separations are necessary if complex matrices, containing several sensitizing and even quenching compounds have to be analyzed. The commonly used phosphorophore in HPLC is biacetyl, a compound with favourable phosphorescence efficiencies in various solvents, used as eluent in HPLC. The concentration of biacetyl is usually not higher than 10^{-4} M in order to avoid background emission due to direct phosphorescence.

312

Compounds measured by sensitized phosphorescence are the well known polychlorinated biphenyls (PCBs) and naphthalenes (PCNs) (ref. 43). The detection limits are in the low nanogram region. Furtheron, ortho-substituted PCBs do not produce any RTPL signal. Therefore the combination of sensitized RTPL and UV absorption detection provides perspectives for the identification of complex PCB mixtures as frequently enountered in industrial samples. In Fig. 21 the UV and sensitized RTPL detected chromatograms of Aroclor 1221 are depicted. All PCBs present in the sample are detected in the UV-chromatogram. As expected the sensitized RTPL chromatogram is more simple since the peaks of the ortho-substituted biphenyls are missing. Of course the excitation wavelength is an additional selectively parameter in sensitized RTPL.

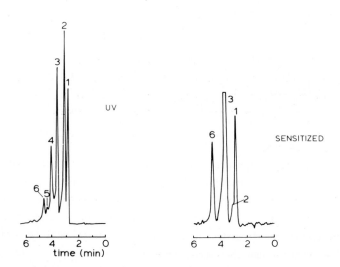

Fig. 21   Reversed-phase chromatograms of Aroclor 1221 obtained with UV detection (λ= 224 nm), concentration 50 ppm and sensitized RTPL detection (λ_{exc} = 265 nm, λ_{em} = 520 nm), concentration 10 ppm. Peaks: 1 = biphenyl, 2 = 2-chlorobiphenyl, 3 = 4-chlorobiphenyl, 4 = 2,2'-dichlorobiphenyl and 6 = 4,4'-dichlorobiphenyl (ref. 43).

A number of PCNs exhibit native fluorescence. Investigations on mixtures of PCNs reveal for most of the compounds a sensitized RTPL detection limit comparable with those obtained with fluorescence detection (ref. 191). Furtheron the linear range is the same as usually found by fluorescence, i.e., 3 to 4 decades. However, the tri- and tetra-substituted compounds have a deviating behaviour. Their triplet state energies are lower than for biacetyl, so that these compounds do not act as sensitizers

but as quenchers of the biacetyl phosphorescence (ref. 43). This aspect
will be discussed thoroughly in the next section.

Another interesting group of compounds which can be detected with RTPL
are the parent compound dibenzofuran and its chlorinated derivatives
(refs. 32, 192). These compounds, known as environmetal hazards have a
high toxicity depending on the numbers and positions of the chloro substituents. Sensitized RTPL detection seems appropriate as the triplet
formation efficiencies are considerable and the triplet energies are high
enough to guarantee an efficient energy transfer to biacetyl. As an
example the chromatogram of a mixture of dibenzofuran (DBF) and 2,8-di-
chlorodibenzofuran (2,8-Cl$_2$DBF) is given in Fig. 22. Various chlorinated
dibenzofurans also exhibit a native fluorescence, a property that has not
received a lot of attention in the literature. Hence, it is interesting to
compare the sensitivities of HPLC combined with fluorescence, sensitized
phosphorescence and UV-absorbance detection. It became clear that there is
a need to increase the sensitivity of the HPLC method by applying a preconcentration procedure. Furtheron, the data revealed that the fluorescence detection of chlorinated dibenzofurans is more sensitive than the
sensitized RTPL mode, if a grating instrument is used, whereas with a
filter instrument the reverse is true (ref. 192). This may be attributed
to the fact that in sensitized RTPL scattering background plays a less
important role.

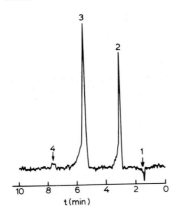

Fig. 22   Liquid chromatogram of a mixture of dibenzofuran (DBF) and 2,8-
          Cl$_2$DBF, separated on LiChrosorb RP-18 column, length 11 cm; flow
          rate 1 ml/min; eluent; 10^{-4}M biacetyl in acetonitrile/water
          83.7/ 16.3 (v/v); sensitized RTPL detection (λ_{exc} = 290 nm, λ_{em}
          = 522 nm); 1 = solvent peak; 2 = DBF (34 ng); 3 = 2,8-Cl$_2$DBF (47
          ng); 4 = unknown (ref. 39).

Another interesting application of sensitized RTPL is the analysis of
biacetyl itself. This determination is important because of the great

influence of biacetyl on the flavour of beer, wine and several dairy products. In these samples biacetyl concentrations in the order of 1 ppb to 1 ppm are relevant. Though biacetyl is a good phosphorophore in fluid solutions, it is very inefficiently excited, because of its extremely low molar absorptivity (above 220 nm $\epsilon < 20$ M^{-1} cm^{-1}). This implies that the detection limits for biacetyl in beer, achieved by UV/Vis absorption and by direct phosphorescence measurements are both very unfavourable. However, sensitized phosphorescence detection can be invoked to improve the excitation of biacetyl in an indirect way (ref. 40, 41). Now biacetyl, acting as acceptor is the analyte and a suitable donor compound has to be found. The appropriate donor should have a high absorptivity, it should guarantee a high energy transfer efficiency to biacetyl, it should be non-phosphorescent itself, but its triplet state T_1 should have a large life-time. For extremely low biacetyl concentrations it is obvious that

$$k_t[A] << (\tau_0^D)^{-1} \tag{37}$$

so that the efficiency of energy transfer (see Eq. 22) can be approximated as

$$\theta_t^{DA} = \tau_0^D k_t[A] \tag{38}$$

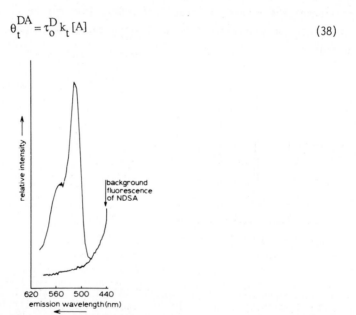

Fig. 23 Sensitized phosphorescence emission spectrum of a mixture of NDSA and biacetyl in a deoxygenated solution of water/acetonitrile 70/30 (v/v) at room temperature. The background fluorescence of NDSA is recorded without deoxygenation; λ_{exc} = 302 nm (a favourable wavelength to excite NDSA) (ref. 35).

Under these conditions the sensitized phosphorescence signal I_p(sens) is linearly dependent on the biacetyl concentration. A number of potential donors has been tested; 1,5-naphthalenedisulfonic acid disodium salt (NDSA) is an appropriate donor. This compound has a favourable solubility in polar solvent mixtures and gives no retention on a reversed-phase column. Fig. 23 shows the emission spectrum of a deoxygenated water/acetonitrile 70/30 (v/v) solution of 1.5×10^{-4}M NDSA and about 5×10^{-8} M biacetyl. A chromatogram for a beer sample containing 14 ppb biacetyl is given in Fig. 24; the corresponding limit of detection is 0.5 ppb.

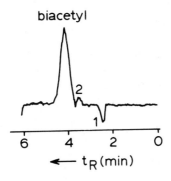

Fig. 24   HPLC chromatogram of a beer sample containing 14 ppb biacetyl with sensitized phosphorescence detection (λ_{exc} = 310 nm; λ_{em} = 516 nm). Column: 25 cm x 4.6 mm I.D. 5 μm Spherisorb; eluent: water/acetonitrile, 70/30 (v/v), pH = 6.5, with 2×10^{-4} M NDSA. 1 = solvent front; 2 = unknown (ref. 35).

3.4.2   QUENCHED PHOSPHORESCENCE
3.4.2.1   INTRODUCTION

Another indirect detection method based on a dynamic principle concerns the quenched phosphorescence in liquid solutions at room temperature. Analytes able to quench the phosphorescence of a compound, for example biacetyl, cause a decrease of the monitored signal. As the amount of quenching is determined by the rate constant of the reaction, between analyte and excited phosphorophore, the quenched phosphorescence method has an inherent selectivity. It is especially useful when one or a few analytes have to be determined in presence of a number of other compounds, for example in complex matrices such as body fluids or environmental samples.

3.4.2.2   THEORETICAL ASPECTS

The reaction scheme showing the reaction pathways for quenched RTPL is given in Table VII. For convenience it is based on biacetyl (B) as the phosphorophore. First biacetyl is excited by means of light absorption directly in an excited singlet state. Subsequently biacetyl falls down to its triplet state T_1. In absence of a quencher, the following step is the return to the ground state by emission of phosphorescence. In presence of a quencher which is able to react rapidly with $^3B*$ the lifetime of biacetyl in the triplet state will decrease, resulting in a reduction of the monitored biacetyl phosphorescence signal (ref. 42).

TABLE VII   Reaction scheme for quenched RTPL of biacetyl by triplet-triplet energy transfer.

---

1. Excitation of biacetyl (B)

$$^1B + h\gamma_{exc}^B \longrightarrow {}^1B*$$

2. Intersystem crossing to the triplet state

$$^1B* \longrightarrow {}^3B*$$

3. Phosphorescence of biacetyl

$$^3B* \longrightarrow {}^1B + h\gamma_p^B$$

4. Quenching of biacetyl phosphorescence by a quencher Q (the analyte)

$$^3B* + {}^1Q \longrightarrow {}^1B + {}^1Q$$

In absence of quencher, the phosphorescence I_o of biacetyl can be expressed as

$$I_o = I_{abs}^B \, \theta_{isc}^B \, \theta_p^B \tag{39}$$

where I_{abs}^B is the rate of light absorption by biacetyl and θ_{isc}^B and θ_p^B represent the respective efficiencies of intersystem crossing and phosphorescence (since only the phosphorescence properties of B play a role, in the following the superscript B will be deleted). The last efficiency is given by

$$\theta_p = k_p \tau_o \tag{40}$$

wherin τ_o is the triplet state lifetime of biacetyl (s). If, however, a quencher is present the lifetime will be reduced, and the following relation holds

$$\tau^{-1} = \tau_o^{-1} + k_Q [Q] \tag{41}$$

k_Q is the bimolecular rate constant of the quenching reaction ($M^{-1}s^{-1}$) and [Q] the concentration of the quencher (M). As a result the phosphorescence signal intensity will decrease from I_o to I. For the ratio I_o/I a relation similar to the well-known Stern-Volmer equation in fluorescence can be derived

$$\frac{I_o}{I} = 1 + k_Q \tau_o [Q] \tag{42}$$

As the quotient $k_Q \tau_o / I_o$ for a chosen phosphorophore and analyte (the quencher) has a constant value it is clear from Eq. (42) that $I^{-1} - I_o^{-1}$ depends linearly from [Q]. This implies that plotting of I^{-1} versus [Q] delivers a straight line with a slope proportional to k_Q and an intercept equal to the inverted intensity of the unquenched signal. In order to realize a linear response between the concentration and the signal height in several experiments an electronic signal inverter has been introduced between the photomultiplier tube and the recorder.

Furtheron it is obvious from Eq. (42) that the sensitivity of the quenched RTPL method for a particular phosphorophore and quencher is determined by the quenching rate constant k_Q. If the lifetime τ_o is known the rate constant k_Q can be calculated from the slope of the Stern-Volmer plot. In static experiments, the value of τ_o in different solvents can be readily estimated in an indirect way from the ratio of the phosphorescence and fluorscence intensities of biacetyl, measured at their respective maxima. Of course, in time-resolved experiments τ_o can be derived directly from the decay of the phosphorescence signal.

Subsequently an estimation of the limits of detection for the analytes can be obtained, since it can be derived that for batch experiments approximately holds

$$\text{l.o.d. (M)} = \frac{10}{k_Q} \text{ (M}^{-1} \text{ s}^{-1})$$

(43)

This implies that the quenched RTPL method would be of interest for analytes quenching the biacetyl phosphorescence with rate constants 10^7 to 10^9 M^{-1} s^{-1}, depending on the availability of alternative detection methods. Of course the sensitivity depends also on τ_0 and thus on the amount of oxygen and impurities present in the solution. Furthermore, the achievable detection limits improve with increasing I_0, since at higher I_0 the signal to noise ratio becomes more favourably. The quenched RTPL is also a selective method, because only analytes with sufficiently high k_Q values can be detected. Otherwise this is the only requirement for detectability. The absorption spectra of the analytes do not play a role. The same holds for the quenching mechanism. Quenched RTPL is not limited to analytes with triplet states lying lower than the T_1 state of the phosphorophore. Other mechanisms, such as electron transfer and proton abstraction may be also involved in the quenched RTPL. Especially the former seems to be important. This explains why for instance some inorganic ions are sensitively detectable with quenched RTPL.

Finally it is emphasized that also fluorescence quenching has been introduced as a detection method in HPLC, i.e., by the group of Winefordner in 1981 (refs. 193, 194). However, it should be realized that this detection principle is essentially different from the quenched phosphorescence method described in the present chapter. Fluorescence quenching is based on the static quenching as a result of the interaction between the fluorophore and the analyte in their electronic ground states and/or on the absorption of fluorescence radiation by the analyte. Dynamic quenching of fluorescence is not effective since the lifetime of the lowest excited singlet states is too short.

3.4.2.3 APPLICATIONS

The potential of quenched RTPL for the detection of a large number of both organic and inorganic compounds has been examined. The relevance of quenched RTPL is obvious since it is applicable for the sensitive detection of various groups of compounds with otherwise poor detection properties.

The chromophoric properties of the analyte do not play a role in quenched RTPL; the analyte needs not to be excited. Also the mechanism of the quenching process is not important. Besides energy transfer reactions, also electron transfer reactions or even proton abstraction reactions may be operative. The only condition is that the quencher reacts rapidly with biacetyl in the triplet state; the rate constant k_Q determines the sensitivity of quenched RTPL for a particular analyte. Generally, quenched RTPL detection is relevant for analytes with rate constants ranging from 10^7 to 10^9 M^{-1} s^{-1} (ref. 42). A screening test of various types of compounds has revealed that quenched RTPL can be used among others for higher chlorinated naphthalenes, aromatic and aliphatic amines, sulphur containing organics such as thioureas and phenothioazines and several inorganic ions (e.g., NO_2^-, SCN^-, $S_2O_3^{2-}$, Sn^{2+}, CrO_4^{2-}) and complexes (e.g., some antitumor agents based on Pt(II)). It is emphasized that quenched RTPL has also an inherent selectivity. This is interesting as in the field of ion chromatography mobile phases with high ionic strengths are used. In general only those ions are observed by quenched phosphorescence that either have a low lying triplet state energy or a low oxidation potential. The others can be used as eluent constituents without inducing any background signal.

In the foregoing section we have seen that PCNs can be detected by sensitized biacetyl phosphorescence though the sensitivity for some higher chlorinated PCNs is low. This has to be ascribed to the triplet energy values of these compounds, which are about the same as, or even lower, than the triplet energy of biacetyl. In this case a reversed energy transfer reaction from biacetyl to the PCN must be taken into account, which leads to a decrease in the sensitized (and the direct) RTPL signal of biacetyl. A reversed energy transfer implies that in addition to UV absorption and sensitized phosphorescence also quenched phosphorescence can be invoked for the analyses and identification of industrial PCN mixtures (ref. 43). It is emphasized that the application of quenched RTPL instead of sensitized RTPL requires only an adjustment of the excitation wavelength. In Fig. 25 the reversed phase chromatograms of Halowax 1099 obtained with UV, sensitized RTPL and quenched RTPL detection are represented.

The sensitivity of the quenched phosphorescence detection is demonstrated for thiourea derivatives. In Fig. 26 a chromatogram detected by quenched RTPL is depicted, using a signal inverter (ref. 42). Quenched RTPL has been used to detect some Pt(II) coordination complexes, which are well known agents with antitumor activity (ref. 46). The k_Q values for the quenching of biacetyl phosphorescence are 7×10^8 and 4×10^8 M^{-1} s^{-1} for CDDP (cisplatin) and CBDCA (carboplatin), respectively; these values give rise to interesting LOD values. On the contrary CHIP, a Pt(IV) derivative, does hardly show any quenching ($k_Q < 10^6 M^{-1}s^{-1}$). This result may be an indication that the quenching is based on electron transfer from the

320

platinum ion to the excited biacetyl molecule. In Fig. 27 a chromatogram of a standard solution of CDDP and CBDCA is presented using quenched phosphorescence as detection (λ_{exc} = 415 mm, λ_{em} = 520 nm). The limits of detection calculated from this chromatogram are 3.0 x 10^{-7} M for CDDP and 3.3 x 10^{-7} M for CBDCA. These data compare favourably with LODs obtained via other detection methods (see Table VIII). Experiments on urine and plasma samples showed interfering compounds co-eluting with the platinum species, so that a clean-up procedure is necessary (ref. 47). In Fig. 28 a chromatogram for a blank plasma sample and a sample spiked with CDDP is depicted, obtained after applying such a sample handling procedure; the chromatography is based on a solvent generated anion exchanger system. It is clear that in this way CDDP can be determined quantitatively in plasma. The same holds for CDDP in urine. The sensitivity of the method is sufficient for the monitoring of therapeutic CDDP levels in clinical samples.

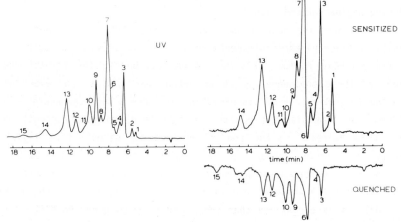

Fig. 25   Reversed-phase chromatograms of Halowax 1099 obtained with UV detection (λ_{exc} = 233 nm; 0.32 a.u.f.s.), sensitized RTPL detection (λ_{exc} = 300 nm; λ_{em} = 520 nm) and quenched RTPL detection (λ_{exc} = 415 nm; λ_{em} = 520 nm); injected amount of sample in all chromatograms: 1 μg. The corresponding peaks in the three chromatograms are indicated by the figs. 1 to 15; it is obvious that combination of these three detection modes is helpful for identification purposes (ref. 43).

Finally, we will consider the determination of chromate as an example of quenched RTPL detection in ion-chromatography (ref. 48). The trivalent cation Cr(III) which is an essential trace element to man, hardly does quench biacetyl phosphorescence. The chromate ion Cr(VI), an enzymatic poison leading to hepatitic and renal damage by exposure, can be determined selectively after separation on a paired-ion reversed phase HPLC system with quenched phosphorescence detection. In these measurements the possibility of time resolution in quenched phosphorescence detection has been invoked. The signal reduction caused by the quencher CrO_4^{2-} depends on both the delay time t_d and the gating time t_g, experimental parameters in a pulsed source detection system. This leads to a modified Stern-Volmer equation:

$$\frac{I_{pulse}}{I'_{pulse}} = (1 + \tau_o k_Q [Q] e^{t_d k_Q [Q]}) \tag{44}$$

wherein the subscription "pulse" is used to denote that the pulsed phosphorescence detection mode is applied. This equation applies provided that t_g is chosen long enough to guarantee that after t_g seconds less than 1 per cent of the signal is recorded. It is obvious that $(I'_{pulse})^{-1}$ is not linearly dependent on [Q]. The influence of the exponential term decreases if t_d is shortened. Experiments on the biacetyl system have shown that for t_g is 0.9-1.0 ms the signal to noise ratio is the most favourable and that the optimum value for t_d is about 0.10 ms. In Fig. 29 a chromatogram for a 1.7×10^{-6} M chromate standard solution is given. The detection limit achieved is 1.4×10^{-7} M, corresponding to 16 ppb, which underlines the relevance of the method as the maximum concentration allowed in drinking water is 50 ppb. The linearity of the method is three orders of magnitude provided that an electronic signal inverter is applied. From the data in Table IX it is clear that the sensitivity of the quenched RTPL method compares reasonably with other detection methods.

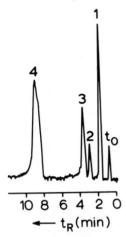

Fig. 26   Quenched RTPL chromatogram of some thiourea derivatives; eluent; 1.0×10^{-4} M biacetyl in water; column RP-18, 12 cm, d_p = 5 μ; flow rate = 1 ml/min; observed t_R (= t_o) for $NaNO_2$ = 45 s; 1 = thourea (41 ng), 2 = thiohydantoine (63 ng), 3 = ethylenethiourea (55 ng), 4 = methimazole (62 ng) (ref. 42).

TABLE VIII   Comparison of l.o.d. values (ng/ml) for CDDP and CBDCA
derived from liquid chromatography with different detection
methods.

Detection method	CDDP	CBDCA	Ref.
UV	20,000	20,000	195
UV after derivatization	40	1200	195
Electrochemicial	15	15	196
		After precolumn derivatization	
Chloride-assisted electrochemical	50	Not measured	197
QRTPL	90	122	46

TABLE IX   Comparison of l.o.d. values (ng/ml) for chromate obtained with
different HPLC methods.

Detection method	l.o.d.	Ref.
Electrochemical (as diethyldithiocarbamate complex)	4	198
Colorimetric (as 1,5-diphenylcarbazide complex)	100	199
Direct current plasma (DCP) emission spectroscopy	7	200
Quenched RTPL	16	48

The chromatogram of tapwater (Fig. 30) and surface water samples
spiked with chromate show the applicability of the quenched phosphorescence method for real samples.

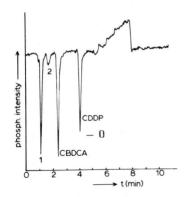

Fig. 27   Quenched phosphorescence chromatogram of CDDP $(6.1 \cdot 10^{-6}$ M) and
CBDCA $(8.3 \cdot 10^{-6}$ M) in 0.15 M aqueous NaCl. 1 = chloride, 2 =
unknown. Column: ODS Hypersil column 5 μm (10 cm x 4.6 mm)
prepared with hexadecyltrimethylammonium bromide (HTAB). Mobile
phase: water/methanol, 99/1 (v/v), pH 5.0 (citrate buffer), 2 x
10^{-5} M HTAB and 10^{-2} M biacetyl. λ_{ex} = 415 nm, λ_{em} = 520 nm
(ref. 46).

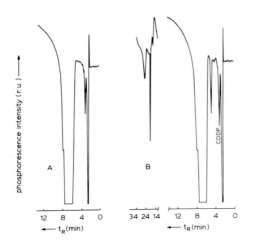

Fig. 28   Chromatogram of a blank plasma sample (A), and of a plasma
sample spiked with 5 x 10^{-6} M CDDP (B). For chromatographic
conditions, see Fig. 27 (ref. 47).

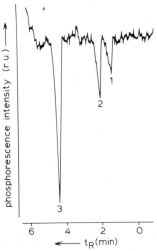

Fig. 29   Chromatogram of a standard solution of chromate (1.7×10^{-6}M). 1 and 2 are quenching impurities, 3 = chromate. Chromatographic conditions: column 12 cm × 4.6 mm I.D. 5 μm ODS Spherisorb; mobile phase: water/acetonitrile 95/5 (v/v), $2 \times_3 10^{-3}$M phosphate buffer, pH = 7.1, 2×10^{-4} M TBACl and 5×10^{-3} M biacetyl; flow rate = 1.9 ml min^{-1}. Time-resolved phosphorescence detection: t_d = 0.01 msec, t_g = 1.00 msec, λ_{exc} = 400 nm (broad band filter), λ_{em} = 515 nm (ref. 48).

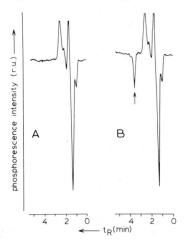

Fig. 30   Chromatogram of a blank tap water sample (A) and of a tap water sample spiked with 1×10^{-5} M chromate (B). The positive peaks are attributed to impurities present in the samples exhibiting native fluorescence or sensitized phosphorescence of biacetyl. The negative peaks are caused by quenching compounds. Chromatographic and detection conditions as in Fig. 29 (ref. 48).

3.5   ALTERNATIVE PHOSPHOROPHORES/LUMINOPHORES

Though the indirect phosphorescence detection, both sensitized and quenched, has proved to give promising results in HPLC experiments with biacetyl as phosphorophore this combination has some inherent disadvantages. First biacetyl is consumed during the experiments as it is part of the eluent. Secondly it is not stable at pH values higher than 7. Thirdly, the measurements have to be done under oxygen free conditions. The first drawback can be raised by using immobilized phosphorophores (ref. 49). In order to eliminate the oxygen removal step rare earth metal ions, which are well known luminophores have been studied in quenched luminescence detection (ref. 50). In this case the transition responsible for the light emission does not involve a triplet and singlet state. Therefore the more general term luminophore is used.

In this section attention will be paid to two alternative phosphorophores/luminophores as detection method in flow systems as HPLC.

3.5.1 IMMOBILIZED PHOSPHOROPHORES

There is a number of advantages in employing immobilized phosphorophores in indirect phosphorescence detection. First there is an extension of the solvent compatability, so that apolar phosphorophores may be combined with aqueous mobile phases and reversed phase HPLC columns. Furtheron, because of immobilization no losses of phosphorophore occur and finally, the localization of the phosphorophore in the cell makes it possible to apply other simpler oxygen removal procedures. Disadvantages are the background scattering caused by the support and the limited transparancy of the cell which limits the excitation range of the phosphorophore down to about 340 nm. Fortunately, the effect of scattering can be efficiently suppressed by using a pulsed source-time resolved luminescence detection system.

Results obtained with 1-bromonaphthalene, a compound known from its RTPL properties (see Table V) have been reported. The phosphorophore is covalently bonded to a solid support via a spacer to create pseudo solution conditions. The immobilization reaction (ref. 49) is schematically represented in Fig. 31. The synthesized batches, which are very stable, could be used under flow conditions at least during six months. They could be applied without problems between pH 2 and 8. The immobilization hardly affects the spectroscopic emission characteristics of the phosphorophore. Although excitation of 1-bromonaphthalene itself is almost impossible at wavelengths higher than 340 nm, introduction of the C=O group on the 4- position results in a low absorptivity in the 340-375 nm region. The possibilities of this system for quenched phosphorescence detection have been tested by injection of a 7×10^{-6} M nitrite solution on a HPLC column. The analytical data are comparable with those obtained with the homogeneous biacetyl system; the detection limit is 2×10^{-7} M.

For I⁻ a lower signal was observed.

According to the Stern-Volmer relation (see Eq. 42) the sensitivity of the quenched phosphorescence detector is determined by the product $k_Q \tau_0$. As the phosphorescence lifetime τ_0 depends mainly on the phosphorophore used (τ_0 is about 0.9 ms) the lower signal for I⁻ has to be described to the relatively low quenching constant k_Q ($k_Q = 1 \times 10^8 M^{-1} s^{-1}$ for I⁻ and $k_Q = 3.3 \times 10^9 M^{-1} s^{-1}$ for NO_2^-). From the data represented, the potential of an immobilized phosphorophore in quenching phosphorescence detection is clear. Until now an example, wherein the immobilized phosphorophore acts in the sensitized mode has not been reported.

Fig. 31 a. Synthesis of 10-(4-bromo-1-naphthoyl) decylamine, intermediate I.
b. Immobilization of intermediate I to silanized silica gel or
CPG (ref. 49).

3.5.2 RARE EARTH METAL IONS AS LUMINOPHORES

The spectroscopic properties of lanthanide ions have been subject of research since many years, but the application of the luminescence of these ions as detection method in liquid chromatography (LC) is still rather restricted. Wenzel at al., have developed an LC method for the determination of polynucleotides and nucleic acids with xanthine, guanine and thioridine unities (ref. 201). After chromatographic separation, a post-column complexation of the analytes with Eu(III) or Tb(III) leads to products that can be sensitively detected by lanthanide luminescence. The use of Eu(III) and Tb(III) as luminophores in LC in a sensitized detection mode has been described by Dibella et al. (ref. 202). The analytes (organic compounds) are excited by UV radiation, energy transfer from the triplet state of the analyte to the lanthanide ion occurs and luminescence of the lanthanide is detected. In this set-up energy transfer is the crucial step starting at the donor in the triplet state, which explains that deoxygenation of the solution is required.

Baumann et al. have examined the use of rare metal ions Eu(III) and Tb(III) as luminophores for quenched detection in LC (ref. 50).

The long living lanthanide luminescence (>0.1 ms) arises from (forbidden) transitions between levels belonging to the 4 f^n electron configuration. As the 4 f electrons are shielded by 5 s and 5 p electrons, the observed transitions are very sharp, even for spectra of liquid solutions. Non-radiative decay usually competes strongly with radiative relaxation, especially in H_2O.

From a practical point of view compared to biacetyl phosphorescence, Eu(III) and Tb(III) luminescence have the strong advantage that oxygen quenching hardly plays any role. This implies that just a common HPLC set-up is compatible with the quenched lanthanide luminescence detection mode; no special experimental requirements need to be fulfilled. On the other hand the fact that oxygen does not induce significant quenching implies that, im comparison with biacetyl, other quenching mechanisms may be operative so that another selectivity should be expected.

As has been pointed out in section 3.4.2.2 the detection limits achievable in dynamically quenched luminescence detection depend on the noise on the luminescence signal and thus on the intensity of this signal. Unfortunately for Eu(III) in practice a relatively low luminescence level is reached because of its low absorptivity ($\epsilon <$ 10 M^{-1} cm^{-1} and the absorption peaks are narrow). For Tb(III) the situation is more favourable: the transition 4 $f^8 \longrightarrow$ 4 $f^7 5d^1$ lying at 220 nm is allowed ($\epsilon \approx$ 300 M^{-1} cm^{-1} in a rather broad band) so that rather efficient excitation is possible provided that a lamp is available with a favourable output at that wavelength.

Preliminary HPLC experiments have been reported for nitrite solutions. The $EuCl_3$ containing mobile water phase was pumped directly through a Li-Chrosorb RP-18 analytical column. The time resolved measurement (t_d = 30 μs, t_g = 2.0 ms) results in a limit of detection of only 2 x 10^{-5} M. As expected for Tb(III) the l.o.d. is more favourable. It was found to be 5 x 10^{-8}M, a quite promising result (ref. 50).

Obviously there are still some aspects to be examined. First of all, it has been shown that the Eu(III) luminescence lifetime is increased about a factor of 30 upon going from H_2O to D_2O; hence such an eluent might be very interesting assuming that microbore HPLC is feasible. Secondly, the choice of buffers, frequently necessary in ion-chromatographic separation seems to be limited; some ions form stable complexes with the lanthanides thus reducing dynamic quenching of luminescence. Thirdly, the excitation efficiency of the lanthanides might be improved via complexation; on the other hand the ligands used in these complexes should not inhibit dynamic quenching processes.

Summarizing this section, we do expect in the near future the publication of some interesting applications of lanthanide luminescence detection in LC especially for the measurements of inorganic ions.

3.6 CONCLUDING REMARKS

From the results described above it is obvious that phosphorescence detection has its particular applicability field in HPLC, despite of the fact that the phenomenon of phosphorescence in normal fluid solutions is quite exceptional. It is important to emphasize that the experimental requirements to be met are not difficult both to realize and to maintain and that the same equipment as applied in fluorescence detection can be utilized. The main requirement is to apply a closed chromatographic system based on stainless steel capillaries and to deoxygenate continuously the eluent by leading nitrogen gas through the eluent vessel.

In sensitized phosphorescence detection, generally the analyte acts as a donor. Hence this detection mode is only appropriate for compounds with an absorption spectrum in the near UV and visible wavelength region. Therefore the results should be considered as additive to those obtained with absorption detection. As has been shown above, the combination of both detection techniques, and if relevant further combintion with fluorescence and / or quenched phosphorescence detection, is quite adequate for the analysis and identification of complex mixtures. The applicability of sensitized phosphorescence wherein the analyte is the acceptor, as for instance utilized for the determination of biacetyl in beer samples, is limited since it requires a high phosphorescence efficiency for the analyte itself.

The quenched phosphorescence detection mode presumably has a wider applicability range than the sensitized one. This indirect method is essentially different from indirect UV absorption- and fluorescence detection known in ion chromatoraphy which both are based on displacement effects. In phosphorescence detection, a dynamic quenching effect is monitored, i.e., a decrease of the phosphorescence efficiency induced by the analytes. This explains the impressive detection limits obtained for efficient quenching analytes. Especially in ion chromatography this detection technique has a good potential.

Nevertheless phosphorescence detection is not widely applied yet. Obviously, the need to deoxygenate the eluent is a serious hindrance for its general acceptance. From this point of view the new developments described in section 3.5, namely the use of immobilized phosphorophores and rare earth luminophores, offer new perspectives. Especially the preliminary results obtained for Eu(III) and Tb(III) luminophores are quite promising since in these systems oxygen removal is not necessary in the quenched mode. It is emphasized that, compared to fluorescence detection, phosphorescence has the great advantage that background radiation due to fluorescent impurities and to scattering can be eliminated quite easily without the need of expensive equipment; a pulsed Xe-lamp instrument suffices.

REFERENCES
1   H. Lingeman, W.J.M. Underberg, A. Takedate and Hulshoff, J. Liq. Chromatogr., 8, 789 (1985).
2   R.W. Frei, N.H. Velthorst and C. Gooijer, Pure & Appl. Chem., 57, 483 (1985).
3   U.A.Th. Brinkman, G.J. de Jong and C. Gooijer, Pure & Appl. Chem., 59, 625 (1987).
4   U.A.Th. Brinkman, Chromatographia, 24, 190 (1987).
5   P. Lindroth and K. Mopper, Anal. Chem., 51, 1667 (1979).
6   J. D'Souza and R.J. Ogilvie, J. Chromatogr., 232, 212 (1982).
7   P. Kucera and H. Umagat, J. Chromatogr., 255, 563 (1983).
8   P. Boehlen and H. Mellet, Anal. Biochem., 94, 313 (1979).
9   J.F. Lawrence and R.W. Frei, Chemical Derivatization in Liquid Chromatography, Elsevier, Amsterdam (1976).
10 J.F. Lawrence, Prechromatographic Chemical Derivatization in Liquid Chromatography, in Chemical Derivatization in Analytical Chemistry, Vol. 2: Separation and Continuous Flow Techniques, R.W. Frei and J.F. Lawrence Eds., Plenum Press, New York (1982).
11 R.W. Frei and A.H.M.T. Scholten, J. Chromatogr. Sci., 17, 152 (1979).
12 R.W. Frei, J. Chromatogr., 165, 75 (1979).
13 J.T. Stewart, Trends Anal. Chem., 1, 170 (1982)
14 G. Schwedt and E. Reh, Chromatographia, 14, 249 (1981).
15 G. Schwedt and E. Reh, Chromatographia, 14, 317 (1981).
16 E.C. Toren Jr. and D.N. Vacik, Anal. Chim. Acta, 152, 1 (1983).
17 R.W. Frei, Chromatographia, 15, 161 (1982).
18 J.F. Lawrence, U.A.Th. Brinkman and R.W. Frei, J. Chromatogr., 185, 473 (1979).
19 R.J. Reddingius, G.J. de Jong, U.A.Th. Brinkman and R.W. Frei, J.Chromatogr., 205, 77 (1981).
20 J.H. Wolf, C. de Ruiter, U.A.Th. Brinkman and R.W. Frei, J. Pharm. Biomed. Anal., 4, 523 (1986).
21 C. de Ruiter, J. Hefkens, U.A.Th. Brinkman, R.W. Frei, M. Evers, E. Matthijs and J.A. Meijer, Int. J. Environ. Anal. Chem., 31, 325 (1987).
22 F. Smedes, J.C. Kraak, C.E. Werkhoven-Goewie, U.A.Th. Brinkman and R.W. Frei, J. Chromatogr., 247, 123 (1982).
23 A.H.M.T. Scholten, U.A.Th. Brinkman and R.W. Frei, Anal. Chim. Acta, 114, 137 (1980).

330

24   M. Uihlein and E. Schwab, Chromatographia, 15, 140 (1982).
25   M.L. Grayeski, Anal. Chem., 59, 1243A (1987).
26   P. van Zoonen, Thesis, Free University Amsterdam, 1987.
27   W.R. Seitz and M.P. Neary in Contemporary Topics in Analytical
     Chemistry, Volume I. D.M. Hercules (Ed.), Plenum, New York (1977).
28   G. Mellbin, J. Liq. Chromatogr., 6, 1603 (1983).
29   S. Kobayashi and K. Imai, Anal. Chem., 52, 424 (1980).
30   P. van Zoonen, D.A. Kamminga, C. Gooijer, N.H. Velthorst and R.W.
     Frei, Anal. Chim. Acta, 167, 249 (1985).
31   P. van Zoonen, D.A. Kamminga, C. Gooijer, N.H. Velthorst , R.W.Frei
     and G.Gübitz,Anal.Chim.Acta, 174, 151 (1985)
32   G. Gübitz, P. van Zoonen, C. Gooijer, N.H. Velthorst and R.W. Frei,
     Anal. Chem., 57, 2071 (1985).
33   P. van Zoonen, I. de Herder, C. Gooijer, N.H. Velthorst, R.W. Frei,
     E. Küntzberg and G. Gübitz, Anal. Letters., 19, 1949 (1986).
34   P. van Zoonen, C. Gooijer, N.H. Velthorst, R.W. Frei, J. H. Wolf, J.
     Gerrits and F. Flengte, J. Pharm. & Biomed. Analysis, 5, 485 (1987).
35   C. Gooijer, R.A. Baumann and N.H. Velthorst, Progr. Anal. Spectrosc.,
     10, 573 (1987).
36   J.J. Donkerbroek, Thesis, Free University Amsterdam, 1983.
37   R.A. Baumann, Thesis, Free University Amsterdam, 1987.
38   J.J. Donkerbroek, C. Gooijer, N.H. Velthorst and R.W. Frei, Anal.
     Chem., 54, 891 (1982).
39   J.J. Donkerbroek, N.J.R. van Eikema Hommes, C. Gooijer, N.H.
     Velthorst and R.W. Frei, Chromatographia, 15, 218 (1982).
40   R.A. Baumann, C. Gooijer, N.H. Velthorst and R.W. Frei, Anal. Chem.,
     57, 1815 (1985).
41   R.A. Baumann, C. Gooijer, N.H. Velthorst. R.W. Frei, J. Strating,
     L.C. Verhagen and R.C. Veldhuyzen-Doorduin, Intern. J. Environ. Anal.
     Chem., 25, 195 (1986).
42   J.J. Donkerbroek, A.C. Veltkamp, C. Gooijer, N.H. Velthorst and R.W.
     Frei, Anal. Chem., 55, 1886 (1983).
43   J.J. Donkerbroek, N.J.R. van Eikema Hommes, C. Gooijer, N.H.
     Velthorst and R.W. Frei, J. Chromatogr., 255, 581 (1983).
44   C. Gooijer, N.H. Velthorst and R.W. Frei, Trends in Anal. Chem., 3,
     259 (1984).
45   C. Gooijer, P.R. Markies, J.J. Donkerbroek, N. H. Velthorst and R.W.
     Frei, J. Chromatogr., 289, 347 (1984).
46   C. Gooijer, A.C. Veltkamp, R.A. Baumann, N.H. Velthorst, R.W. Frei
     and W.J.H. van der Vijgh, J. Chromatogr., 312, 337 (1984).
47   R.A. Baumann, C. Gooijer, N.H. Velthorst, R.W. Frei, I. Klein and
     W.J.F. van der Vijgh, J. Pharm. Biomed. Anal., 5, 2, 165 (1987).
48   R.A. Baumann, M. Schreurs, C. Gooijer, N.H. Velthorst and R.W. Frei,
     Can. J. Chem., 65, 965 (1987).
49   R.A. Baumann, C. Gooijer, N.H. Velthorst, R.W. Frei, I. Aichinger and
     G. Gübitz, Anal. Chem., 60, 1237 (1988).
50   R.A. Baumann, D.A. Kamminga, H. Derlagen, C. Gooijer, N.H. Velthorst
     and R.W. Frei, J. Chromatogr., 439, 165 (1988).
51   W.P. Bostick, M.S. Denton and S.R. Dinsmore in Bioluminescence and
     Chemiluminescence, Instruments and Applications, K. van Dijke (Ed.),
     CRC, Boca Raton, (1985) Vol. II.
52   J.W. Birks and M.C. Kuge, Anal. Chem., 52, 897 (1980).
53   B. Shoemaker and J.W. Birks, J. Chromatogr., 209, 251 (1981).
54   H. Schaper, J. Electroanal. Chem., 129, 335 (1981).
55   R.C. Massey, C. Crews, D.J. McWeeny and M.E. Knowles, J. Chromatogr.,
     236, 527 (1982).
56   W.C. Yu and E.U. Golf, Anal. Chem., 55, 29 (1983).
57   A.V.Hartkopf and R. Delumya, Anal. Lett., 7, 79 (1974).
58   M.P. Neary, W.R. Seitz and D.M. Hercules, Anal. Lett., 7, 583 (1974).
59   R. Delumya and A.V. Hartkopf, Anal. Chem., 18, 1402 (1976).

60   J.L. Burguera, M. Burguera and A. Townshend, Anal. Chim. Acta, 127, 199 (1981).
61   T.Hara, M. Toriyama and K. Tsukagoshi, Bull. Chem. Soc. Japan, 56, 1382 (1983).
62   T. Hara, M. Toriyama and T. Ebuchi, Bull. Chem. Soc. Japan, 58, 109 (1985).
63   A. MacDonald and T.A. Nieman, Anal. Chem. 57, 936 (1985).
64   J.P. Auses, S.L. Cook and J.T. Maloy. Anal. Chem., 47, 244 (1975).
65   D.T. Bostick and D.M. Hercules, Anal. Chem., 47, 447 (1975).
66   C. Ridder, E.H. Hansen and J. Ruzicka, Anal. Lett., 15, 1751 (1982).
67   C.A. Koerner and T.A. Nieman, Anal. Chem., 58, 116 (1986).
68   R. Fagerstroem, P. Seppanen and J. Janne, Clin. Chim. Acta, 143, 45 (1984).
69   U. Bachrach and Y.M. Plesser, Anal. Biochem., 152, 423 (1986).
70   M.P. Neary, R.W. Seitz and D.M. Hercules, Anal. Lett., 7, 583 (1974).
71   M.L. Gandelman and J.W. Birks, J. Chromatogr., 242, 21 (1982).
72   T. Kawasaki, M. Maeda and A. Tsuji, J. Chromatogr., 328, 121 (1985).
73   R.L. Veazey and T.A. Nieman, J. Chromatogr., 200, 153 (1980).
74   R.A. Steen and T.A. Nieman, Anal. Chim. Acta, 155, 123 (1983).
75   M. Maeda and A. Tsuji, J. Chromatogr., 352, 213 (1986).
76   C.L.R. Catherall, T.F. Palmer and R.B. Cundall, J. Chem. Soc. Faraday Trans. 2, 80, 823 (1984); 80, 837 (1984).
77   F.J. Alvarez, N. J. Parekh, B. Matuszweski, R.S. Givens, T. Higuchi and R.L. Showen, J. Am. Chem. Soc., 108, 6435 (1986).
78   P. van Zoonen, D.A. Kamminga, C. Gooijer, N.H. Velthorst, R.W. Frei and G. Gübitz, Anal. Chem., 58 1245 (1986).
79   P. van Zoonen, H. Bock, C. Gooijer, N.H. Velthorst and R.W. Frei, Anal. Chim. Acta, 200, 131 (1987).
80   K.W. Sigvardson and J.W. Birks, Anal. Chem., 55, 432 (1983).
81   K.W. Sigvardson, J.M. Kennish and J.W. Birks, Anal. Chem., 56 1096 (1984).
82   M.L. Grayeski and A.J. Weber, Anal. Lett., 17, 1539 (1984).
83   K. Miyaguchi, K. Honda and K. Imai, J. Chromatogr., 316, 501 (1984).
84   K. Miyaguchi, K. Honda and K. Imai, J. Chromatogr., 303, 173 (1984).
85   K. Miyaguchi, K. Honda, T. Toyo'oka and K. Imai, J. Chromatogr., 352, 255 (1986).
86   G. Mellbin and B.E.F. Smith, J. Chromatogr., 312, 203 (1984).
87   T. Koziol, M.L. Grayeski and R. Weinberger, J. Chromatogr., 317, 355 (1984).
88   K. Kobayashi, J. Sekino, K. Honda and K. Imai, Anal. Biochem., 112, 99 (1981).
89   K. Honda, K. Miyaguchi and K. Imai, Anal. Chim. Acta, 177, 111 (1985).
90   B. Mann and M.L. Grayeski, J. Chromatogr., 386, 149 (1987).
91   L.D. Bowers, Anal. Chem., 58, 513A (1986).
92   L. Dalgaard, Trends in Anal. Chem., 5, 185 (1986).
93   K. Honda, K. Miyaguchi, H. Nishiro, H. Tanaka, T. Yao and K. Imai, Anal. Biochem., 153, 50 (1986).
94   F.M. Freeman and W. R. Seitz, Anal. Chem., 50, 1242 (1978).
95   K. Hool and T.A. Nieman, Anal. Chem., 59, 869 (1987).
96   R.D. Lippman, Anal. Chim. Acta, 116, 181 (1980).
97   B.R. Branchini, F.G. Salituro, J.D. Hermes and N.G. Post, Biochem. Biohpys. Res. Comm., 97, 334 (1980).
98   J.R. Poulson, J.W. Birks, G. Gübitz, P. van Zoonen, C. Gooijer, N.H. Velthorst and R.W. Frei, J. Chromatogr., 360, 371 (1986).
99   J.R. Poulson, J.W. Birks, P. van Zoonen, C. Gooijer, N.H. Velthorst and R.W. Frei, Chromatographia, 21, 587 (1986).
100   C.E. Goewie, M.W.F. Nielen, R.W. Frei and U.A.Th. Brinkman, J.Chromatogr., 301, 325 (1984).
101   H. Jansen, R.W. Frei, U.A. Th. Brinkman, R.S. Deelder and R.P.J. Snellings, J. Chromatogr., 325, 255 (1985).

332

102 D. Pilosof and T.A. Nieman, Anal. Chem., 54, 1698 (1982).
103 G. Scott, W.R. Seitz and J. Ambrose, Anal. Chim. Acta, 115, 221 (1980).
104 J.L. Meek and C. Eva, J. Chromatogr., 317, 343 (1984).
105 H.H. Weetall in "Immobilized Enzymes" (U. Moshbach Ed.), Academic Press, New York (1976).
106 G. Damsma, B.H.C. Westerink and A.S. Horn, J. of Neurochem., 45, 1649 (1985).
107 P. van Zoonen, D.A. Kamminga, C. Gooijer, N.H. Velthorst and R.W. Frei, J. Liq. Chromatogr. 10, 819 (1987).
108 K. Honda, J. Sekino and K. Imai, Anal. Chem., 53, 940 (1983).
109 P. van Zoonen, D.A. Kamminga, C. Gooijer, N.H. Velthorst and R.W. Frei, Anal. Chem., 58, 1245 (1986).
110 P. van Zoonen, H. Bock, C. Gooijer, N.H. Velthorst and R.W. Frei, Anal. Chim. Acta, 200, 131 (1987).
111 K. Honda, K. Miyaguchi and K. Imai, Anal. Chim. Acta, 177, 103 (1985).
112 C. Gooijer, P. van Zoonen, N.H. Velthorst and R.W. Frei, J. of Bioluminescence and Chemiluminescence, in press (1989).
113 G.N. Lewis and M. Kasha, J. Am. Chem. Soc., 66, 2100 (1944).
114 N.J. Turro "Modern Molecular Photochemistry", The Benjamin/Cummings Publishing Co., CA, (1978).
115 J.B.F. Lloyd and J.N. Miller, Talanta, 26, 180 (1980).
116 J.D. Brown, J. Chem. Soc., (1958).
117 M. Roth, J. Chromatogr., 30, 276 (1967).
118 E.M. Schulman and C. Walling, Science, 178, 53 (1972).
119 E.M. Schulman and C. Walling, J. Phys. Chem., 77, 902 (1973).
120 R.T. Parker, R.S. Freedlander and R.B. Dunlap, Anal. Chim. Acta, 119, 189 (1980).
121 R.T. Parker, R.S. Freedlander and R.B. Dunlap, Anal. Chim. Acta, 120, 1 (1980).
122 S.L. Wellons, R.A. Paynter and J.D. Winefordner, Spectrochim. Acta, 30A, 2133 (1974).
123 T. Vo-Dinh, E. Lue Yen and J. D. Winefordner, Talanta, 24, 146 (1977).
124 J.S. McHale and P.G. Seybold, J. Chem. Ed., 53, 654 (1976).
125 E. Lue Yen Bower and J.D. Winefordner, Anal. Chim. Acta, 101, 319 (1978).
126 G.J. Niday and P.G. Seybold, Anal. Chem., 50, 1577 (1978).
127 C.G. de Lima and E.M. de Nicola, Anal. Chem., 50, 1658 (1978).
128 R.T. Parker, R.S. Freedlander, E.M. Schulman and R.B. Dunlap, Anal. Chem., 51, 1921 (1979).
129 M.L. Meyers, R. Zellmer, R.K. Sorrell and P.G. Seybold, J. Lum., 20, 215 (1979).
130 C.D. Ford and R.J. Hurtubise, Anal. Chem., 51, 659 (1979).
131 D.L. McAleese and R.B. Dunlap, Anal. Chem., 56, 836 (1984).
132 D.W. Abbott and T. Vo-Dinh, Anal. Chem., 57, 41 (1985).
133 M.M. Andino and J.D. Winefordner, J. Pharm. Biomed. Anal., 4, 317 (1986).
134 C.D. Ford and R.J. Hurtubise, Anal. Chem. 50, 610 (1978).
135 C.D. Ford and R.J. Hurtubise, Anal. Chem., 52, 656 (1980).
136 W. Honner, G. Krabichler, S. Uhl and D. Oelkrug, J. Phys. Chem., 87, 4872 (1983).
137 R.M.A. von Wandruszka and R.J. Hurtubise, Anal. Chem., 48, 1784 (1976).
138 R.M.A. von Wandruszka and R.J. Hurtubise, Anal. Chem., 49, 2164 (1977).
139 R.M.A. von Wandruszka and R.J. Hurtubise, Anal. Chim. Acta, 93, 331 (1977).

140 S.M. Ramasamy and R.J. Hurtubise, Anal. Chem., 59 432 (1987).
141 V.P. Senthilnathan and R.J. Hurtubise, Anal. Chem., 57, 1227 (1985).
142 R.A. Dalterio and R.J. Hurtubise, Anal. Chem., 54, 224 (1982).
143 R.A. Dalterio and R.J. Hurtubise, Anal. Chem. 55, 1084 (1983).
144 V.P. Senthilnathan, S.M. Ramasamy and R.J. Hurtubise, Anal. Chim. Acta, 157, 203 (1984).
145 R.P. Bateh and J.D. Winefordner, Talanta, 29, 713 (1982).
146 J.J. Dekkers, G.Ph. Hoornweg, K.J. Terpstra, C. MacLean and N.H. Velthorst, Chem. Phys., 34, 253 (1978).
147 J.L. Ward, E. Lue Yen-Bower and J.D. Winefordner, Talanta, 28, 119 (1981).
148 D.L. McAleese and R.B. Dunlap, Anal. Chem., 56, 600 (1984).
149 E.M. Schulman and R.T. Parker, J. Phys. Chem., 81, 1932 (1977).
150 D.L. McAleese and R.B. Dunlap, Anal. Chim. Acta, 162, 431 (1984).
151 P.G. Seybold and W. White, Anal. Chem., 47, 1199 (1975).
152 W. White and P.G. Seybold, J. Phys. Chem., 81, 2035 (1977).
153 E. Lue Yen-Bower and J.D. Winefordner, Anal. Chim. Acta, 102, 1 (1978).
154 M.L. Meyers and P.G. Seybold, Anal. Chem., 51, 1609 (1979).
155 T. Vo-Dinh and J.R. Hooyman, Anal. Chem., 51, 1915 (1979).
156 R.J. Hurtubise, Talanta, 28, 145 (1981).
157 R.J. Hurtubise and G.A. Smith, Anal. Chim. Acta, 139, 315 (1982).
158 D.L. McAleese and R.B. Dunlap, Anal. Chem., 56, 2244 (1984).
159 S.M. Ramasamy and R.J. Hurtubise, Anal. Chem., 54 2477 (1982).
160 S.M. Ramasamy and R.J. Hurtubise, Anal. Chim. Acta, 152, 83 (1983).
161 C.M. O'Donnell and J.D. Winefordner, Clin. Chem. 21, 285 (1975).
162 R.P. Bateh and J.D. Winefordner, Anal. Lett., 15, 373 (1982).
163 J.N. Miller, Trends Anal. Chem., 1, 31 (1981).
164 T. Vo-Dinh, G.L. Walden and J.D. Winefordner, Anal. Chem., 49, 1126 (1977).
165 J.B.F. Lloyd, Analyst, 103, 775 (1978).
166 R.J. Hurtubise, Anal. Chem., 55, 669A (1983).
167 G.D. Boutilier and J.D. Winefordner, Anal. Chem., 51, 1384 (1979).
168 G.D. Boutilier and J.D. Winefordner, Anal. Chem., 51, 1391 (1979).
169 R.M. Wilson and T.L. Miller, Anal. Chem., 47, 256 (1975).
170 T. Vo-Dinh, Anal. Chem., 50, 396 (1978).
171 T. Vo-Dinh and R.B. Gammage, Anal. Chim. Acta, 107, 261 (1979).
172 T. Vo-Dinh and P.R. Martinez, Anal. Chim. Acta, 125, 13 (1981).
173 T. Vo-Dinh, R.B. Gammage and P.R. Martinez, Anal. Chim. Acta, 118, 313 (1980).
174 L.J. Cline Love, M. Skrilec and J.G. Habarta, Anal. Chem., 52, 754 (1980).
175 M. Skrilec and L.J. Cline Love, Anal. Chem., 52, 1559 (1980).
176 L.J. Cline Love and R. Weinberger, Spectrochim Acta, 38B, 1421 (1983).
177 L.J. Cline Love, J.G. Habarta and J.G. Dorsey, Anal. Chem., 56, 1132A (1984).
178 R. Weinberger, P. Yarmchuk and L.J. Cline Love, Anal. Chem., 54, 1552 (1982).
179 F.J. DeLuccia and L.J. Cline Love, Anal. Chem., 56, 2811 (1984).
180 F.J. DeLuccia and L.J. Cline Love, Talanta, 32, 665 (1985).
181 R.A.Femia and L.J.Cline Love, Anal.Chem., 56, 327 (1984).
182 H.L.J. Bäckström and K. Sandros, Acta Chem. Scand., 12, 823 (1958).
183 M. Algrem, Photochem. and Photobiol., 6, 829 (1967).
184 C.A. Parker and T.A. Joyce, Trans. Faraday Soc., 65, 2823 (1969).
185 N.J. Turro, L. Kou Chiang, C. Ming-Fea and P. Lee, Photochem. and Photobiol., 27, 523 (1978).
186 N.J. Turro, Mol. Photochem., 4, 213 (1972).
187 J.J. Donkerbroek, J.J. Elzas, C. Gooijer, R.W. Frei and N.H. Velthorst, Talanta, 28, 717 (1981).

334

188 T. Vo-Dinh, "Room Temperature Phosphorimetry for Chemical Analysis",
     Wiley-Interscience, New York (1984).
189 See e.g., J.A. Barltrop and J.D. Coyle, "Principles of
     Photochemistry", Wiley Interscience, New York, Chapter 4 (1978).
190 N.J. Turro, Modern Molecular Photochemistry, Chapter 9,
     Benjamin/Cummings, CA (1978).
191 J.J. Donkerbroek, N.J.R. van Eikema Hommes, C. Gooijer, N.H.
     Velthorst and R.W. Frei, Applied Spectr., 37, 188 (1983).
192 E. Blanco-Gonzalez, R.A. Baumann, C. Gooijer, N.H. Velthorst and
     R.W. Frei, Chemosphere, 16, 1123 (1987).
193 S.Y. Su, A. Jurgensen, D. Bolton and J.D. Winefordner, Anal. Lett.,
     14, A1, 1 (1981).
194 S.Y. Su, E.P.C. Lai and J.D. Winefordner, Anal. Lett., 15, A5, 439
     (1982).
195 K.C. Marsh, L.A. Sternson and A.J. Repta, Anal. Chem., 56, 491
     (1984).
196 I.S. Krull, X.-D. Ding, S. Braverman, C. Selevka, F. Hochberg and
     L.A. Sternson, J. Chromatogr. Sci., 21, 166 (1983).
197 W.N. Richmond and R.P. Baldwin, Anal. Chim. Acta, 154, 133 (1983).
198 A.M. Bond and G.G. Wallace, Anal. Chem., 54, 1706 (1985).
199 J. Rüter, U.P. Fislage and B. Neidhart, Chromatographia, 19, 62
     (1985).
200 I.S. Krull, D. Bushee, R.N. Savage, R.G. Schleicher and S.B. Smith,
     Anal. Lett., 15 (A3), 267 (1982).
201 R.J.Wenzel and L.M. Collette, J. Chromatogr., 436, 299 (1988).
202 E.E. Dibella, J.B. Weissman, M.J. Joseph, J.R. Schultz and T.J.
     Wenzel, J. Chromatogr., 328, 101 (1985).

CHAPTER VII

CONTINUOUS SEPARATION TECHNIQUES IN
FLOW-INJECTION ANALYSIS

M. VALCÁRCEL and M. D. LUQUE DE CASTRO

1.   INTRODUCTION

Chromatographic and non-chromatographic continuous separation processes are currently among the most relevant aspects of analytical chemistry. Both are characterized by the continuous motion of one or both liquid or gas phases involved, accomplished by means of a propelling system
(e.g. a peristaltic or piston pump, pressurized gas, etc.). The liquid
(gas) sample can be introduced into the system either by injection or insertion, or by continuous aspiration. These systems generally accommodate
a continuous detector allowing identification and/or quantitation of the
analytes concerned. One of their most interesting features is the
possibility to decrease to a greater or lesser extent human participation
in the analytical process (automatization) thanks to their inherent
dynamism. This trend towards automatization has been reinforced with the
increasing use of the currently irreplaceable microprocessors, used both
to control the process and for data acquisition and treatment purposes
(ref. 1).

Gas chromatography is the obvious continuous separation technique to
be chosen whenever the carrier or the sample itself is a gas. There are
three analytical hydrodynamic techniques of widespread use, namely high

performance liquid chromatography (HPLC) (ref. 2), field flow fractionation (FFF) (refs. 3, 4) and continuous flow analysis (CFA) in its two chief variants: segmented flow analysis (SFA) (ref. 5) and flow injection analysis (FIA) (refs. 6, 7). The first two methodologies are based on the continuous separation of the analytes and interferents present in a chromatographic column or in the application of a centrifugal force or a thermal or electric gradient. On the other hand, continuous flow analysis methods are not primarily intended for separation purposes, but for automatization of analytical determinations of one or more analytes; yet, they do allow for realization of continuous separations, although these are markedly less efficient - particularly as regards discrimination between several analytes.

FIA is a CFA mode developed along the past twelve years and characterized by a number of features, namely: (a) the flow is not segmented by air bubbles; (b) the liquid sample is directly injected or inserted into a fluid stream; (c) the sample-reaction zone is dispersed in a partial, controlled fashion in the flow and (d) neither physical (homogenization) nor chemical equilibrium has been reached by the time detection is performed. FIA recordings resemble typical chromatograms.

On comparing the instrumental schemes in Fig. 1 one may conclude that HPLC and FIA are relatively similar: in fact, both techniques use liquid reservoirs, pumps, injectors and continuous systems. However, there are also a number of differences between both, chiefly as regards working pressure, use of an interface, versatility, type of analytical problems dealt with, cost, etc. In any case, the greatest difference lies in the continuous separation carried out in the chromatographic column, which is essential to HPLC and only occasionally employed in FIA (ref. 8).

Interfaces are used in FIA for one of two chief purposes. On the one hand, they can be employed to develop continuous separation processes improving on others formerly carried out manually (liquid-liquid extraction, ion-exchange, adsorption, etc.). On the other hand, they can be used for non-separative purposes (e.g. to improve or facilitate the analytical determination). In this case, advantage is taken of the chemical reaction taking place between a solid phase and a liquid flowing through it. Thus, redox columns have been used to handle reagents sensitive to atmospheric agents (ref. 9) and to perform multideterminations (e.g. of nitrite and nitrate). The use of enzymes immobilized on packed columns or tube walls is also a very interesting alternative (ref. 11). Voltammetric and potentiometric stripping techniques performed in a continuous fashion can also be included here insofar as they pursue the two aforesaid objectives in their two essential steps: preconcentration and determination (refs. 12 - 14).

HIGH PERFORMANCE LIQUID CHROMATOGRAPHY

FLOW INJECTION ANALYSIS

Fig. 1 Basic components of HPLC and FIA.

The different types of interfaces used in FIA and the corresponding continuous separation techniques employed are summarized in Table I. The second phase, which acts as an analyte or interference collector, can be continually introduced into the separation system (liquid-liquid extraction, dialysis and gas diffusion). Alternatively, it can be a permanent part of the FIA system (e.g. an adsorption or ion-exchange minicolumn) - in this case, FIA and HPLC are nearly identical and only differ appreciably in that retention-elution processes in the former are one-stage rather than multi-step. The second phase can also be created in the separation system itself through a physical change (distillation) or a chemical reaction (precipitation). Whenever an analytical reaction is required, this can be developed in the course of the separation process (e.g. metal chelate or ion-pair extraction) or at a later stage - by merging the line carrying the isolated analyte(s) with one or several reagent lines prior to the detector, in much the same way as in post-column HPLC reaction detectors.

This chapter deals with the most relevant features and applications of continuous separation techniques used in FIA configurations. Especial emphasis is placed on the coupling of FIA systems to liquid chromatographs. The objectives pursued and advantages offered by the incorporation of non-chromatographic separation techniques into unsegmented flow configurations are also critically discussed.

TABLE I   Continuous separation techniques used in FIA

Interface	Technique
gas-liquid	gas-diffusion distillation hydride generation
gas-solid	
liquid-liquid	extraction dialysis
liquid-solid	ion-exchange adsorption precipitation others

2.   GAS-LIQUID INTERFACES

The gas-liquid separation systems typically used in FIA can be classified into three groups, namely (a) gas-diffusion, in which a gas present in the donor phase or generated in a chemical reaction diffuses to the other phase - also liquid -, acting as acceptor; (b) distillation, where the gas phase is formed by heating, condensation at a suitable temperature and collection into a second liquid phase and (c) hydride generation, in which the gas phase is the result of a chemical reaction and the second one is a gas driving the sample to the detection system.

2.1 GAS-DIFFUSION

The transfer of a gas between two donor streams, still not too widely applied in FIA, can be used with a large variety of analytes, matrices and detection systems (Table II).

The analyte making the gas phase can be as a gas in the donor (O_2, O_3, Cl_2) or alternatively be generated in it by a straightforward chemical reaction induced by an acid (formation of SO_2, CO_2 or HCN) or a base (formation of NH_3), or through the action of relatively high temperatures (formation of acetone from oxidized ketone bodies). Photometry has been so far the detection technique most frequently used in conjunction with these systems on account of its suitability for analytes with acid-base properties, which usually diffuse to a solution containing an indicator whose colour change is a function of the amount of analyte diffused.

As a rule, phase separation is isotermal and is effected by passage through a suitable membrane - usually PTFE. Occasionally, there is no separation membrane, but this is produced between two parallel rubber sheets supported by Perspex[R] plates. The stream containing the sample spreads throughout the lower sheet and yields a film which traverses the entire length of the rubber before going to waste. During the transport, if the partial pressure of the gas in the liquid is higher than that in the surrounding atmosphere, it volatilizes and diffuses to and is absorbed by the stream where its partial pressure is lower (ref. 15).

The chief analytical purpose of the FIA/gas-diffusion association is interference removal from complex matrices (biological fluids, foods, vegetable tissues, environmental samples, etc.); yet, enhanced selectivity can also be obtained - as demonstrated by Pacey et al. (ref. 16) - by incorporating kinetic discrimination in the system timing or the reagent concentrations and experimental conditions of a given method and using a gas diffusion unit. These authors describe two examples of kinetic discrimination, namely the determination of ozone with Indigo Blue and that of chlorine dioxide based on luminol chemiluminescence. Both compounds react with chlorine.

In the batch determination of ozone with Indigo Blue, equilibrium measurements do not allow discrimination between ozone, chlorine and manganese (VII) (ref. 17); yet, the differences between the rates of reaction of these analytes with the reagent permit their discrimination by using a single-channel FIA system which increases the selectivity for ozone over chlorine by a factor of about 3, and that over Mn (VII) by a factor of about 2 - these factors can be further increased by incorporating a gas-diffusion unit into the manifold (Fig. 2). The diffusion membrane completely overcomes the interference by Mn (VII), while that posed by chlorine is significantly decreased from 1 mg chlorine (corresponding to an apparent ozone concentration of 0.36 mg/ml) in the manual method to an apparent 0.008 mg O_3/ml in the gas-diffusion method. Selectivity is enhanced by a factor of 2.5 as a result of the kinetic discrimination introduced by the reagent, and by a further factor of 45 resulting from the incorporation of the gas-diffusion unit, so that the overall selectivity enhancement factor amounts to 112. Such a selectivity level is adequate to safely determine residual ozone in the presence of chlorine in disinfected water samples.

TABLE II   Features of FIA methods involving gas-diffusion processes

Analyte	Matrix	Gas-phase formation	Phase separation	Analytical purpose	Detection	Special features	Ref.
NH_3	whole blood	pH change	membrane	I.R.	P	use of an indicator	20
	blood	=	=	=	pot		23
	plant	=	none	=	P, pot		15
NH_3, urea		chemical reaction	membrane	=	opt	integrated microconduits	22
NO_x, NO_2^-		pH change	membrane		pot	ISE	19
SO_2	wine, beer, fruit juice	=	=	=	P		24
		chemical reaction			chem.		28
CO_2	wine, food, plasma	pH change	=	=	amp.		29
			=	=	P		21
ClO_2			=	=	"		25
ClO_2, O_2			=	discrimination-separation	chem.		16
			=	I.R.	P		26
CN^-	waste water,	=	=	=	Pot		30
Oxidized ketone bodies	milk	heat	=	=	P		27

P: photometry   pot: potentiometry   chem.: chemiluminescence
amp.: amperometry   opt: optosensor   I.R.: interference removal

The differences in the signal produced by Cl_2 and ClO_2 in the chemi-luminescence determination of chlorine dioxide with luminol are time and pH-dependent. Chlorine dioxide reacts extremely quickly with luminol, while chlorine reacts more slowly and with longer lifetimes. By use of a flow-injection system, both the timing and the chemistry can be controlled in such a way as to minimize the signal yielded by chlorine. The added use of pH 13 makes the reagent even more selective - under the conditions typically used in FIA - towards chlorine dioxide. The observed overall selectivity factor is over 500. The incorporation of a gas-diffusion unit into the system overcomes the interference posed by both ionic and organic materials absorbing light in the ultraviolet region (ref. 18). The selectivity enhancement between ClO_2 and Cl_2 resulting from the use of the membrane is 3.1; this, multiplied by the enhancement factor resulting from the difference in molar absorptivity between both species at the wavelength used, 175, yields the overall factor of 500, which can be further increased to over 5,000 thanks to the masking effect of oxalic acid on chlorine (ref. 18).

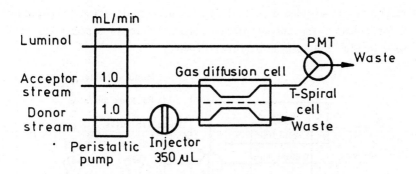

Fig. 2   Manifold for the determination of chlorine dioxide based on luminol chemiluminescence. Both the donor and the acceptor stream are water at pH 2, adjusted with sulphuric acid. The reagent, $1 \cdot 10^{-3}$M luminol, is merged at the T-cell prior to the photomulplier tube (PMT). The membrane consists of 0.45 μm PTFE and the tubing (0.5 mm i.d.) is also PTFE. All flow-rates are 1 ml/min. (Reproduced from (ref. 16) with permission of Elsevier Science Publishers).

Martin and Meyerhoff (ref. 19) have developed a general procedure for enhancing the selectivity of anion-responsive liquid-membrane electrodes based on the use of an acceptor channel receiving the flow through a gas-diffusion unit as the injection loop of an FIA system. A suitable membrane prevents ionic interferents from reaching an ion-selective

342

electrode in the final flow stream. The authors apply their highly
selective semi-automated method to the determination of dissolved
nitrogen oxides or nitrite at levels above 5 µM. Nitrogen dioxide is
trapped across a PTFE membrane in the separation unit and converted to
nitrate by a buffered peroxide recipient solution that is injected and
carried to a tubular nitrate electrode. As noted by its proponents, the
method excels the selectivity and detection capabilities of the nitrate
electrode alone and of conventional sensing systems based on pH
electrodes.

Gas-liquid interfaces have been used in a number of interesting
clinical determinations such as that of ammonia in whole blood and plasma
(ref. 20) or that of carbon dioxide in plasma (ref. 21). The former,
proposed by Evensson and Anfält, uses the configuration depicted in Fig.
3, where the sample is injected into a distilled water stream later
merged with a 0.5N NaOH solution converting ammonium ion to ammonia,
which in turn diffuses across the PTFE membrane to a stream of Phenol Red
in 10^{-3}N NaOH, the resultant process being monitored at 540 nm. The
method is quite convenient, uses small sample volumes (90 µl) and
features a relatively high sampling frecuency (60 h^{-1}). According to its
proponents, it yielded excellent results upon application to samples from
17 individuals. The use of plasma instead of whole blood makes coils A
and B in Fig. 3 dispensable (refs. 15-30).

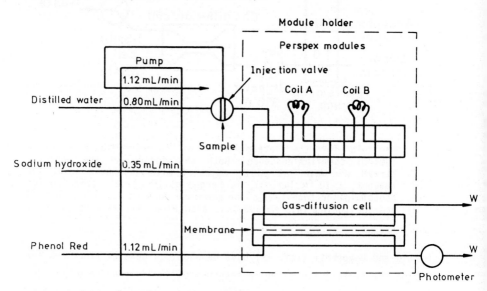

Fig. 3   Manifold for determination of ammonia in blood accommodating a
          gas-diffusion system. (Reproduced from (ref. 20) with per-
          mission of Elsevier Biomed. Press).

A similar configuration is used for the determination of CO_2 in plasma, an acid stream acting as donor and one of Red Cresol as acceptor.

Ruzicka and Hansen (ref. 22) have applied reflectance spectrophotometry to flow-injection measurements of pH and ammonia and urea assays with the aim of demonstrating the principle behind and testing the performance of optosensors integrated into microconduits. With pH measurements, detection is based on commercial non-bleeding acid-base indicator papers placed in the flowing stream at the tip of the fibre optic. The determination of ammonia and urea (via urease) involves the use of a Bromothymol Blue stream and a miniature gas-diffusion device (ref. 8).

2.2 DISTILLATION

Continuous microdistillation systems have long been used in classical continuous air-segmented configurations for determination of volatile analytes (e.g. volatile acidity in wine), and are commercially available from a number of companies such as Technicon or Skalar. On the other hand, there is only one FIA method using distillation with the chief purpose of interference removal from such a complex matrix as waste water usually is for the determination of cyanide. The method in question, proposed by Pihlar and Kosta (ref. 31), uses a distillation system consisting of a distillation and an absorption unit. The former, borosilicate glass half-packed with glass helices and wrapped in a heating wire, is entered by the nitrogen stream at the bottom of the distillation column and carries hydrogen cyanide through the condenser into the absorption unit. An 0.1M solution of sodium hydroxide is pumped to the top of the absorption column. A debubbler prior to the voltammetric detector removes all gas from the system. Differentiation between total and strongly bound metal cyanide complexes is achieved by UV photodecomposition of the complexes. Cyanide is thereby recovered quantitatively, except when its concentration is beyond the operating range of the distillation and/or the absorption unit. Either a calibration graph or the standard-addition method can be used, though the latter is to be preferred when the sample viscosity is markedly different from that of the standards. Up to 60 samples per hour can be thus assayed by alternating samples and washing solutions every 30 sec.

2.3 HYDRIDE GENERATION

This is a sui generis example of a gas-liquid separation process calling for a chemical agent - usually sodium borohydride - to yield a volatile compound which is separated from the solution by a gas acting as the second phase and transporting the analyte to an atomic optical detector. Hydride generation is usually aimed at interference removal (refs. 32 - 38) and, secondarily, to speciation based on the different rate of formation of the hydrides of the different chemical forms in which a given analyte occurs (ref. 37, Table 3).

The FIA/hydride generation association has materialized in two types of generic configuration differing in the way gas-separation is effected, namely by means of a conventional debubler from which the generated hydride is driven to the spectrophometer quartz cell by a gas stream (Fig. 4) or by use of one of the above-mentioned gas-diffusion units where the stream receiving the gas (hydride) transferred across the membrane is another gas which functions to drive it to the detector (Fig. 4). The former type of configuration was used in 1982 for determiantion of bismuth by Aström (ref. 32), who emphasized the promising advantages of FIA in controlling interference effects in hydride generation systems. The configuration (Fig. 4), designed to avoid void volumes as far as possible, involves an acid stream into which the sample (700 μl) is injected and later mixed with sodium borohydride and sprayed with nitrogen or argon into the gas-liquid separator. The gas hydride is swept into the electrically heated tube furnace and the concentration is then measured at 223.1 nm on an atomic absorption spectrophotometer. Standard bismuth solutions are always injected before and after each interfering test solution to check for potential changes in sensitivity. Consultation of the literature, prompted the author to use the integral rather than the peak height as a measure of the hydride concentration - the peak height is dependent upon the oxidation state of the element in question because of kinetic effects involved, while the integral is not (ref. 34). However, as bismuth does not pose that oxidation state problem, and on account of the ease with which analysis, detection and recording are performed in FIA, the authors opted for making peak height measurements, which in addition made the method more widely applicable insofar as many instruments are not equipped with integrating facilities. The results obtained show the efficiency of the FIA system in minimizing the severe interference from copper and nickel (refs. 35, 36) - the interferent concentrations tolerated are 100 to 1000 higher than those afforded by conventional hydride generation systems for bismuth. Except for the ideal reagent concentrations used, the improvement is the result of keeping the reaction time as short as possible in order to favour the main reaction.

The latter type of configuration is represented by a manifold designed by Pacey et al. (ref. 33) for determination or arsenic, which is injected into a water stream merging with an acid and an IK stream - intended to remove interferents formed in the hydride generation and to improve the arsenic signal - prior to mixing with the sodium borohydride stream. The gas passing across the membrane of the gas-diffusion unit is driven to the detector by a hydrogen stream. This dual-phase gas-diffusion system provides remarkably better results than a conventional configuration (refs. 37 - 41).

3.   GAS-SOLID INTERFACES

There are few methodologies involving gas-solid interfaces in general and only one direct determination of chlorine and bromine in FIA (ref. 42). This is based on the monitoring of the transient signal resulting from two consecutive reactions taking place at a gas-solid interface. The reaction giving rise to the increased signal is that of the halogen and α-naphthoflavone, yielding a red-brown complex. The baseline restoration is due to a slower, simultaneous combination of two processes, namely the spontaneous decomposition of the transient complex and the reaction of the coloured complex with As(III), in which α-naphthoflavone is regenerated and the halogens are reduced to their corresponding halides. All these reactions take place at the surface of a piece of filter paper on which a layer of dried reacting mixture has been deposited. The course of the reaction is monitored by transmittance spectrophotometry. The method involves no separation process indeed.

TABLE III   Features of FIA methods involving hydride generation

Analyte	Matrix	Analytical purpose	Detection	Special features	Ref.
As (III), As(V)		I.R.	A.A.S	use of gas-diffusion membrane	33
Bi		I.R	A.A.S		32
As, Sb, Bi, Se, Te	termal water	I.R	A.A.S	speciation	37
As	standard reference	I.R.	A.A.S		38
	NBS orchard leaves	I.R.	MECA		39
	glycerine	I.R.	ICP-AES		41
Hg		I.R.	AAS		40

I.R.: interference removal
MECA: molecular emission cavity analysis

AAS: atomic absorption spectrometry
ICP-AES: inductively coupled plasma-atomic emmission spectrometry

a)

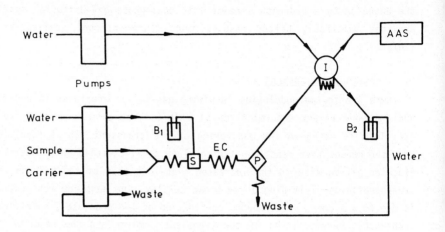

b)

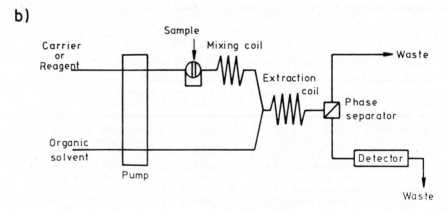

Fig. 4   (a) Flow-injection manifold for determination of bismuth by
            atomic absorption with hydride generation (S, injection port -
            sample loop --; a, b and c, coils of i.d. 0.7, 0.5 and 0.5 mm,
            respectively; d, gas-liquid separator; W, waste). (b) Manifold
            for determination of As(III) and As(V) using gas diffusion.
            (Reproduced from (refs. 19, 20) with permission of the American
            Chemical Society).

4.   LIQUID-LIQUID INTERFACES

The way in which the transfer of matter between two liquid phases takes place in continuous separation systems depends essentially on the type of contact between the phases involved and on their miscibility. Liquid-liquid extraction relies on the immiscibility of the two phases and the establishment of a dynamic contact zone between both facilitating the transfer of matter. On the other hand, in dialysis, the phases concerned - generally aqueous - are miscible and the transfer of matter takes place across a semi-permeable membrane separating both liquid streams and accommodated in the separation unit (dialyser).

These two separation techniques have been used to a different extent in FIA. Thus, while extraction has been employed relatively frequently (refs. 43, 44), dialysis has been applied to a somewhat lesser extent.

4.1   EXTRACTION

The on-line coupling of a liquid-liquid extraction system to an FIA configuration was simultaneously reported by Karlberg and Thelander (ref. 45), and Bergamin et al. (ref. 46). In both cases, the extraction system was located behind the injection valve. Currently, the separation unit is also occasionally placed prior to the injection system, so that the separation process results in a continuous stream of organic phase containing the analyte and filling the injection valve (refs. 46, 47). Less common in FIA is the off-line positioning of the system (refs. 48 - 50). As a result of this coupling, a number of configurations of variable complexity, suited to specific needs, are now available, namely:

(a)   Without phase separation. In this mode - the simplest -, the aqueous sample is injected into a single-channel manifold carrying the organic stream extractant, which flows through the extraction coil. This is where the formation of an extractable complex between the analyte and the reagent dissolved in the organic phase - measured as it passes through the fluorimetric flow-cell (ref. 51) takes place. The application of ultrasounds to the extraction coil results in the formation of a microemulsion that increases contact between the phases and hence the process yield and measurement reliability (ref. 52). Microemulsion formation prior to introduction into an FIA system has been investigated by Worsfold (ref. 53). This author uses a special propelling system capable of reversing the direction of the flow in order to introduce an immiscible organic phase plug into the sample stream; the plug is passed alternately in both directions through the detector, which allows control over the yield of the extraction process and its extraction kinetics to be studied (ref. 54).

(b)   Single extraction with the separation system located prior to or after the injection unit.

(c)   Multi-extraction, where the separation process is repeated several times by using the same or a different extractant at each stage (refs. 55, 56), thereby increasing the selectivity, sensitivity and efficiency of the overall process.

(d)   Back-extraction. This is a multi-stage extraction mode in which the aqueous sample is first extracted into the organic medium and then back-extracted into an aqueous phase where measurements are performed (ref. 57).

The presence of an organic phase in an FIA system requires a series of cautions to be taken on account of its corrosive properties. Thus, the transport tubing, connectors and extraction system must be steel, platinum, glass or PTFE. An organic stream can be set in motion by: (a) a peristaltic pump - the flexible PVC tubing commonly employed in other systems is completely useless here and is to be replaced with inert materials such as modified PVC, silicon rubber or fluoroplasts; (b) by the displacement technique, which involves pumping an aqueous stream into a closed container - this can be achieved with a peristaltic pump and ordinary tubing - that is filled with the organic solvent and fed at a constant flow-rate into the FIA system; (c) by setting a constant pressure with the aid of an inert gas forcing the extractant to circulate along the FIA manifold.

The most serious shortcoming arising from the use of FIA/liquid-liquid extraction is currently the lack of an elaborate theory. The studies carried out so far in this area have only dealt with specific aspects of the subject - yet, the number of papers reporting new contributions is fortunately increasing (refs. 58 - 63).

In their work on the hydrodynamic and interfacial origin of phase segmentation , Sweileh and Cantwell (ref. 60) developed a semi-quantitative physicochemical model for the process whereby alternating segments of aqueous and immiscible organic phases are produced on merging of both phases. A growing "drop" of one phase is dislodged to produce a segment when the hydrodynamic force exerted by it as a result of the flow of the other solvent equals the interfacial force holding it in place. Hydrodynamic forces are expressed by Poiseuille's and Bernouilli's laws, while the interfacial force is expressed by a form of the Tate equation (refs. 64, 65) in terms of liquid-liquid interfacial tension and solid-liquid contact angle. These authors also derived a series of equations for calculation of the extracted analyte fraction, the dependence of the peak area on the flow-rate, the distribution ratio and a proportionality constant characteristic of the chemical system, the peak height as a function of the analyte concentration in the injected sample, the flow-rate ratio, the extracted analyte fraction, spectrophotometer sensitivity and a factor similar to that of dispersion reported by Ruzicka (ref. 5) and adapted to systems accommodating liquid-liquid extraction.

The efficiency of a separation process has been studied in depth by Rossi et al. (ref. 58), who determined the optimum characteristics of the extraction coil for different types of phases and the influence of FIA variables on the efficiency.

Nord and Karlberg (ref. 58) have investigated the mechanism of dispersion in flow-injection extraction systems and determined its influence on the peak width in terms of related variables. In addition, Backström et al. (ref. 62) have evaluated the efficiency achieved and the dispersion involved in various phase separators.

Laser-induced excitation (ref. 63) and the alternate passage through the detector in both directions (ref. 54) will foreseeably aid in establishing the theoretical background of this technique.

Every automatic solvent extraction FIA system has three essential components, namely:

(a)   Segmentor, in which the streams of the two phases involved merge. Its chief aim is to obtain identical alternate segments of both immiscible liquids reaching the extraction coil.

(b)   Extraction coil, where the transfer of matter between the segments of both phases is effected. PTFE coils repel the aqueous phase, which is carried as bubbles; conversely, the walls of glass coils are wetted by the aqueous phase, so that the organic phase is transported by the former as bubbles. Selley et al. (ref. 55) defined some criteria for coil selection. Thus, coils should ideally be made of materials allowing the analyte to pass into the bubble phase. In addition, the ratio between the interfacial area and the initial analyte volume should be as high as possible and the analyte motion should be facilitated to achieve maximum efficiency.

(c)   Phase separator, which receives the segmented flow from the coil and continuously splits it into two separate streams of both phases.

Of all three elements, the most complex and interesting is no doubt the phase separator, of which a variety of models have been designed with the aim to improve on earlier ones (refs. 44, 46, 63, 66 - 69). There are three chief types of continuous separator: (a) devices using a chamber relying on gravity to separate the phases; (b) gravity-based devices with a T-shaped separator and with or without a sort of phase guide made of material wetted selectively by one of the phases; (c) devices with a membrane separator based on the selective permeability of a microporous membrane towards the phase wetting it. This last type of separator features a number of advantages over the other two, namely: a smaller inner volume, which lessens the dispersion or dilution of the analyte or its reaction product and hence results in lesser band broadening and increased sensitivity; greater reliability in separating the phases at higher flow-rates, which redounds to shorter analysis times; flexibility for use with a variety of water-immiscible solvents as a result of no difference in density between the aqueous and the organic phase being required; greater separation efficiency. A recent publication (ref. 70a) has demonstrated that many of these above - mentioned advantages can also be achieved with a new generation of sandwich - type separators based on wetting.

The joint use of FIA and liquid-liquid extraction has aided in solving a number of analytical problems in various areas - particularly environmental, clinical and pharmaceutical chemistry -, where this association has been chiefly applied for separation and - occasionally - preconcentration of the analyte. In Table IV are summarized the applications reported so far, classified according to the type of analyte determined (inorganic or organic) and, within each group, according to the detection system used.

Fig. 5 shows the two basic types of FIA/liquid-liquid extraction assembly, namely with the extraction unit located prior to (a) or after (b) the injection system.

The configuration including an extraction system prior to the injection valve is the most commonly used in the FIA/extraction/AAS association for analyte preconcentration and separation. The advantages offered by the joint use of this triad have been emphasized by several authors (refs. 6, 7, 70) (e.g. in the determination of copper, nickel, zinc and cadmium proposed by Nord and Karlberg (ref. 71), in that of zinc in biological and environmental samples (ref. 72) or iron matrices (ref. 59), or in that of perchlorate in serum and urine, based on the formation of an ion-pair between this anion and the $Cu(I)$/6-methylpicolinaldehyde azine complex (ref. 47). The sequential determination of nitrate and nitrite in foodstuff, also based on the formation of an ion-pair between nitrate and the $Cu(II)$/neocuproine complex which is extracted into MIBK is illustrative of the potential of this association for simultaneous analyses. In this determination, the samples are spiked with oxidant, $Ce(IV)$, or inhibitor (amidosulphonic acid), for determination of the sum of both anions and nitrate alone, respectively (ref. 73).

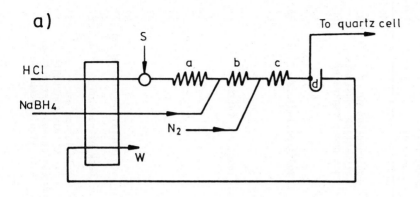

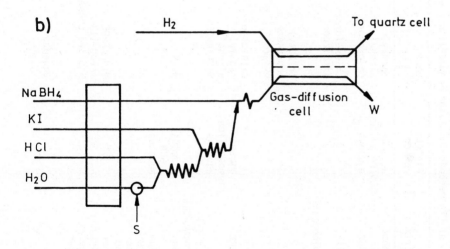

Fig. 5   Generic types of FIA/liquid-liquid extraction assemblies. (a)
With the extraction system before the injection system (B_1 and
B_2, displacement bottles; S, segmentor; EC, extraction coil;
P, phase separator; I, injection system; AAS, atomic absorption photometer). (b) With the extraction system behind
the injection system.

TABLE IV   Features of FIA methods involving liquid-liquid extraction

Analyte	Matrix	Analyte phase	Second phase	Detection	Special features	Ref.
Mo	plant	aqueous SCN/Fe system	isoamyl alcohol	P		46
Pb, Cd	water	aqueous	chloroform	"	calculation of extraction constants	83
Cd	urine	"	chloroform/dithizone	"	new phase separator	66
PO_4^{3-}	biological	"	tributyl phosphate			79
U	"	"	CCl_4/		modified T-piece	80
Zn	soil	"		"		81
Ga	water	aqueous/lumogallion	dithizone isoamyl alcohol	F	laser excitation	63
Zn, Cu, Pb, Ni	"	aqueous	MIBK/APDC	AAS		71
Zn	biological, environmental iron	"		"		72
ClO_4^-	urine, serum	aqueous/SCN⁻	MIBK	"		59
NO_2^-, NO_3^-	water food	aqueous	"	"		47
			"	"		82
Cu	water		MIBK/APDC	FAAS	sequential determination	73
Cd, Cu, Co, Pb, Ni	diluted samples		freon-113		back extraction	46
			H_2O 3rd phase			57
Amines	water	"	aqueous	P	calculation of extraction constants	87
	"	"	aqueous	"	aqueous/aqueous extraction with liquid membrane	84

Compound	Sample	Aqueous	Organic	P	Remarks	Ref.
Caffeine	tablets	aqueous	chloroform	P	aqueous/aqueous extraction with liquid membrane	44
Codeine, acetyl-salicylic acid	"	SCN⁻/ aqueous	"	=	aqueous/aqueous extraction with liquid membrane	85
Anionic surfactants	industrial water	"	"	=	with liquid membrane	68
	sewage water	"	toluene	=		86
	waste water	"	MIBK	=	increased sensitivity	95
Cationic surfactants	water	"	chloroform	=	use of various segmentors	48
Non-ionic surfactants	water	"	1,2-dichloro-ethane	=		87
Caffeine, surfactants	water	"	chloroform	=		88
8-Dichlorotheo-phylline, diphenyldramamine	tablets	"	cyclohexane	=	measurements in both phases	67
Procyclidine	"	"	chloroform	=	theoretical quantitativity studies	89
Bittering compounds	beer, malt	"	iso-octane	=		74
Terodiline	serum	"	n-heptane	=	gas chromatography	90
Enalapril	tablets	"	dichloro-methane	=		91
Drugs	biological	"	chloroform	F	adsorption problems	92
Vitamin B₁	tablets	heptane	aqueous	=		93
Alkylamines	organic	"		=	microemulsions without phase separation	53
Steroids	water	aqueous/ lucigenin	1,2-dichloro-ethane	chem.		94

MBIK: methyl isobutyl ketone   APDC: 1-pyrrolidinecarbodithioic acid

The manifold depicted in Fig. 5b was developed and used by Ishibashi et al. (ref. 62) for the determination of gallium based on the formation of a fluorescent complex with lumogallion which is extracted into isoamyl alcohol. A similar configuration has been proposed by Karlberg et al. (ref. 70) for adaptation of the standard manual liquid-liquid extraction method for determination of bitterness by the FIA technique. The bittering compound is extracted into iso-octane and its absorption in the organic medium measured at 275 nm, thus making the separate solvent blank extraction, required in the batch procedure, unnecessary. The injected sample volume used in 100 μl and up to 60 samples can be assayed per hour with as little iso-octane consumption as 1 ml per sample. The method has been recently applied to the determination of this type of compounds in must as a means of on-line monitoring of their evolution in the course of beer making (ref. 74). An interesting method for simultaneous determination of two organic compounds (diphenyldramamine and 8-chloro-theophylline) has been proposed by Fossey and Cantwell (ref. 67). These authors use a dual-membrane separator (lipophilic -PTFE- and hydrophilic -paper-) to obtain clear aqueous and organic phases, each being led to a different spectrophotometer. The aqueous portion, of pH 10, contains 8-chlorotheophylline, while the organic phase (cyclohexane) contains dyphenyldramamine.

The inherent versatility of FIA allows for adaptation of the con-figurations typically used to the particular characteristics of the system involved. Thus, the extraction efficiency can be improved through a salting-out effect by using further streams merging with the two-phase system (ref. 45). Also, the system can be adapted for changes in viscosity or pH scanning (ref. 78). The kinetic nature of this technique - measurements are made under non-equilibrium conditions, which allows FIA methods to be classed as fixed-time kinetic (ref. 72) - is increased by joint use with liquid-liquid extraction, which further increases its selectivity (refs. 78, 80). This powerful association provides a number of valuable advantages such as lower sample, reagent and organic solvent consumption, higher determination rate, greater instrumental simplicity and reproducibility and less expensive instrumentation.

The FIA/extraction association has also been applied for non-ana-lytical purposes such as the calculation of the extracted analyte fraction (ref. 59) or that of the peak area (ref. 61) and height (ref. 59) -based on the use of a dual-membrane phase separator - as a function of other parameters typical of the chemical system such as acidity constants (ref. 78). The still small number of applications in this area

(Table IV) will predictably be increased by the use of laser-induced excitation (ref. 62), multi-extraction systems (ref. 55) and fast-scan detectors (refs. 55, 56) for studying reaction mechanisms and kinetics (refs. 79 - 95).

4.2 DIALYSIS

Membrane-based FIA/liquid separation systems have so far been almost exclusively used in clinical analysis. Their chief use is interference removal - only once have these been used for dilution purpose (ref. 96). Table V lists the most significant achievements of the FIA/dialysis association, of which the work by Gorton and Ögren (ref. 97) is a typical example. These authors determine glucose and urea in serum with suitable immobilized enzymes. In the configuration used for determination of urea, depicted in Fig. 6a, the sample is injected into a donor buffer which is driven to waste once the analyte has passed through the membrane, wherefrom it is led by the acceptor stream (phosphate buffer of pH 6) to the tubing zone containing the reactor, packed with urease immobilized on controlled pore glass. The detection system (ammonia-selective electrode) calls for the use of a basic stream merging with the main line after the enzymatic reactor in order to obtain a sample plug of a pH adequate for the release of the monitored product (ref. 97). In their report, these authors evaluate the effect of the dialyser, enzyme reactor and detector on the dispersion.

Chang and Meyerhoff (ref. 98) used a membrane-dialyser injection loop to enhance the selectivity of anion-responsive liquid-membrane electrodes in FIA (Fig. 6b) and applied to the determination of salicylate. The system consists of a tubular polymer membrane electrode based on manganese (III) tetraphenylporphyrin chloride to sense salicylate ions formed in a recipient buffer solution held within the upper channel of the flow-through membrane dialyser assembly. Samples containing salicylic acid are manually introduced into the lower channel of the dialysis unit, where a thin silicone rubber membrane separates the two channels. The analyte is trapped across the membrane as salicylate ions within a static layer of a suitable recipient buffer. After a preselected trapping time, the recipient plug is flushed to the electrode in the typical flow-injection fashion. The peak potentials obtained are logarithmically related to the salicylic acid concentrations in the original samples. A near-Nernstian response is obtained in the range 10^{-4}-10^{-2}M salicylate for a trapping time of 2 min. The detection limits can be modified by changing this trapping time. The resultant system is highly selective towards salicylate (as salicylic acid) over most inorganic and organic anions commonly found in blood.

TABLE V   Features of FIA methods involving dialysis

Analyte	Matrix	Analyte phase	Second phase	Detection	Special features	Ref.
Li	serum	aqueous ligand solution	borax ligand solution	P	ISE	104
Various metal ions	water				theoretical	105
Zn	=	aqueous	reagent solution	volt		103
Ca^{2-}	milk	=		P		106
SO_4^{2-}	urine			=		107
Cl^-, PO_4^{3-}	serum	acid solution		=		96
Glucose	milk, waste water, fermentation broth			=	new dialysis probe	99
Glucose, urea	urine, serum	buffer	basic medium	P, pot	enzyme reactor	97
Galactose	serum	aqueous sample	sample	P	=	108
Glucose	=			chem.	=	109
	plasma	phosphate buffer	phosphate buffer	=		110
	=			volt		112
Galactose, lactose, di-hydroxylactose	urine, milk	aqueous	aqueous	amp.	enzyme reactors	111
Salicylic acid	serum	biological fluid	aqueous buffer	pot	ISE	98
Sulphite, sulphide, methanethiol	serum	aqueous	aqueous	F	reagent introduction by dialysis. speciation	100
H_2O_2, CH_3OH_2	snow	aqueous	aqueous	=	reagent introduction by dialysis. speciation	101
Amines		aqueous	aqueous	P	liquid membrane for cleanup and preconcentration	84
Sulphonamides	serum	biological fluid	aqueous	=	calculation of	102

a)

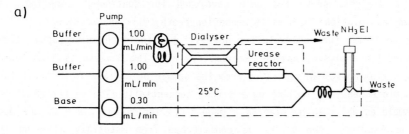

b)

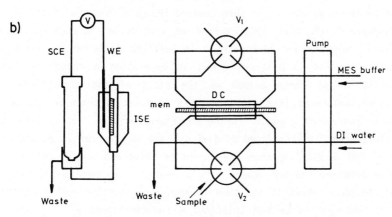

c)

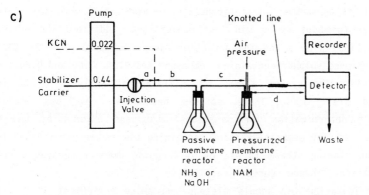

Fig. 6 (a) Configuration for determination of glucose in serum with sample dialysis. (Reproduced from (ref. 97) with permission of Elsevier Science Publishers). (b) Scheme of the dialyser/flow-injection set-up used for determination of salicylic acid (ISE, tubular PVC ion-selective membrane electrode; WE, working electrode for potentiometric measurements; SCE, saturated calomel electrode; V, pH/mV meter; DC, dialysis chamber; mem, silicone rubber membrane; V_1 and V_2, flow-injection valves). (Reproduced from (ref. 98) with permission of Elsevier Science Publishers). (c) Membrane-based FIA system for determination of sulphite, sulphide and methanethiol in water. A KCN stream is used when formaldehyde is also present. a, b, c, and d are 24 cm, 15 cm - reactor included -, 39 cm - reactor included - and 148 cm in length, respectively (Reproduced from (ref. 100) with permission of the American Chemical Society).

A dialysis probe has been developed for continuous sampling from complex solutions such as fermentation broth, milk and waste water, aimed at rendering them suitable for analysis by liquid chromatography, flow-injection analysis, enzyme calorimetry, etc. The analyte is transferred to a flowing stream separated from the sample by a dialysis membrane that is protected from fouling by a strong tangentially flowing stream of the sample established by placing a magnetic stirring bar close to the membrane surface. The device is constructed from materials allowing the probe to be steam-sterilized when mounted inside a fermentor (ref. 99).

Dasgupta et al. (ref. 100) carried out a study on the preservation of sulphite, sulphate and methanethiol in buffered formaldehyde and oxaldihydroxamic acid stabilizers aimed at developing a method for fast determination of these anions on the basis of their reaction with N-acridinylmaleimide (NAM) in a water/N, N-dimethylformamide medium to yield a fluorescent product. In the configuration used, depicted in Fig. 6c, the reagent is introduced by dialysis and the carrier is pumped at a rate of 440 μl min^{-1} through the injection valve and sample loop into a passive membrane reactor immersed in a concentrated ammonia solution to raise the pH to about 10 and then through a pressurized porous membrane reactor immersed in an NAM solution. A superincumbent air pressure of 11.5 psi is adequate for introduction of NAM at a suitable rate, equivalent to about 44 μl min^{-1} of conventional addition. The typical reaction time is 50 s and the detection limits achieved for the three above-mentioned sulphur-containing compounds are 0.04, 0.60 and 0.80 μM, respectively. The system allows for differential analysis to be implemented (ref. 100). These authors have also developed another speciation determination for peroxides (H_2O_2 and CH_3OH_2) by use of a pressurized PTFE membrane reactor containing the enzyme peroxidase. The pH of the flowing stream is set by introducing ammonia through a non-porous cation-exchange reactor (ref. 101).

A configuration for sample cleanup and amine enrichment in a flow system recently reported involves passing the sample through a liquid membrane whereupon the analyte is released and subsequently trapped by a stagnant acceptor phase on the other side. The resultant analyte plug is then swept from the membrane separator to the detection system The proponents provide a theoretical discussion of the mass transfer across the membrane and the influence of the transport on the acceptor concentration profile. Strictly, this is an example of mixed extraction-dialysis. The enrichment factor achieved with this configuration, whose results compare well with theoretical predictions, is dependent upon the sample volume, supporting matrix, type of immobilizing solvent used, donor flow-rate and coefficient of partition of the analyte between the donor and the membrane phase (ref. 84).

Other non-determinative applications of dialysis have been reported by Macheras and Koupparis (ref. 102) and by Bernhandsson et al. (ref. 103). The former authors used an automated flow-injection analyser interfaced to a dialysis unit to study drug-protein binding interactions (e.g. between some sulphonamides and bovine serum albumin), with results similar to those obtained by other procedures. A complete run, including calibration, takes about 100 min. The dialysable sulphonamides are quantitated spectrophotometrically by a modification of the Bratton-Marshall method. The system also allows calculation of dialysis rate constants.

Berhandsson et al. (ref. 103) have discussed the transfer of mass in infinite parallel plate dialysers with co-flow between sample and detector streams by applying three theoretical models. These authors derived analytical expressions for the coupled diffusion and transfer phenomena occurring in both channels and obtained numerical solutions for a laminar flow regime by the finite-difference approximation method. They also considered the results obtained by a mixing-cup model applied under steady-state conditions. With the dimensions typical of analytical dialysers there were only small differences between the results provided by the laminar-flow and plug-flow models. The mixing-cup model predicted higher fluxes through the membrane than the other two, particularly with increased channel heights. The theoretical results were consistent with those experimentally obtained in the dialysis of Zn(II) ions and the flow dependence also agreed reasonably well with theory provided that the hydrostatic pressures were equal on both sides and that stresses potentially resulting in membrane bulging were kept low (see Table V) (refs. 104 - 112).

5. SOLID-LIQUID INTERFACES

Separation techniques involving solid and liquid phases have been used in conjunction with FIA almost since this technique was introduced. While ion exchange was originally the separation technique most frequently used with FIA, adsorption and preconcentration are gradually becoming commonplace in this context.

5.1 ION EXCHANGE

The FIA/ion exchange association has been preferentially used for preconcentration of minor species from complex samples (e.g. industrial, rain and sea water, soldering smokes, biological fluids, etc.), though it has also been employed for separation purposes and to facilitate the determination of different analytes in the same sample by sequential elution of these, kept in a suitable active agent (ref. 113). Insofar as the analytes most frequently determined are metal cations, the commonest active agents used are different types of chelating resin. Table VI lists the major species determined by methods involving the FIA/ion exchange association, classified according to the type of analyte concerned into cationic species (individually and in mixtures), anionic species and conjugate acid-base pairs. A representative example of the versatility of this association and its ease of adaptation to different problems lies in the AAS determination of heavy metals in sea water proposed by Ruzicka et al. (ref. 114). The metals are preconcentrated in a chelating-resin microcolumn incorporated into one of three configurations of differing compexity that these authors designed in order to overcome the problems successively encountered in their experiments. A single-channel manifold featuring two series of injection valves located prior to the column (Fig. 7a) is the simplest alternative for implementation of the pre-concentration step. The propelling system is gas pressure-based. The carrier, ammonium acetate, drives the sample injected by means of valve I_1 to the microcolumn through a coil and a by-pass of valve I_2, where the analytes are retained. The second step involves injection of the eluant through valve I_2. This configuration poses a series of problems such as the appearance of a prepeak due to the sample matrix, disturbances arising from changes in the resin compactness in changing from the ammonium to the proton from and the lack of homogenization between sample and carrier in the central zone ot the sample plug, which is very acidic and hinders retention. These shortcomings were circumvented by using the configuration depicted in Fig. 7b, with a merging point for the ammonium acetate stream and elution of the analytes in the direction opposing the retention path by means of selecting valves whose operation is illustrated in the figure. The analytical procedure was automated by its proponents by using a manifold with a single injection valve and a system consisting of two pumps and a timer for synchronization of the operation of the pumps ($pump_1$ acted during the preconcentration step, while $pump_2$ worked during the elution). The sample matrix never reached the detector in either case and the microcolumn was regenerated in the elution step.

Townshend et al. (refs. 115 - 116) have shown the vast potential of application of ion-exchange microcolumns in FIA systems. These authors have developed a determination for Zn and Cd by using an exchange column

where both cations, present in the same sample, are retained to be subsequently eluted sequentially and their concentration determined indirectly through inhibition of the cobalt-catalysed chemiluminescence
generation from luminol (ref. 115). These authors use the displacement of
thiocyanate from a strongly basic ion-exchange resin by other anions to
determine comon anions with spectrophotometric detection of the iron-
(III)-thiocyanate complex formed. Nitrate can be determined in the
presence of chloride and sulphate, which are removed by a precolumn
packed with a cation-exchange resin in silver form followed by a zinc
reductor. Binary mixtures (e.g. chloride and nitrate) can be determined
simultaneously by splitting the sample in the flowing system so that part
of it goes through the chloride suppressor (yielding a response
corresponding to nitrate alone) while the rest by-passes it and gives a
response corresponding to the sum of chloride and nitrate.

The joint use of ion exchange and conversion techniques with FIA has
materialized in the on-line conversion of soluble species to insoluble
compounds by means of a tag material which is subsequently determined.
This approach (Fig. 7) has been used to determine sulphide by flame
atomic absorption spectrometry with the aid of cadmium(II) as precipitation tag reagent. Excess Cd(II) is collected on a chelating ion
exchanger and later eluted. The detection limit for sulphide is 10 μg/l
and the sampling rate achieved is 100 samples hr^{-1} , the typical standard
deviation being 1.2%. Of all potential interferents, only phosphate has
any effect on the determination (ref.117).

Hwang and Dasgupta (refs. 101, 118) use ion exchange in a rather
uncommon fashion in their peroxide determination - the required pH change
is effected by introducing ammonia through a non-porous cation-exchange
membrane reactor.

A novel, highly interesting contribution to this area is represented
by the use of integrated microconduits (refs. 119 - 137) (see Table VI).

5.2 ADSORPTIVE PRECONCENTRATION

The adsorptive preconcentration of analytes in FIA has so far been
tackled with two chief active agents, namely activated alumina and
electrode surfaces (carbon paste or platinum). Adsorption on aluminia has
been used for preconcentration of chromic ion in biological samples
(urine) prior to its determination by ICP-AES (ref. 138) in the
speciation of chromium (ref. 139), as well as for that of oxyanions such
as arsenate, borate, chromate, molybdate, phosphate, selenate and
vanadate with the same detection system. Preconcentration on a carbon

paste electrode prior to the voltammetric determination of the analyte has been used in the analysis for drugs such as chlorpromazine (ref. 140) and doxorubicin (ref. 141) in urine. The pulsed amperometric determination of electroinactive adsorbates such as chloride and cyanide at platinum electodes (ref. 142) is but another proof of the FIA/adsorptive preconcentration association.

An activated aluminia microcolumn has been used for separation and preconcentration of Cr(VI) from Cr(III) in synthetic aqueous solutions prior to ICP detection at 267.72 nm, yielding a linear calibration graph between 0 and 1 000 μg/l of Cr(VI) or Cr(III), with relative standard deviations at the 10 μg/l level of 2.2% and 1.1% for Cr(III) and Cr(VI), respectively, for a 2-ml sample, the corresponding detection limits being 1.4 and 0.20 μg/l, respectively. The procedure has been applied to the determination of both chromium forms at the μg/l level in reference NBS water (ref. 139).

Preconcentration and quantitation of doxorubicin, a cancer chemotherapy drug, are accomplished by a flow-injection approach involving adsorption of the drug onto a carbon paste electrode, medium exchange and differential voltammetry on the adsorbing surface (ref. 141). Linear response is obtained for concentrations from 10^{-6}M to the detection limit (10^{-9}M). No preliminary steps are required for determination of the drug in urine by direct injection. Chlorpromazine can be determined under similar conditions in the presence of a tenfold excess of non-adsorbable species with similar redox potentials. The preconcentration step also results in increased sensitivity.

A special type of adsorptive presoncentration is that of electroinactive species on Pt electrodes (ref. 142). Thus, chloride and cyanide modify the rate of surface oxide formation following a positive potential step. Hence, triple-step potential waveforms similar to those used successfully for pulsed amperometric detection of electroinactive adsorbates (e.g. alcohols, carbohydrates and amino-acids) can also be applied to electroinactive adsorbates injected into an electrolyte stream. Depending of the wave form, the overall anodic current at the detection peak will be greater or less than the baseline signal corresponding to oxide formation in the absence of the adsorbate. The sensitivy achieved is very high indeed (ref. 143 and Table VII).

TABLE VI   Features of FIA methods involving ion-exchange processes

Analyte	Matrix	Analytical purpose	Active agent	Detection	Special features	Ref.
NH_3	rain water	separation	Amberlite- R120	P	column in sample loop. Alternating flow	121
Ni	river water	preconcentration separation	Amberlite IRA-400 chelating resin	F	multi-function valve	122
Ca		=	"	AAS	"	123
Cu		preconcentration separation	"	=		124
		preconcentration separation	8-quinolinol	pot		125
Mn, Pb, Cu	soldering smokes	preconcentration	Dowex A-1 chelating resin	AAS	series injection valves (sample-eluent)	126
Zn, Cd		=	chelating resin selectivity	P	study of resin	49
		simultaneous determination	Amberlite IRA-400	chem.	sequential elution	115
Pb, Cd, Cu		preconcentration	Chelex-100	AAS	integrated microconduits simultaneous determination	119
Ba, Be, Cd, Co				ICP-AES		129
Cu, Mn, Ni, Pb	tap, sea, polluted water	=	chelating resin	AAS	two alternate columns	128
Ni, Cu, Pb, Cd	sea water	=	Chelex-100	"	series injection and selecting valves	114
Cd, Pb, Cu, Zn	"	=	Chelex-100 and 8-quinolinol	"	series injection and selecting valves	120
	"	=	Chelex-100	AAS		129
Cd	reference materials	=	muromac A-1	ICP-AES		130
Ti, V, Al, Cr, F, Cd, Co		=	triPEN	FAA		131
S^{2-}, CN^-		indirect determination	8-quinolinol	AAS	continuous precipitation	117
$P_2O_7^{2-}$, PO_4^{3-}		simultaneous determination	TKS-gel SAX	P		132

Analyte	Sample	Purpose	Material	Method	Ref.
Various anions		simultaneous determination	basic resin	"	116
Br⁻	sea water, serum, chlorinated reagents	separation	Amberlite XAD-2	"	133
Acid-base pairs		salt removal determination	resin	cond.	134 135
Polyphosphates		separation	TXK-gel SAX	P	136
H_2O_2, CH_3OH_2	rain water	NH_3 introduction	Nafion 811X cation-exchange membrane	F	hydrolytic catalysis enzyme membrane 118
H_2O_2	water	separation	Diaion SK-1	P	137

F: fluorimetry   cond.: conductimetry   ICP-AES: inductively coupled plasma-atomic emission spectroscopy
triPEN: N, N, N-tri(2-pyridylmethyl)ethylene diamine

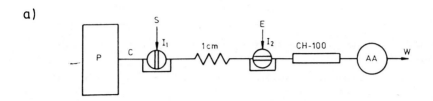

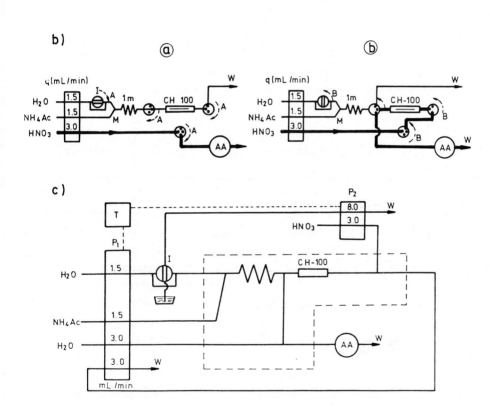

Fig. 7 FIA/ion-exchange configurations of variable - increasing - complexity. (a) Single-channel manifold. (b) With elution [a] in the reverse direction of retention [b]. (c) Automated manifold. (Reproduced from (ref. 114) with permission of the Royal Society of Chemistry).

TABLE VII   Features of FIA methods involving adsorptive preconcentration

Analyte	Matrix	Analytical purpose	Active agent	Detection	Special features	Ref.
Cr(III)	urine	separation	activated alumina	ICP-AES		138
Cr(III), Cr(VII)	reference water	separation, preconcentration	activated alumina	"	speciation	139
Oxyanions		preconcentration, separation	activated alumina	"		140
Chlorpromazine	urine	preconcentration	carbon paste electrode	volt		141
Doxorubicin	"		carbon paste electrode	"		142
CN^-, Cl^-		determinatin	Pt electrode	amp.		143

TABLE VIII   Features of FIA methods involving precipitation-dissolution

Analyte	Matrix	Analytical purpose	Active agent	Detection	Special features	Ref.
NH_3, Cl^-, $C_2O_4^{2-}$		theoretical	Fe^{3+}, Ag^+, Ca^{2+}	AAS	with and without continuous precipitate dissolution	145
Cl^-	waters	determination	Ag^+	AAS	normal and reversed FIA	147
Cl^-, I^-	foodstuff	simultaneous determination	"	AAS	mixture resolution	148
Pb^{2+}	waters	preconcentration	NH_3	AAS	sample aspiration	149

5.3   PRECIPITATION AND DISSOLUTION

This separation technique, of widespread use in classical analytical chemistry applications, has been scarcely automatized owing to the intrinsic difficulties involved.

Filtration through a piece of paper moving at right angles to the flow has been used in air-segmented methods for cleanup of samples prior to introduction into the system and in some methods involving the analyte precipitation (e.g. the determination of potassium in fertilizers with tetraphenyl borate and photometric monitoring of the precipitating reagent (ref. 144)), though with not too brilliant results.

Recently the authors have shown the possibility to obtain satisfactory results from precipitation/dissolution processes implemented in continuous unsegmented systems (Table VIII). In Fig. 9 is shown the operational scheme of an FIA configuration for indirect AA determination of anionic species incorporating a conventional HPLC cleanup filter. The system uses three valves; one for injection (IV) of the sample containing the anion; another for alternating introduction of washing and precipitate solvent solutins (SV$_1$) and a third, four-way diverting one (SV$_2$) for directing streams to the detector or to waste.

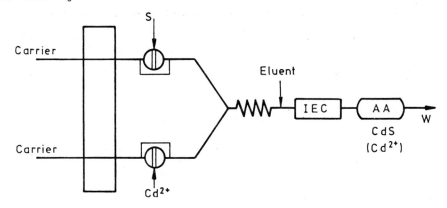

Fig. 8 Configuration for precipitation of the analyte (sulphide) with cadmium (II). The precipitate circulates freely along the system and is detected by AAS. Excess Cd(II) is removed by the ion-exchange column (IEC) and subsequently eluted. (Reproduced from (ref. 117) with permission of Elsevier Science Publishers).

In the precipitation-filtration step (Fig. 9a), the sample is injected into the reagent-carrier. After precipitation, the flowing stream is passed by SV$_2$ through the filter, where the precipitate is retained. Since the cation-reagent stream continuously reaches the detector in this first stage, the signal (baseline) is quite high. On passage through the detector, the precipitate fluid zone yields a negative signal (peak) proportional to the analyte concentration and corresponding to the reagent disappearance. Thus, no additional step is required - and valves SV_1 and SV_2 can be dispensed with - when the precipitate is relatively pure. A diluent stream later merged at the filter is occasionally needed when the reagent concentration in the carrier required to ensure precipitation is too high to be directly introduced into the atomic absorption spectrophotometer. In the wash step (Fig. 9b), valve SV_1 introduces the washing solution, which is passed through the filter via SV_2 and later led to the detector, where it gives a parasitic, non-analytical signal corresponding to the adsorption (contamination) of the cation-reagent on the precipitate surface. Valve SV_2 then sends the flow emerging from the precipitation coil to waste. In the dissolution step (Fig. 9c), valve SV_1 is switched to introduce a - generally acidic - solvent. The remaining components act as in the previous operation. On passage of the stream through the filter, the freshly precipitated small mass is rapidly dissolved. A flow plug contains the stoichiometric amount of cation-reagent present in the precipitate. On passing through the detector, this zone yields a transient signal (peak) obviously proportional to the amount of precipitated anion-analyte.

These continuous precipitation/filtration systems have been used with three types of precipitate: gelatinous ($Fe_2O_3 \cdot xH_2O$, with Fe^{3+} as carrier and NH_3 samples), curdy (AgCl, with Ag^+ as carrier and Cl^- samples) and crystalline ($CaC_2O_4 \cdot 2H_2O$, using Ca^{2+} as carrier and $C_2O_4^{2-}$ samples). Recoveries close to 100% are achieved in every case, even at low analyte concentrations (ref. 145). Interferents are much less disturbing than in the classical precipitation-filtration procedure, probably as a result of the decreased precipitation and digestion times involved (ref. 146).

By use of two FIA configurations (normal and reversed), continuous precipitation has been applied to the determination of chloride in different types of water with Ag(I) as reagent (ref. 147). Chloride and iodide have also been determined in various foods and drinks by using a configuration involving washing and sequential dissolution of the precipitates with ammonia and nitric acid for Cl^- and I^-, respectively (ref. 148).

These assemblies also allow the implementation of on-line preconcentration with the measuring device used. Such is the case with the determination of lead traces in various types of water (ref. 149). The direct aspiration of the sample into the flame of the atomic absorption spectrophotometer yields no signal. However, the set-up depicted in Fig. 10

allows preconcentrating to the extent required by continuously pre-cipitating the lead as a basic salt. Only pump 2 works in the pre-concentration step. The sample stream - aspirated rather than injected -, is merged with a precipitating reagent stream (NH_3) and the mixture is led by SV_2 to the precipitation coil and onto the filter, where the precipitate formed is retained and trough which the flow goes to waste. In the dissolution step, pump 2 is stopped and a nitric acid stream is introduced into the system by pump 1, valve SV_2 leading the flow to the detector. The nitric acid rapidly dissolves the precipitate built up on the filter and a positive peak proportional to the amount of analyte contained in the aspirated sample volume is yielded. Lead can thus be determined over a wide concentration range (1.2 - 1 500 ng/ml), with a maximum preconcentration factor of 10^3. The potential interference from other transition metal ions also precipitated and dissolved in the process is overcome by using the characteristic spectrum line of lead.

6.   HPLC-FIA ASSOCIATION

Though a number of papers in the literature have HPLC and FIA among their keywords, few of them actually report on the on-line coupling of both techniques. The development of novel continuous hydrodynamic detection systems or the improvement of those already existing involves both techniques on account of their common features (ref. 150). Also, parts of a liquid chromatograph - separation column excluded - have been used in FIA determinations, although this alternative is not re-commendable because of the increased cost of components designed to withstand high pressures and of the need for one high-pressure pump per channel (ref. 151).

Strictly, only when there is a complementary pre or post-column in-jection should the possibility of an FIA system being coupled to a liquid chromatograph be considered. Thus, the basic components of an on-line coupled HPLC-FIA system are two injection valves, two pumps, a chroma-tographic column, a reactor and a continuous detector, in addition to the usual reservoirs for the eluent(s), carrier(s) and reagent(s). Restrictor coils are also frequently used to prevent the formation of air bubbles. In Fig. 11 are illustrated the three general manners in which this association can be experimentally implemented according to the position of the injection valve within the FIA subsystem: (a) precolumn; (b) post-column with injection prior to merging of the carrier or reagent stream with the chromatographic effluent and (c) post-column with the FIA valve placed at the merging point itself.

370

a)

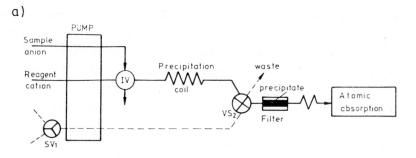

b)

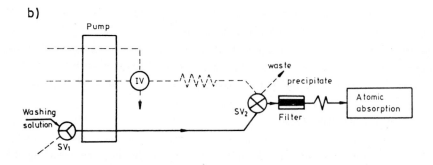

c)

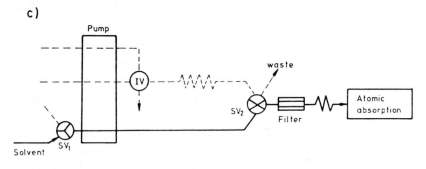

d)

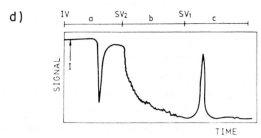

Fig. 9   Continuous precipitation/dissolution configurations for indirect
         determination of anions by AAS. (a) Precipitation; (b) washing;
         (c) dissolution; (d) transient signal obtained in each step. For
         further details, see text.

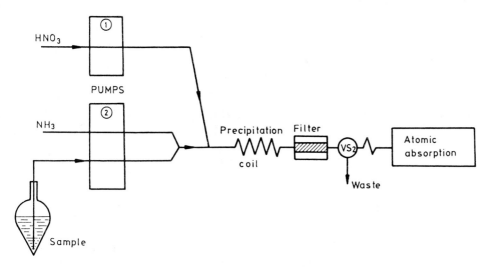

Fig. 10 Continuous on-line system for preconcentration of lead traces in
water prior to its determination by AAS with continuous precipitation (pump 2 in operation) and precipitate dissolution
(pump 1 in operation). (Reproduced from (ref. 149) with
permission of the Royal Society of Chemistry).

6.1 PRE-COLUMN ASSEMBLIES

Continuous pretreatment (sample conditioning, preconcentration and
interference removal) and derivatization systems applied prior to sample
introduction in HPLC are of special relevance whenever the possibility of
automatization is involved. The most outstanding advances in this area
have arisen from the use of pre-columns packed with an active material
and coupled to assemblies consisting of several rotary valves for sample
cleanup and trace enrichment (refs. 152, 153).

Liquid-liquid extraction has occasionally been applied prior to liquid
(ref. 154) or gas chromatography (ref. 155). Indeed few FIA-HPLC systems
use the FIA valve prior to the chromatographic column (Fig. 11a). A
typical example is the determination of zinc in the range $2 \ 10^{-7}$-$20 \ 10^{-7}$M
based on its activating effect on metal-free carboxypeptidase A
immobilized in a reactor (ref. 157) (Fig. 12). A diverting valve allows
switching between water and regenerating solution (1,10-phenanthroline)
streams, where 500-μl samples containing the analyte and the substrate
(hippuryl-L-phenylalanine, the decomposition products of which are sensed
by reversed-phase liquid chromatography) are injected. The two injection
valves used are connected on-line to each other and a volume of 10 μl is
injected into the chromatograph 30 sec after sample injection (carrier
flow-rate 1 ml/min).

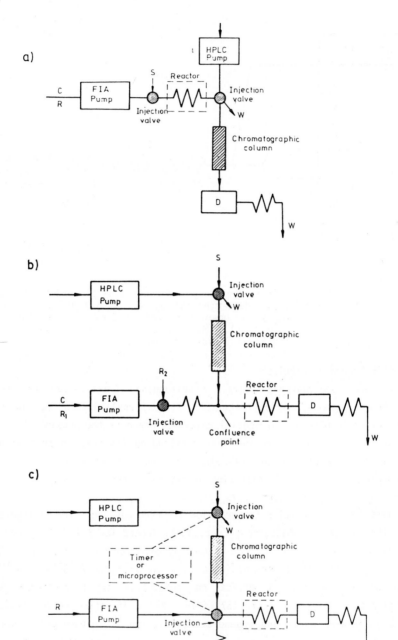

Fig. 11 Generic types of HPLC-FIA assemblies. Precolumn flow injection
(a); Post-column flow injection with valve located prior to (b)
or at the point of merging (c) of the chromatographic eluate and
the carrier or reagent. C, carrier; R, reagent; S, sample; D,
continuous detector; W, waste.

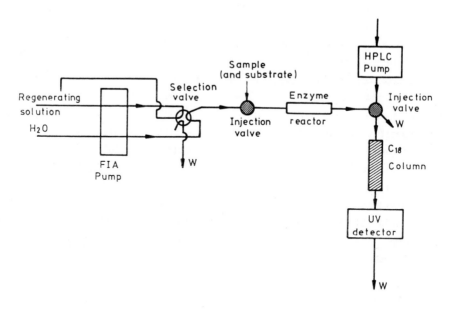

Fig. 12 Operational scheme of an FIA system used prior to a chromatographic separation (determination of zinc through its activating
effect on an enzyme immobilized in the reactor. For details, see
text. (Reproduced from (ref. 156) with permisssion of Elsevier
Science Publishers).

6.2 POST-COLUMN ASSEMBLIES

Post-column reaction detectors are the commonest way of implementing
on-line derivatization in HPLC aimed to improve or facilitate detection
(ref. 157). According to Frei et al. (ref. 158), one of the chief
shortcomings of these configurations is the need for post-column reagent
addition. The reagent can be introduced in three different manners,
namely (a) in a continuous stream merged with the effluent from the
chromatographic column; (b) by injection into a carrier later merging
with the effluent and (c) by means of a solid-phase reactor where the
reagent - generally a catalyst - is immobilized. A further pump is needed
in the first two cases to set the reagent or carrier flow. Other
post-column pumpless reaction units include electrochemical,
photochemical and thermal sensing.

Several FIA systems in which injection can be substituted by merging
with the chromatographic effluent have been described (Fig 13). The
overall analyte concentration is determined by injecting the sample
through the FIA valve whereas discrimination between different analytes

(multidetermination) is accomplished by acommodating the effluent in the post-column system, which uses no injection valve and thus acts as an open-tube reaction detector. Inorganic polyphosphates (ref.159), poly-phosphoric acids in phosphorous smokes (ref. 160), phosphate and phos-phonate (with two parallel (ref. 161) or series (ref. 162) photometric detectors) and the complexing abilities of ligand for metal ions (refs. 163, 164) have been determined with these dual configurations. A real post-column on-line HPLC-FIA configuration is only justified when specific problems are involved. Such is the case with the determination of phosphinate, phosphonate and phosphate (ref. 165), in which sodium bisulphite is previously required to oxidize P(I) and P(III) to P(V), the species ultimately responsible for the analytical reaction with the chromogenic reagent - Mo(V)-Mo(IV). As the sulphite solution tends to corrode stainless steel and disturb the flow-rate of the reciprocating pump, it is introduced with a loop-valve injector to avoid contact with the pump.

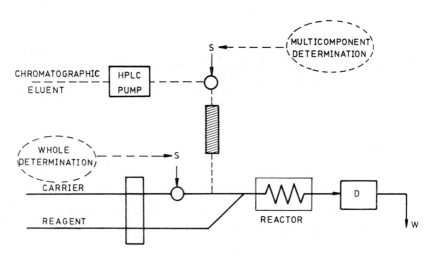

Fig. 13 Alternate use of a flow-injection valve or the chromatographic effluent for individual and multi-determinations, respectively, with the same flow-injection reaction/detection system.

Another interesting way to couple an FIA system after a liquid chroma-tograph involves filling the flow-injection valve with the chromato-graphic effluent and introducing microvolumes of this into a reagent or carrier stream at regular intervals (Fig 11c). The automatic functioning of the valve is obviously mandatory in this use. Mixtures of reducing sugars (ref. 116) and amino-acids have been resolved photometrically and amperometrically, respectively, with configurations involving the synchronized operation of the two injection valves. In this respect it is worth noting the possibility to use an FIA assembly as the interface between a liquid chromatograph and an atomic absorption spectrophotometer (ref. 168). In Fig 14 is shown the scheme of the configuration developed for studying metal-ligand binding in clinical samples. It involves dual continuous detection; photometric of molecular species (citrate, albumin) and atomic absorption spectrophotometric of metal ions (Ca^{2+} and Mg^{2+}). This configuration allows for individual optimization of both integrated processes (HPLC-photometric detection and FIA-atomic absorption detection). The serum samples used (250 μl) are manually introduced into the HPLC injector, while the flow-injection valve sequentially and automatically introduces 11 μl of the effluent into the carrier with a delivery time of 5 sec (6 cycles per minute). The atomic spectroscopic detector gives a chromatographic 'peak' whose profile is formed by the maxima of the FIA peaks. Both chromatograms are recorded by a dual-channel recorder.

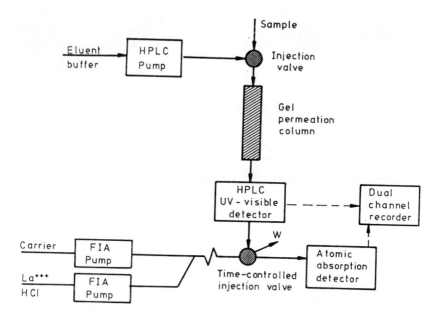

Fig. 14   Use of a - dual-detection - FIA system as interface between a liquid chromatograph and an atomic spectroscopic detector in the study of metal-ligand binding interactions in clinical samples. For details, see text.

7.   FINAL REMARKS

Continuous separation techniques have so far been used only occasionally in FIA - only in about 10% of all instances as can be seen from Fig. 15a. Liquid-liquid extraction is the separation technique most commonly used with this methodology. Roughly 20% of the work dealing with the joint use of FIA and separation techniques involved ion-exchange microcolumns. Gas-diffusion was used in a similar proportion, while dialysis was employed to a lesser extent.

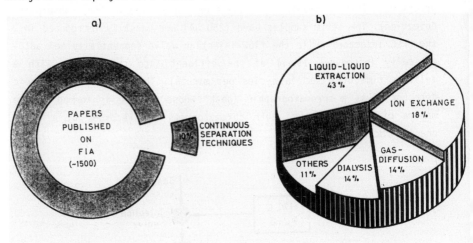

Fig. 15   Statistics on the use of separation techniques with FIA.

Membrane separation (Fig. 16) of molecules (dialysis), gases (gas diffusion) and immiscible liquids (extraction) was the foundation of over 60% of the continuous separation processes developed by FIA to date (ref. 169).

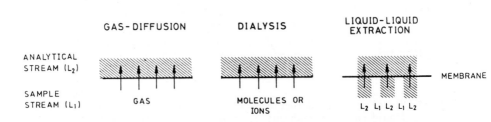

Fig. 16   Membranes used in continuous non-chromatographic separation techniques.

In addition to the use of solid-liquid interfaces for implementation of some analytical procedures, the chief purpose of the joint use of separation techniques and segmented flow systems is to improve sensitivity (preconcentration) and selectivity (sample cleanup, multi-determinations), and - in some cases - to improve or facilitate the analytical reaction and/or detection, otherwise unfeasible. Another advantage of this association over batch non-chromatograpic separation techniques lies in the higher sampling rates achieved, which is of great relevance to routine determinations.

These separation processes carried out in a continuous fashion are intermediate, both kinetically and thermodynamically, between batch processes, where equilibrium is reached once or several times, and chromatographic processes, in which equilibrium is attained many times.

It is interesting to note the decisive role played by kinetics in these continuous separation processes. As a rule, physico-chemical equilibrium is not reached by the time detection is performed, in contrast with batch and air-segmented continuous flow methods. This should result in decreased precision; yet, the relative standard deviations obtained by batch and air-segmented methods and by FIA are comparable. On the other hand, the kinetic discrimination afforded by the continuous methodology results in enhanced selectivity (ref. 149), to the detriment of sensitivity - this should not be too much of a worry if the separation process involved is intended for preconcentration purposes.

Despite their proven advantages, few FIA systems have been used in conjunction with continuous separation techniques so far. This can be attributed to the occurrence of a number of deterrent experimental factors influencing these dynamic systems. Nevertheless, it would suffice to test any of the above-described configurations to immediately realize the scarce technical and instrumental difficulties involved. Applications in this field will no doubt increase significantly in the years to come, particularly in clinical, food and environmental analysis, where the sample matrix and low analyte concentrations usually dealt with are decisive factors.

8. ACKNOWLEDGEMENT

The authors wish to acknowledge the support of the comision Inter-ministerial de Ciencia y Technologia (Grant no. PA 86-0146) for research on this topic.

REFERENCES

1   M. Valcárcel and M.D. Luque de Castro, Automatic Methods of
    Analysis, Elsevier, Amsterdam, 1988.
2   C.F. Simpson, Ed., Techniques in Liquid Chromatography, J. Wiley,
    New York, 1982.
3   J.C. Giddings, K.A. Graff, K.D. Caldwell and M.N. Myers, Advances
    in Chemistry Series, C.D. Craver, Ed., Amer. Chem. Soc.,
    Waschington D.C., 1983.
4   J.C. Giddings, Anal. Chem., 53 (1981), 1170A.
5   W.B. Furman, Continuous Flow Analyis. Theory and Practice, Marcel
    Dekker, New York, 1976.
6   J. Ruzicka and E.H. Hansen, Flow Injection Analysis, J. Wiley, New
    York, 1981.
7   M. Valcárcel and M.D. Luque de Castro, Flow Injection Analysis:
    Principles and Applications, Ellis Horwood, Chichester, 1987.
8   J. Ruzicka and E.H. Hansen, Anal. Chim. Acty, 179 (1986) 1.
9   G. Den Boef and R.C. Schorthorst, Anal. Chim. Acta, 180 (1986) 1.
10 L. Anderson, Anal. Chim. Acta, 110 (1979) 123.
11 H.A. Mottola, Anal. Chim. Acta, 145 (1983) 27.
12 D. Jagner, M. Josefson and K. Aren, Anal. Chim. Acta, 141 (1982)
    147.
13 J.A. Wise and W.R. Heineman, Anal. Chim. Acta, 172 (1985) 1.
14 J. Wang and H.D. Dewald, Anal. Chim. Acta, 162 (1984) 189.
15 E.A.G. Zagatto, B.F. Reis, H. Bergamin F^0 and F.J. Krug, Anal.
    Chim. Acta, 109 (1979) 45.
16 G.E. Pacey, D.A. Hollowell, K.G. Miller, M.R. Straka and G. Gordon,
    Anal. Chim. Acta, 179 (1986) 259.
17 H. Bader and J. Joigne, Water Res., 15 (1981) 71.
18 D.H. Hollowell, Diss., Miami Univ., Oxford, OH, (1985).
19 G.B. Martin and M.E. Meyerhoff, Anal. Chim. Acta, 186 (1986) 71.
20 G. Svensson and T. Anfält, Clin. Chim. Acta, 119 (1982) 7.
21 H. Baadenhuijsen and H.E.H. Seuren-Jacobs, Clin. Chem., 25 (1979)
    443.
22 J. Ruzicka and E.H. Hansen, Anal. Chim. Acta, 173 (1985) 3.
23 M.E. Meyerhoff, and Y.M. Fraticelli, Anal. Lett., 14 (1981) 415.
24 J. Möeller and B. Winter, Fresenius Z. Anal. Chem., 320 (1985) 451.
25 D.A. Hollowell, G.E. Pacey and G. Gordon, Anal. Chem., 57 (1985)
    2851.
26 M.R. Straka, G. Gordon and G.E. Pacey, Anal. Chem., 57 (1985) 1799.
27 P. Marstorp, T. Anfält and L. Anderson, Anal. Chim. Acta, 149
    (1983) 281.
28 D.A. Hollowell, J.R. Gord, G. Gordon and G.E. Pacey, Anal. Chem.,
    58 (1986) 1524.
29 M. Granados, S. Maspoch and M. Blanco, Anal. Chim. Acta, 179 (1986)
    445.
30 C. Okumoto, M. Nagashima, S. Mizoiri , M. Kazama and K. Akiyama,
    Eisei Kagaku, 30 (1984) 7.
31 B. Pihlar and L. Kosta, Anal. Chim. Acta, 114 (1980) 276.
32 O. Aström, Anal. Chem., 54 (1982) 190.
33 G.E. Pacey, M.R. Straka and J.R. Gord, Anal. Chem., 58 (1986) 504.
34 D.D. Siemer, P. Koteel and V. Jariwala, Anal. Chem., 48 (1976) 836.
35 F.D. Pierce and H.R. Brown, Anal. Chem., 48 (1976) 693.
36 F.D. Pierce and H.R. Brown, Anal. Chem., 49 (1977) 1417.
37 M. Yamamoto, M. Yatsuda and Y Yamamoto, Anal. Chem., 57 (1985)
    1382.
38 R.R. Liversage and J.C. van Loon, Anal. Chim. Acta, 161 (1984) 275.
39 M. Burguera and J.L. Burguera, Analyst, 111 (1986) 171.
40 J.C. De Andrade, C. Pasquini, N. Baccan and J.C. van Loon,
    Spectrochim. Acta, Part B, Oct. 38 (1983) 1329.
41 N.H. Tioh, Y. Israel and R.M. Barnes, Anal. Chim. Acta, 184 (1986) 205.

42 S.M. Ramasamy, M.S.A. Jabbar and H.A. Mottola, Anal. Chem., 52 (1980) 2062.
43 M.D. Luque de Castro, J. Autom. Chem., 8 (1986) 56.
44 B. Karlberg, Anal. Chim. Acta, 180 (1986) 16.
45 B. Karlberg and S. Thelander, Anal. Chim. Acta, 98 (1978) 1.
46 H. Bergamin F⁰, J.X. Medeiros, B.F. Reis and E.A.G. Zagatto, Anal. Chim. Acta, 101 (1978) 9.
47 M. Gallego and M. Valcárcel, Anal. Chim. Acta, 169 (1985) 161.
48 J. Kawase, Anal. Chem., 52 (1980) 2124.
49 A. Deratini and B. Sebille, Anal. Chem., 53 (1981) 1742.
50 J.L. Burguera, M. Burguera, L. Cruz and O.R. Naranjo, Anal. Chim. Acta, 186 (1986) 273.
51 K. Kina, T. Shiraishi and N. Ishibashi, Talanta, 25 (1978) 295.
52 P. Linares, F. Lázaro, M.D. Luque de Castro and M. Valcárcel, Anal. Chim. Acta, 200 (1987).
53 M.H. Memon and P.J. Worsfold, Anal. Chim. Acta, 183 (1986) 179.
54 A. Ríos, M.D. Luque de Castro and M. Valcárcel, Anal. Chem. (in press).
55 D.C. Shelly, T.M. Rossi and I.M. Warner, Anal. Chem., 54 (1982) 87.
56 T.M. Rossi, D.S. Shelly and I.M. Warner, Anal. Chem., 54 (1982) 2056.
57 M. Bengtsson and G. Johansson, Anal. Chim. Acta, 158 (1984) 147.
58 L. Nord and B. Karlberg, Anal. Chim. Acta, 164 (1984), 233.
59 J.A. Sweileh and F.F. Cantwell, Anal. Chem., 57 (1985) 420.
60 F.F. Cantwell and J.A. Sweileh, Anal. Chem., 57 (1985) 329.
61 K. Backstrom, L.G. Danielsson and L. Nord, Anal. Chim. Acta, 169 (1985) 43.
62 L. Nord and B. Karlberg, Anal. Chim. Acta, 125 (1981) 199.
63 T. Imasaka, T. Harada and N. Ishibashi, Anal. Chim. Acta, 129 (1981)195.
64 J.T. Davies and E.K. Rideal, Interfacial Phenomena, 2nd edn., Academic Press, New York, 1963, Chapter 1.
65 A.W. Adamson, Physical Chemistry of Surfaces, 2nd edn., Interscience, New York, 1967, Chapters 1 and 7.
66 J.L. Burguera and M. Burguera, Anal. Chim. Acta, 153 (1983) 207.
67 L. Fossey and F.F. Cantwell, Anal. Chem., 55 (1983) 1882.
68 J. Kawase, S. Nakae and M. Yamanaka, Anal. Chem., 51 (1979) 1640.
69 K. Ogata, K. Taguchi and T. Imanari, Anal. Chem., 54 (1982) 2127.
70 B. Karlberg in Chemical Derivatization in Analytical Chemistry Vol. 2, Separation and Continuous Flow Techniques, R.W. Frei and J. F. Lawrence, Plenum Press, New York, 1982.
70a C. de Ruiter, J.H. Wolf, K.A.Th. Brinkman and R.W. Frei, Anal. Chim. Acta 192 (1987) 267.
71 L. Nord and B. Karlberg, Anal. Chim. Acta, 145 (1983) 151.
72 K. Ogata, S. Tanabe and T. Imanari, Chem. Pharm. Bull, 31 (1983) 1419.
73 M. Gallego, M. Silva and M. Valcárel, Anal. Chim. Acta, 179 (1986) 439.
74 Y. Sahlestöm, S. Twengström and B. Karlberg, Anal. Chim. Acta, 187 (1986) 339.
75 O. Klinghofer, J. Ruzicka and E.H. Hansen, Talanta, 27 (1980) 169.
76 F. Lázaro, M.D. Luque de Castro and M. Valcárcel, Anal. Chim. Acta, 169 (1985) 132.
77 P. Linares, M.D. Luque de Castro and M. Valcárcel, Anal. Chim. Acta, 161 (1984) 257.
78 L. Fossey and F.F. Cantwell, Anal. Chem., 57 (1985) 992.
79 K. Ogata, K. Taguchi and T. Imanari, Bunseki Kagaku, 31 (1982) 641.
80 K. Ogata, K. Taguchi and T. Imanari, Bunseki Kagaku, 31 (1982) 89.
81 L. Sun, L. Li and Z. Fang, Fenxi Huaxue, 13 (1985) 447.
82 M. Gallego, M. Silva and M. Valcárel, Fresenius Z. Anal. Chem. 323 (1986) 50.

380

83   P. A. Johansson, B. Karlberg and S. Thelander, Anal. Chim. Acta, 114 (1980) 215.
84   G. Audunsson, Anal. Chem., 58 (1986) 2714.
85   B. Karlberg, P.A. Johansson and S. Thelander, Anal. Chim. Acta, 104 (1979 21.
86   Y. Hirari and K. Tomokuni, Anal. Chim. Acta, 167 (1985) 409.
87   M.J. Whitaker, Anal. Chim. Acta, 179 (1986) 459.
88   Y. Sahlestöm and B. Karlberg, Anal. Chim. Acta, 179 (1986) 315.
89   L. Fossey and F.F. Cantwell, Anal. Chem., 54 (1982) 1693.
90   L. Nord, S. Johansson and H. Brötell, Anal. Chim. Acta, 175 (1985) 281.
91   T. Kato, Anal. Chim. Acta, 175 (1985) 339.
92   S. Johansson and H. Brötell, Flow Analysis II, Lund (Sweden) 1982.
93   B. Karlberg and S. Thelander, Anal. Chim. Acta, 114 (1980) 129.
94   M. Maeda and A. Tsuji, Analyst, 110 (1985) 665.
95   M. Gallego, M. Silva and M. Valcárcel, Anal. Chem., 58 (1986) 2265.
96   E.H. Hansen and J. Ruzicka, Anal. Chim. Acta, 87 (1976) 363.
97   L. Gorton and L. Ögren, Anal. Chim. Acta, 130 (1981) 45.
98   Q. Chang and M.E. Meyerhoff, Anal. Chim. Acta, 186 (1986) 81.
99   C.F. Mandenius, B. Danielsson and B. Mattiasson, Anal. Chim. Acta, 163 (1984) 135.
100 P.K. Dasgupta and H.C. Yang, Anal. Chem., 58 (1986) 2839.
101 H. Hwang and P.K. Dasgupta, Anal. Chem., 58 (1986) 1521.
102 P.E. Macheras and M.A. Koupparis, Anal. Chim. Acta, 185 (1986) 65.
103 B. Bernhardsson, E. Martins and G. Johansson, Anal. Chim. Acta, 167 (1985) 111.
104 R.Y. Xie and G.D. Christian, Anal. Chem., 58 (1986) 1806.
105 E. Martins, M. Bengtsson and G. Johansson, Anal. Chim. Acta, 169 (1986) 31.
106 W.D. Basson and J.F. van Staden, Analyst, 104 (1979) 419.
107 J.F. van Staden and W.D. Basson, Lab. Pract., 29 (1980) 1279.
108 B. Olsson, H. Lundback and G. Johansson, Anal. Chim. Acta, 167 (1985) 123.
109 D. Pilosof and T.A. Nieman, Anal. Chem., 54 (1982) 1698.
110 P.J. Worsfold, J. Farrelly and M.S. Matharu, Anal. Chim. Acta, 164 (1984) 103.
111 H. Lundback and B. Olsson, Anal. Lett., 18(B7) (1985) 871.
112 M. Masoom and A. Townshend, Anal. Chim. Acta, 166 (1984) 111.
113 A. Townshend, Anal. Chim. Acta, 180 (1986) 49.
114 S. Olsen, L.C.R. Pessenda, J. Ruzicka and E.H. Hansen, Analyst, 108 (1983) 905.
115 J.L. Burguera, M. Burguera and A. Townshend, Anal. Chim. Acta, 127 (1981) 199.
116 A.T. Faizullah and A. Townshend, Anal. Chim. Acta, 179 (1986) 233.
117 B.A. Petersson, Z. Fang, J. Ruzicka and E.H. Hansen, Anal. Chim. Acta, 184 (1986) 165.
118 H. Hwang and P.K. Dasgupta, Anal. Chem., 58 (1986) 1521.
119 J. Ruzicka and E.H. Hansen, Anal. Chim. Acta, 161 (1984) 1.
120 Z. Fang, J. Ruzicka and E.H. Hansen, Anal. Chim. Acta, 164 (1984) 23.
121 H. Bergamin F^{o}., B.F. Reis, A.O. Jacintho and E.A.G. Zagatto, Anal. Chim. Acta, 117 (1980) 81.
122 H. Mikasa, S. Motomizu and K. Toei, Bunseki Kagaku, 34(8) (1985) 518.
123 Z. Fang, S. Xu and S. Zhang, Fenxi Huaxue, 12 (1984) 997.
124 O.F. Kamson and A. Townshend, Anal. Chim. Acta, 155 (1985) 253.
125 L. Risinger, Anal. Chim. Acta, 179 (1986) 509.
126 P. Hernández, L. Hernández, J. Vicente and M.T. Sevilla, Anal. Quím., 81 (1985) 117.
127 S.D. Hartenstein, J. Ruzicka and G.D. Christian, Anal. Chem., 57 (1985) 21.

128   Z. Fang, S.Xu and S. Zhang, Anal. Chim. Acta, 164 (1984) 41.
129   S. Hirata, Y. Umezaki and M. Ikeda, J. Flow Inj. Anal., 3 (1986) 8.
130   S. Hirata, Y. Umezaki and M. Ikeda, Anal. Chem., 58 (1986) 2602.
131   M. Bengtsson, F. Malamas, A. Torstensson, O. Regnell and G. Johansson, Mikrochim. Acta, III (1985) 209.
132   H. Hirano, Y. Baba, N. Yoza and S. Ohashi, Anal. Chim. Acta, 179 (1986) 209.
133   P.I. Anagnostopoulou and M.A. Koupparis, Anal. Chem., 58 (1986) 322.
134   T.S. Stevens and T.E. Miller, Patent US 4290775.
135   T. Miller and T. Stevens, Adv. Instrum., 35 (1980) 21.
136   N. Yoza, H. Hirano, Y. Baba and S. Ohashi, J. Chromat., 325 (1985) 385.
137   T. Yamane, Bunseki Kagaku, 33 (1984) E203.
138   A.G. Cox, and C.W. McLeod, Anal. Chim. Acta, 179 (1986) 487.
139   A.G. Cox, I.G. Cook and C.W. McLeod, Analyst, 110 (1985) 331.
140   I.G. Cook, C.W. McLeod and P.J. Worsfold, Anal. Proc., 23 (1986) 5.
141   J. Wang and B.A. Freiha, Anal. Chem., 55 (1983) 1285.
142   E.N. Cheney and R.P. Baldwin, Anal. Chim. Acta, 176 (1985) 105.
143   J.A. Polta and D.C. Johnson, Anal. Chem., 57 (1985) 1373.
144   J.M. Skinner and A.C. Docherty, Talanta, 14 (1967) 1393.
145   P. Martínez, M. Gallego and M. Valcárcel, Anal. Chem., 59 (1987) 69.
146   M. Valcárcel, Analyst, 112 (1987), 729.
147   P. Martínez, M. Gallego and M. Valcárcel, J. anal. Atom. Spec., 2 (1987) 211.
148   P. Martínez, M. Gallego and M. Valcárcel, Anal. Chim. Acta, 193 (1987), 127.
149   P. Martínez, M. Gallego and M. Valcárcel, Analyst, 112 (1987) 1233.
150   C.E. Lunte, T.H. Ridgway and W.R. Heineman, Anal. Chem., 59 (1987) 761.
151   F.P. Bigley, R.L. Grob and G.S. Brenner, Anal. Chim. Acta, 181 (1986) 241:
152   R.W. Frei, Swiss Chem., 6 (1984) 55.
153   M.W.F. Nielen, R.W. Frei and U.A.Th. Brinkman, Selective sample handling and detection in liquid chromatography, Vol. I, (Editors Frei and Zech) Elsevier, Amsterdam, 1987.
154   J.N. Dolan, J.R. Grant, N. Tanaka, R.W. Gise and B.L. Karger, J. Chromat. Sci., 16 (1986) 616.
155   E. Fogelquist, M. Krysell and L.G. Danielsson, Anal. Chem., 58 (1986) 1516.
156   L. Risinger, L. Ögren and G. Johansson, Anal. Chim. Acta, 154 (1983) 251.
157   I.S. Krull, Ed., Reaction Detection in Liquid Chromatography, Marcel Dekker, New York, 1986.
158   R.W. Frei, J. Hansen and V.A.Th. Brinkman, Anal. Chem., 57 (1985) 1529A.
159   Y. Hirai, N. Yoza and S. Ohashi, Anal. Chim. Acta, 115 (1980) 269.
160   R.S. Brazell, R.W. Holmberg and J.H. Moneyhun, J. Chromatogr., 290 (1984) 163.
161   Y. Baba, N. Yoza and S. Ohashi, J. Chromatogr., 295 (1984) 153.
162   Y. Baba, N. Yoza and S. Ohashi, J. Chromatogr., 318 (1985) 319.
163   N. Yoza, T. Miyaji, Y. Hyrai and S. Ohashi, J. Chromatogr., 283 (1984) 89.
164   N. Yoza, T. Shuto, Y. Baba, A. Tanaka and S. Ohashi, J. Chromatogr., 298 (1984) 419.
165   Y. Hirai, N. Yoza and S. Ohashi, J. Chromatogr., 206 (1981) 501.
166   D. Betteridge, N.G. Courney, T.J. Sly and D.G. Porter, Analyst, 109 (1984) 91.

382

167   J.B. Kafil and C.O. Hubber, Anal. Chim. Acta, 139 (1982) 347.
168   B.W. Renoe, C.E. Shideler and J. Savory, Clin. Chem., 27 (1984) 1546.
169   M. Valcárcel and M.D. Luque de Castro, J. Chromatogr., 393 (1987) 3.

SUBJECT INDEX

JOURNAL OF CHROMATOGRAPHY LIBRARY

A Series of Books Devoted to Chromatographic and Electrophoretic
Techniques and their Applications

Although complementary to the *Journal of Chromatography*, each volume in the Library Series
is an important and independent contribution in the field of chromatography and
electrophoresis. The Library contains no material reprinted from the journal itself.

Other volumes in this series

392

394